Dear MyCopy Customer,

This Springer book is a monochrome print version of the eBook to which your library gives you access via SpringerLink. It is available to you at a subsidized price since your library subscribes to at least one Springer eBook subject collection.

Please note that MyCopy books are only offered to library patrons with access to at least one Springer eBook subject collection. MyCopy books are strictly for individual use only.

You may cite this book by referencing the bibliographic data and/or the DOI (Digital Object Identifier) found in the front matter. This book is an exact but monochrome copy of the print version of the eBook on SpringerLink.

R. Bausière · F. Labrique · G. Séguier

Power Electronic Converters

DC-DC Conversion

With 268 Figures

Springer-Verlag Berlin Heidelberg GmbH

Professor Dr. Robert Bausière

Université des Sciences et Techniques de Lille
59655 Villeneuve d'Ascq Cedex
France

Professor Dr. Francis Labrique

Université Catholique de Louvain
1348 Louvain-la-Neuve
Belgique

Professor Dr. Guy Séguier

Université des Sciences et Techniques de Lille
59655 Villeneuve d'Ascq Cedex
France

Translated from the French by the authors
with the help of N. Quayle

This is the third volume in a series. The first two were published in English by McGraw Hill
the first–on AC/DC conversion, in 1986,
the second–on AC/AC conversion, in 1987.
These two volumes are often referred to in the present work.

DOI 10.1007/978-3-642-52454-7

Library of Congress Cataloging-in-Publication Data

Bausière, R. (Robert), 1947
[Convertisseurs de l'électronique de puissance. English]
Power electronic converters: DC–DC conversion/R. Bausière, F. Labrique, G. Séguier.
p. cm. — (Electric energy systems and engineering series).
Translation of: Les convertisseurs de l'électronique de puissance.
Includes bibliographical references and index.

1. Electric current converters. I. Labrique, F. (Francis), 1946– II. Séguier, Guy. III. Title.
IV. Series.
TK7872.C8B39 1993 92-10524
621.3815'322—dc20 CIP

Typesetting: Macmillan (India) Ltd., Bangalore, India
61/3020 - 5 4 3 2 1 0 – Printed on acid-free paper

www.springer.com/mycopy

Series Editors:

Prof. J.G. Kassakian

Massachusetts Institute of Technology,
77 Massachusetts Ave., Cambridge, MA 02139, USA

Prof. D.H. Naunin

Institut für Elektronik, Technische Universität Berlin,
Einsteinufer 19, W-1000 Berlin 10, FRG

Introduction to the Electric Energy Systems and Engineering Series

Concerns for the continued supply and efficient use of energy have recently become important forces shaping our lives. Because of the influence which energy issues have on the economy, international relations, national security, and individual well-being, it is necessary that there exists a reliable, available and accurate source of information on energy in the broadest sense. Since a major form of energy is electrical, this new book series titled *Electric Energy Systems and Engineering* has been launched to provide such an information base in this important area.

The series coverage will include the following areas and their interaction and coordination: generation, transmission, distribution, conversion, storage, utilization, economics.

Although the series is to include introductory and background volumes, special emphasis will be placed on: new technologies, new adaptations of old technologies, materials and components, measurement techniques, control–including the application of microprocessors in control systems, analysis and planning methodologies, simulation, relationship to, and interaction with, other disciplines.

The aim of this series is to provide a comprehensive source of information for the developer, planner, or user of electrical energy. It will also serve as a visible and accessible forum for the publication of selected research results and monographs of timely interest. The series is expected to contain introductory level material of a tutorial nature, as well as advanced texts and references for graduate students, engineers and scientists.

The editors hope that this series will fill a gap and find interested readers.

John G. Kassakian · Dietrich H. Naunin

Preface

This book is the third in a series of four devoted to *POWER ELECTRONIC CONVERTERS*:

The first of these concerns AC to DC conversion.
The second concerns AC to AC conversion.
This volume examines DC to DC conversion.
The fourth is devoted to DC to AC conversion.

Converters which carry out the DC–DC conversion operate by chopping the input voltage or current: they are called choppers or switch-mode power converters. Their operating frequency is not imposed by either the input or the output, both of which are at zero frequency. A frequency which is much greater than that of the industrial network can be chosen, provided that suitable configurations and semiconductor devices are used. This is the first difference compared to the rectifiers and AC–AC converters, analyzed in the previous volumes and which often operate at the industrial network frequency.

The second difference concerns the commutation mode. Choppers operate in forced commutation. The beginning of an operating phase does not automatically turn off the semiconductor devices which were conducting during the previous phase and which have to be brought to the blocking state. This turn-off must be carried out autonomously.

These two differences – the higher frequency of commutations and, especially, the different mode of commutation – justify the first two chapters in this work:

- Chapter 1 examines general notions concerning converters, supplies and loads, and more especially, how they can be characterized with regard to commutations.
- Chapter 2 presents semiconductor devices during commutation. The notions concerning the diode and the thyristor outlined at the beginning of Vol. 1 are insufficient, when commutations become an essential phenomenon. Moreover, other components, specially adapted to forced commutation, have to be presented.

These two chapters provide an introduction to the analysis of choppers as well as of inverters. The latter are the subject of the fourth volume in this series.

The following four chapters are devoted to the detailed analysis of DC–DC converters.

- Chapter 3 gives a general presentation of chopper structures.
- Chapter 4 provides a quantitative analysis of the most common types of choppers with direct energy transfer.
- Chapter 5 examines the most commonly used procedures employed for turning off thyristors in choppers.
- The most widely used configurations of switch-mode power supplies form the basis of the analysis in Chap. 6.

We thank Robert Bausière and Francis Labrique for their valuable contributions to the preparation of this volume. R. Bausière was responsible for Chaps. 2 and 6, as well as for the Appendix on snubbers. F. Labrique was responsible for Chaps. 4 and 5.

As we noted at the beginning of the previous volume, it is somewhat artificial to divide the whole group of converters into four groups according to the conversions each carries out. Different conversions can be obtained with the same structure and can be analyzed together.

This is especially true in the case of choppers and inverters. When controlled in a different way, reversible choppers can operate as inverters. In the part concerning the thyristor turn-off circuits, the analysis of circuits common to two thyristors, connected in series under the same voltage will not be analyzed in detail. Nor will the study of the switching DC power supplies consisting of the cascade association of an inverter, a transformer and a rectifier. Both topics will be taken up in the following volume.

The guiding principle of this collection is to provide a useful and practical tool to power electronics specialists – whether they are designers or users of power converters, in activity or graduating. We have tried to remain faithful to this principle.

The chapter on semiconductor devices in no way lays claim to being a summary of power components physics. It only uses the aspects of this topic which are indispensable for explaining the operation of semiconductor devices in the circuits into which they are embedded, for understanding how they are characterized and the way in which they can be most efficiently used.

Before developing the quantitative analysis of DC–DC converters, we give a very brief description of their basic principles in order to focus on the topics which are to be dealt with; this is the main aim of Chap. 3.

In the detailed further analysis of the most common types of choppers and of switch-mode power supplies, we have developed the computations as far as necessary to obtain the characteristics which are of most immediate interest to users.[1]

Lille, August 1992 G. Séguier

[1]We are very grateful to Mrs. Reine Del Vitto for typing the manuscript and to Mr. Paul Leroy for drawing the figures of this book.

Notes

In order to limit the bibliography to a reasonable length, we have only quoted publications from 1970 onwards.

Insofar as articles are concerned, we have limited the references to those published in the *Proceedings* and *Transactions of the Institute of Electrical and Electronic Engineers* (U.S.A.), the *Proceedings of the Institute of Electrical Engineers* (G.B.), and the *Revue Générale de l'Electricité* (France).

The bibliography concerning:

- Power Semiconductor devices is to be found at the end of Chap. 2.
- Choppers, at the end of Chap. 4.
- Switch-Mode Power Supplies, at the end of Chap. 6.

Articles concerning commutation circuits of thyristors are to be found at the end of chapter 5 and those concerning snubbers at the end of Appendix.

Contents

Chapter 1
Converters, Supplies and Loads: Introductory Remarks 1

Chapter 2
Switching Power Semiconductor Devices 17

Chapter 5
Forced Commutation of Thyristors

Appendix
Snubbers

Nomenclature

Prime are used to distinguish quantities relating to the output from the corresponding quantities relating to the input

at	magnetomotive force
AT	average value of a magnetomotive force
A	anode
B	base
BV	breakdown voltage
C	collector
C	capacitor
D	drain
D	diode
D_{AR}	anti-return diode
e	electromotive force
E	constant electromotive force
E	emitter
f	frequency
G	gate
i	instantaneous current
I	special values of a current through a device
I	average value of a direct current
I_{H}	holding current
I_{L}	latching current
I^2t	overcurrent factor
J	junction
j	positive integers
k	positive intergers or zero
k	ratio
K	cathode
K	switch
ℓ, L or $\mathscr{L}$	inductance
m	related value of an e.m.f.
n	number of paralled choppers
N	N-type layer
N	speed

p	instantaneous power
P	power
P	P-type layer
Q	transistor
Q	ratio $L\omega/R$
r or R	resistance
Rh	rheostat
t	time
t_c	commutation time
t_d	delay time
t_f	fall time
t_{fr}	forward recovery time
t_{gq}	gate recovery time
t_p	reverse-bias time
t_q	turn-off time
t_r	rise time
t_{rr}	reverse recovery time
T	cycle
T	transistor, thyristor
T	temperature
TC	controlled turn-on/off switch
T_{off}	turn-off time
T_{on}	turn-on time
u	voltage
U	average voltage or constant voltage
v	voltage
V	special value a voltage across a device
V_0 or V_T	threshold voltage
W	energy
z	damping ratio
α	switch duty ratio
α	current gain
β	current gain
β	turn-on relative value
γ	capacitor
Δat	magnetomotive force ripple
Δi	current ripple
Δu	voltage ripple
ζ	damping factor
λ	inductance ratio
φ	flux
ω	angular chopping frequency
ω_O	angular resonant frequency

Subscripts

a	of rotor armature
a	of damping circuit
AV	average
B	of base
BE	between base and emitter
BR	breakdown
c	of commutation
c	of case
C or γ	of capacitor C or γ
C	of collector
CE	between collector and emitter
D	of diode
D	of drain
DS	between drain and source
f	of filter
f	of field winding
f or F	forward
G	of gate
GS	between gate and source
j	of junction
K	of switch K
max or M	maximum
min or m	minimum
mean	average
nom	rated
off or OFF	off-state
on or ON	on-state
o	at initial instant
O	with zero current base
P	peak
R or r	reverse
R	repetitive
RMS	root mean square
s	of smoothing inductor
S	surge
T	of thyristor or transistor T
Q	of transistor Q

Chapter 1

Converters, Supplies and Loads: Introductory Remarks

As was the case with rectifiers and AC regulators in the previous volumes, choppers and inverters are static converters using semiconductor devices operating in the switching mode, i.e. commutating from ON state to OFF state or vice versa.

However, the supply of a chopper or an inverter is no longer AC but DC. When a semiconductor switch is in the ON state, it tends to conduct permanently. It is no longer automatically switched off at the end of its normal conducting interval. The majority of DC–DC converters and inverters operate in a *forced commutation mode* and not in natural commutation mode, as do rectifiers and AC regulators.

- In a DC–DC converter, the semiconductor switches are still controlled periodically, but the frequency of their switching cycle is no longer imposed by the supply, since the converter is supplied from a DC source. As the output frequency is also zero, the internal switching frequency can be chosen freely. It is normally given a value considerably higher than that of the industrial AC mains. Operating at a *high frequency* has two advantages:
  - it allows for a size reduction of the inductors and of the capacitors needed to filter the input and output waveforms;
  - it allows for a reduction of the converter response time.

  In an inverter, the output frequency is imposed. If it is of low value, the converter switches are often operated at a much higher frequency; during each of its half-cycle, the output waveform is chopped by introducing supplementary commutations. This is known as the Pulse Width Modulation technique, which enables filtering to be reduced.

- The change from natural commutation to forced commutation, together with the increase in switching frequency, leads to
  - the need to further the study of semiconductor devices and, more precisely, their switching behaviour. This will be done in Chap. 2;
  - the need to clarify some general notions concerning converters.

  Clarifying such notions concerning
    - commutations
    - sources

will indicate that time intervals of considerable difference must be taken into account in the analysis of converters, which, incidentally, makes this analysis much easier.[1]

1.1 Commutation Modes

Static converters are based on the use of semiconductor devices which act as switches, opening and closing paths at a given frequency. They often need to open when a current is flowing through them, i.e. to perform the commutation of a current. *The commutation of a current is its transfer from one path to another.*

In practice, commutation problems are linked to the current turn-off in the path to be opened, especially when this path is inductive. It thus follows that converters are classified according to the way their semiconductor devices are switched off.

1.1.1 Natural Commutation, Forced Commutation

- *Natural* commutation occurs when the turn-off of the current in the path to be opened requires no specific action on the switch which closes this path,
  - either because the current goes by itself to zero,
  - or because the turn-on of another switch leads automatically to the turn-off of the current.

- In AC regulators and frequency multipliers considered in Vol. 2, the current flowing through a path which has been closed falls to zero by itself. Moreover, one path may not be closed before the preceding one has been opened. In fact, there is no current commutation and the converters are thus said to operate in a *free* natural commutation mode.

- In rectifiers operating in continuous conduction mode, the current in a path is different from zero when this path is to be opened. But closing the following path brings the current to zero in the first one.

Commutation in rectifiers has been thoroughly studied (Vol. 1, Chap. 2 § 3.1, Chap. 3, § 1.3 and 2.3, Chap. 4, § 3.2, . . .). It depends on the parameters of the AC mains. When a path is closed, the AC mains applies a voltage across the inductive circuit formed by the path being closed and by the one through which the current previously flowed. This voltage increases the current in the first path and reduces it in the second. Commutation is ended when the current in the path to be opened becomes zero, i.e. when the current previously flowing through it has been entirely transferred to the path which has just been closed.

---

[1] This Chapter, as well as Chap. 3, owes a great deal to the results of work carried out by a certain number of French university research teams.

The turn-off of a semiconductor switch results automatically from the firing of the next one; the voltage necessary to perform the commutation is supplied by the mains. This type of commutation is often referred to as *line commutation* to distinguish it from free natural commutation.

– Some inverters operate in natural commutation.
When the current becomes zero before the end of each half-cycle, free natural commutation occurs.

When the turn-off of a semiconductor switch results from the turn-on of another one, the voltage necessary to perform the commutation is then supplied by the load, *load-assisted commutation* occurs. For this to happen, the load must be capacitive or contain an EMF

- Most choppers and inverters operate in a *forced commutation* mode. In a semiconductor switch with controlled turn-off, the current reverts to zero neither by itself nor by the turn-on of another switch. Current flow is interrupted by a specific action:
  - an action on the control electrode, in the case of a switch that can be turned on and turned off (transistor, GTO thyristor, . . .);
  - or the action of an auxiliary circuit when the semiconductor switch can only be turned on (standard thyristor). This auxiliary circuit is referred to as a commutation circuit.

1.1.2 Nature of the Path to be Opened

1.1.2.1 The Semiconductor Switch

In describing the transient behaviour of a semiconductor switch (as described in the next chapter), two switchings are studied:

- turn-on switching: voltage across the terminals falls and current rises;
- turn-off switching: current falls and voltage rises.

Ideal environmental conditions are assumed (supplies without any internal impedance, paths which in no way limit current and voltage variations) so that the transient behaviour of the semiconductor switch depends on no external factors.

The semiconductor switch behaviour can be modified by its *own environment*, by adding protective circuits. To reduce the current rate of rise at turn-on (limitation of $\mathrm{d}i/\mathrm{d}t$) and to limit the voltage rate of rise at turn-off (limitation of $\mathrm{d}v/\mathrm{d}t$), "*snubbers*" are added where necessary:

- turn-on snubber: small inductor, usually saturable, connected in series with the semiconductor switch and ensuring its protection when conduction starts;
- turn-off snubber: small capacitor connected in parallel with the device and ensuring its protection when conduction is interrupted.

1.1.2.2 The Semiconductor Switch Incorporated in the Converter

During switching, as well as in steady state, the behaviour of semiconductor switches is affected by the circuits in which they are embedded.

In particular, the opening of a path depends not only on the semiconductor switch which ensures the opening, but also on the nature of this path (as well on the nature of the one to which the current must be transferred).

Two basic cases are to be distinguished, concerning how the current falls to zero in the path to be turned off.

- *First case: switching of an inductive path*

If the path to be opened has a noticeable inductance L (i.e. in practice, an inductor forming an element of the basic diagram of the converter, its supply, or its load), this inductance contains an energy $1/2\ Li^2$ when the current i flows through it. The opening of this path results from turning off of the semiconductor switch and diverting the energy stored in the inductor path into another path.

This phenomenon of energy evacuation is most important: it determines the duration of current fall time and the difficulty in turning it off. The dynamic characteristics of the semiconductor switch only come into play later, when the voltage is restored across the terminals.

If the operation of the converter requires commutations between highly inductive paths (e.g. rectifiers), its operating frequency must be low.

- *Second case: switching of a semiconductor switch alone*

Topologies in which there is no inductor to load or unload at each commutation are used in converters which have to operate at high frequency (DC–DC converters, voltage inverters with a high output frequency or a low output frequency but with Pulse Width Modulation). If there is an inductor and if a series-connected semiconductor device is switched off, the current flowing through the inductor is thus transferred to another semiconductor.

When commutations do not involve any energy change in an inductive path, their time duration is only a function of the turn-off characteristics of the semiconductors which are governed by the carriers recombination. So the transient characteristics of only these semiconductor switches (taking possibly, eventually, into account the effects of snubbers) are relevant. These characteristics – and notably the switching losses – are only to limit the possible operating frequency.

1.2 Sources

In power electronics, a generator or a load is defined in a different way than in electrotechnics. This particular description of the supply and load – on either

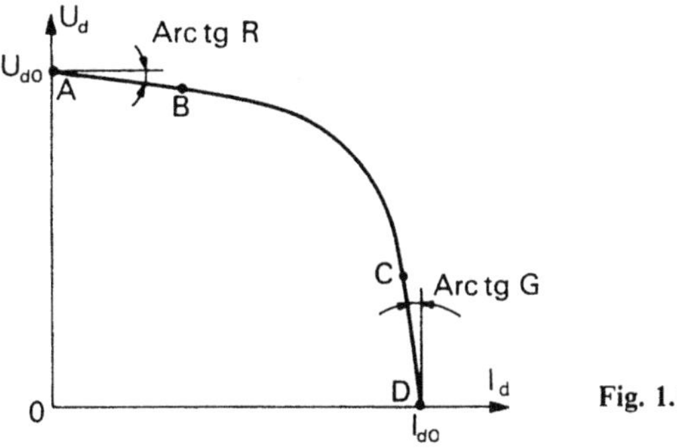

Fig. 1.1

side of a DC–DC converter or of an inverter – must be clearly defined since they determine the converter structure.

- *In conventional electrotechnics*, a DC generator is characterised by curves as in Fig. 1.1 showing the voltage across its terminals U_d as a function of output current delivered I_d. U_d and I_d are the mean steady-state values.

For operation on section AB of the curve, the generator is said to be a voltage supply. The equation

$$U_d = U_{d0} - R\, I_d$$

is used where

U_{d0} = supply voltage produced by an ideal voltage source,
R = internal resistance connected in series.

For operation on section DC of the curve, the generator is said to be a current supply. The equation

$$I_d = I_{d0} - G U_d$$

is used where

I_{d0} = supply current produced by an ideal current source,
G = parallel internal conductance.

- *In power electronics*, and in the particular case of DC–DC converters and inverters working by chopping the input DC voltage or current, the generators and loads are to be characterised by their behaviour in the presence of pulsed signals, i.e. signals with fast variations corresponding to *switchings*.

1.2.1 Voltage/Current Generator or Load

In describing a generator or load, it should be indicated if it is of voltage type or of current type and its voltage or current waveform should be given.

1.2.1.1 Definitions and Representations

- A *voltage* generator or load imposes the voltage at its terminals independently of the current flowing through it. In particular, the voltage cannot be affected by any discontinuity of this current.

In a *current* generator or load, the waveform of the current flowing through it cannot be modified by the circuit to which it is connected.

- The voltage waveform imposed by a voltage generator (or load) and the current waveform imposed by a current generator (or load) may be of any type. However, in studying the functional principles of DC–DC converters and inverters, only DC and AC generators and loads are generally considered.

The generator (or load) voltage (or current) is said to be a DC one, if it undergoes no change of polarity; as a first approximation, it is assumed to be constant.

The voltage or current is said to be an AC one, if it is cyclical and has an average value equal to zero; as a first approximation, it is assumed to be sinusoidal.

- A voltage generator or load is represented by a circle in which there is the symbol $=$ if it is a DC voltage or the symbol $\sim$ if it is an AC voltage (Fig. 1.2).

A current generator or load is represented by two overlapping circles; in one of them will be found the symbol $=$ or $\sim$ depending on whether it is a DC or an AC one.

Examples of waveform for *perfect* generators or loads are shown by full line in Fig. 1.2.

- If they are of voltage type, the voltage across their terminals remains constant or sinusoidal, even if there are sharp variations in current (e.g. current of a square shape is shown).
- For current generator (or load) the current remains constant or sinusoidal, whatever the voltage across its terminals (e.g. voltage of a square shape is shown).

1.2.1.2 Real Generators and Loads

An ideal *voltage* generator or load has an internal impedance equal to zero. A real generator or load will be said to be a voltage one, if its *internal inductance is very low*, i.e.

- if it allows for fast variations of i;
- if these fast variations do not induce important spikes in the voltage waveform u.

An ideal *current* generator or load has an infinite internal impedance. A real generator or load is still referred to as current, if its *internal inductance is*

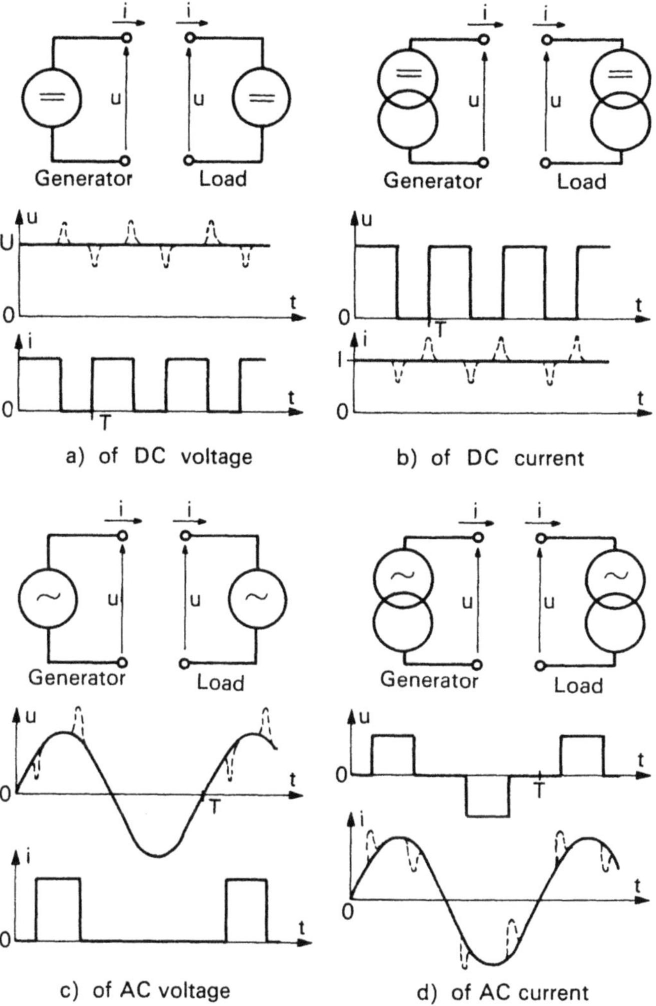

Fig. 1.2

sufficiently high, with respect to the other inductances in the circuit, so that the current flowing through it is hardly affected by fast cyclical voltage variations at its terminals.

In Fig. 1.2, the *broken lines* indicate changes in voltage across the terminals of voltage generators, due to their internal inductance. For current generators, the changes in current caused by the finite value of their inductance are represented in the same way.

1.2.1.3 Operating Conditions

When a converter is used to connect a generator or a load to the rest of the circuit, the basic principles of circuit theory (unicity of the potential in one point, sum of the currents at a node equal to zero) must not be invalidated.

It must be remembered that

- an *inductor*, which opposes current discontinuities through it, acts as a current supply;
- a *capacitor*, which opposes voltage discontinuities across its terminals, acts as a voltage supply.

- A *voltage* generator (or load)
  - cannot be directly connected to another of different value or to a capacitor loaded to another voltage;
  - can be placed in open circuit;
  - must not be short-circuited as its current would then become very high, and theoretically infinite.

- Similarly, a *current* generator or load
  - cannot be directly connected to another of different value or to another inductor with a different current flowing through it;
  - can be short-circuited;
  - must not be placed in open circuit as it would lead to high, and theoretically infinite, overvoltage.

1.2.2 Generator and Load Reversibility. Sources. Examples

1.2.2.1 Instantaneous Reversibility

When the functional characteristics of converters are studied, interest is focused on the mean value of the power supplied by a generator or received by a load. There is said to be reversibility when inversion of energy flow becomes possible.

Instantaneous reversibility has to be considered, when studying the internal operation of converters, in order to determine the characteristics to be attributed to switches and their control laws:

- A generator or a load is *voltage-reversible* if the instantaneous value of the voltage across it can change of polarity.
- It is *current-reversible* if the instantaneous value of the current flowing through it can be reversed.

When the mean power can be reversed, the terms "generator" or "load" will no longer be used; "*source*" shall be used whether the mean power is generated or received.

This use will be extended to cases where the instantaneous power can be reversed. If a generator or a load shows at least one possibility of reversibility, it will be called a "source".

1.2.2.2 DC Source Examples

The mean value of a DC voltage source and the mean current of a DC current source remain constant, for a given operational mode. These values can only be reversed if the operational mode is modified: e.g. current source changing from generator mode to load mode.

However, in steady state, within a chopping period, the current delivered by a voltage source or the voltage across the terminals of a current source can, in certain cases, be reversed.

- *A storage battery* behaves as a generator during the intervals when it generates a current with the same polarity as its internal EMF; it behaves as a load when the current changes its polarity. As its internal inductance is usually negligible, it can be described as a current-reversible DC voltage source.

(At high frequencies, variations in electrochemical balance produce an inductive effect. A capacitor must be connected in parallel with the battery for the latter to remain a voltage source).

- *A DC machine* has an armature circuit similar to an EMF connected in series with a resistance and an inductance. The voltage across the circuit can be reversed as can the current flowing through it. On account of its inductance, the DC machine will be described as a current-reversible and voltage-reversible DC current source.

The effects of imperfections in this "source", i.e. the finite value of its inductance, has been studied at length in Vol. 1 (Chap. 6, § 2; Chap. 7, § 6 and Chap. 8, § 2.6).

1.2.2.3 AC Source Examples

An AC voltage source is necessarily voltage-reversible; an AC current source is necessarily current-reversible.

However, an AC voltage source is not necessarily current-reversible. The waveforms of Fig. 1.2c correspond to a current non-reversible AC voltage source.

Similarly, an AC current source may or may not be voltage-reversible. The waveforms shown in Fig. 1.2d require this reversibility.

- *The industrial AC mains* (possibly via a transformer) is current- and voltage-reversible. Is it a voltage source or a current source?

In studying rectifiers, the AC mains reactances (and the transformer leakage reactances) are not taken into account at first, and the mains is considered as a voltage source. However, further study of line commutation process leads to consider the AC mains as a current source.

In studying AC regulators, the AC mains can be considered as a voltage source since, strictly speaking, no commutation occurs (these converters are in fact operating in a free commutation mode).

If the AC mains supplies a converter operating in forced commutation mode, the mains must be considered as a current source, e.g. in the case of certain special rectifiers operating in Pulse Width Modulation.

- *A synchronous or induction motor* is a current- and voltage-reversible current source.

  It will occasionally be considered as a voltage source as far as concern the fundamental components of the waveform imposed by the converter feeding it (induction motor fed by a current source inverter, self-commutated synchronous motor). But its inductances must be taken into account during commutations.

1.2.3 Improving or Changing the Nature of a Source

The former examples show that it is often difficult to determine whether a source is a voltage source or a current source. If a source differs greatly from an ideal one, it must be improved or "corrected".

If the nature of a source is not adapted to the converter to which it is to be connected, this nature can be changed by adding passive energy-storage elements such as inductors or capacitors.

1.2.3.1 DC Sources

- *Improving a source*

In order to improve a DC *current source*, its inductance must be increased by the addition of another inductance connected in series.

The greater the total inductance L, the stronger the induction EMF, $L\,di/dt$, will oppose current variations under the effect of voltage variations.

- In order to improve a DC *voltage source*, its apparent impedance must be reduced. This can be done by adding a capacitor across its terminals.

  This capacitor prevents discontinuities in voltage u and reduces the effects of variations of i on u, by supplying current when u tends to decrease and by absorbing current when u tends to increase.

- *Changing the nature of a source*

- In order to obtain a DC *current source* from a DC voltage source, a sufficiently high *inductance L* has to be connected in series with the latter (Fig. 1.3a).

  Inductance L charges and discharges under the effect of variations of u. If its value is high enough the variations of i will be negligible.

- To transform a DC current source into a DC *voltage source*, a capacitor with a sufficiently high capacitance C has to be connected across the former (Fig. 1.3b).

  Voltage u variations will be negligible if the energy, $(1/2)\,C\,u^2$, stored in the capacitor is high compared to the energy received or supplied by this voltage source during an operation period.

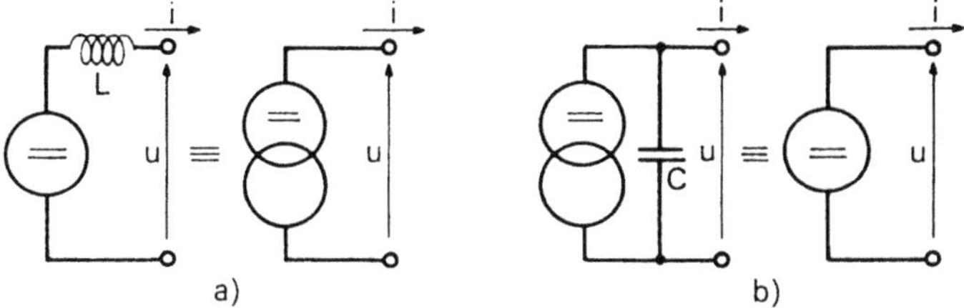

Fig. 1.3

- *Consequences on the reversibilities*

Adding an energy storage element to a DC source does not modify its functional reversibilities. If a high inductance is connected in series with a storage battery, it changes from a current-reversible voltage source to a current-reversible current source.

However, such an addition may lead to an *instantaneous* reversibility. If a capacitor is connected in parallel across a current-non-reversible current source, it may store the energy corresponding to transient reverse currents: the voltage source thus obtained is current reversible concerning the instantaneous value of the current but not concerning its mean value.

- *Example*

A thyristor rectifier bridge, fed by an AC voltage supply of negligible or non-negligible internal impedance, is neither a good DC voltage source nor a good DC current source when considered from its output terminals. The rectified voltage waveform shows important variations and can be influenced by the nature of the load. The rectified current has a waveform closely dependent on the load.

The rectifier can become a DC *current source* by connecting a high inductance in series with it. This source is voltage-reversible but not current-reversible.

In order to obtain a DC *voltage source*, a capacitor must be connected in parallel, in addition to the inductance in series: the LC filter at the rectifier output allows for the DC component of the rectified voltage to pass through and stops the harmonics. The rectifier–filter set is *voltage and current* reversible: although the mean value of the output current may not change polarity, its instantaneous value may reverse during certain intervals; in such a case, the current charges the capacitor.

- *Remarks*

- Depending on the problem under consideration, the study of improving or changing the nature of a source is made easier by considering either the energy or the frequency aspect.

The *energy* aspect consists of deducing the voltage or current variations from the variations of the energy stored in an inductor or a capacitor.

The *frequency* aspect consists of comparing the effects of additional elements on the mean value of the signal and its harmonics.

- In DC–DC converters with an intermediate energy storage, an inductor or a capacitor is used for this purpose. This storage element is connected alternately to the input and the output, in order to ensure energy transfer between input and output.

1.2.3.2 AC Sources

An AC source can similarly be improved or its nature changed by adding an inductor or a capacitor. However the fundamental component of the current or the voltage will induce an excessive voltage drop or current consumption across these elements, if they are of high value. Therefore a *filter* is normally used, enabling a greater difference between the effects produced on the fundamental and these produced on the harmonics to be obtained.

Let us take the *example* of an imperfect sinusoidal voltage source, feeding a converter which takes a square-wave current from it. Such is the case of the single-phase AC mains supplying, directly or via a transformer, a four-thyristor bridge rectifier feeding a highly inductive load.

Due to the internal reactance of the AC source, the voltage u across its terminals is affected by the current i taken by the rectifier (Fig. 1.4a).

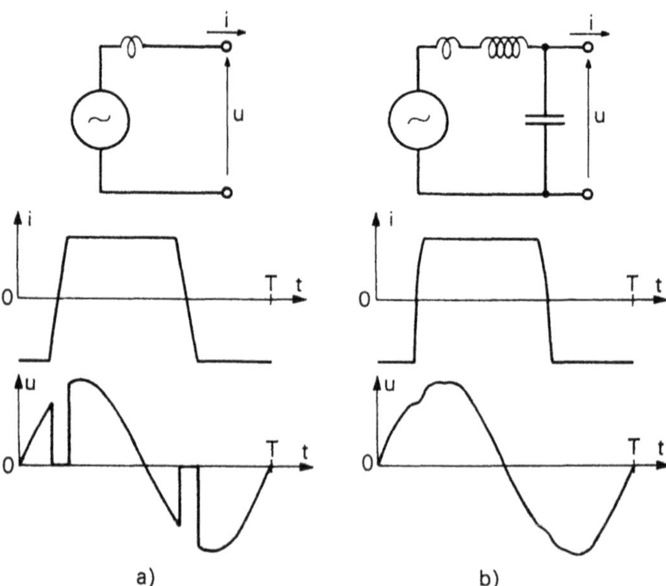

Fig. 1.4

To improve this source a filter made up of a series-connected inductor and a parallel-connected capacitor may be added (Fig. 1.4b). The flow of i has less perturbing effects on the output voltage u of the source improved by the filter.

1.2.3.3 Frequency Increase: Advantages

The preceding short overview of how to improve a source clearly shows the advantages in increasing the operating frequency of converters.

1. The first advantage is the size *reduction of additional elements* required for improving sources or for changing their nature.

   For a DC source, the inductor or capacitor to be added must be such that the energy stored is greater than the energy exchanged with the external circuits during a chopping period. For a given level of power, this energy is all the smaller as the chopping period is short. By increasing the frequency, the additional inductor or capacitor can be reduced or even eliminated.

   For AC sources, increasing the converter switching frequency enables the filter cut-off frequency to be raised and the elements of the latter to be reduced.

   The other two advantages result from the first.

2. For steady-state operation, voltage u of a DC voltage source or current i of a DC current source must show a low ripple around their mean value. However, during transients, the mean value of u or i needs to vary rapidly. The higher the chopping frequency, the greater the possible reduction of energy stored in L or C, the greater the possible reduction of the response time of the generator – converter – load set.

   Similarly, with AC sources the response time of a system including a filter is all the shorter as the cut-off frequency is high.

3. Reduction of inductors and capacitors size (and consequently of the energy stored in these elements) is beneficial when *short-circuit occurs.* Indeed, this energy is then dissipated, either wholly or partly, in a semiconductor switch which must be rated in accordingly.

1.3 Time Intervals

The various time intervals to be considered in the study of a converter are normally very different. This makes that study much easier.

1.3.1 Example

- Consider, for example, a step-down chopper feeding a DC motor, the speed of which is to be varied.

The chopper, fed by a constant voltage (U) source, consists of a controlled turn-on and turn-off switch, a transistor T for instance, and a diode D. Its output voltage u' is applied to the motor which takes a current i' (Fig. 1.5).

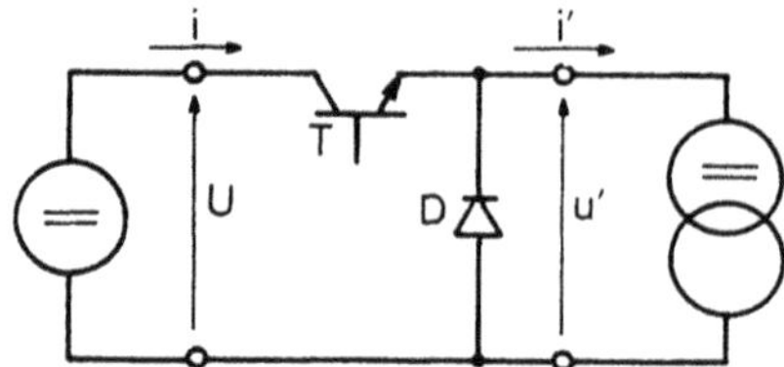

Fig. 1.5

By assuming ideal semiconductor devices (zero voltage drop in the ON state, negligible commutations duration), and denoting as f the chopper switching frequency and as α $(1/f)$ the on-time duration of T in each cycle,

when T is conducting, $u' = U$, $i' = i$;

when T is blocked, i' flows through D, $u' = 0$.

The mean value of u', equal to $\alpha\, U$, can be varied from zero to U by acting on α.

(Current i' in the load flows alternatively through the diode and through the transistor. These switchings – from a diode to a transistor, or from a transistor to a diode type – are very frequently met in force commutated converters. In the following chapter, the step-down chopper topology will thus be frequently used in the study of semiconductor devices switching characteristics).

- Let us examine an increase in the speed of the motor if the mean value of current i' is kept constant.
  - If we concentrate on the speed increase, interest will be on the evolution of u'_{mean} concerning the converter output. Variations in u' (and in i') at the chopper operating frequency f may be neglected (Fig. 1.6a).
  - In studying chopper operation (i.e. relationships set up between its input and output), the variations of u' (and i') during a switching period $1/f$ are considered. Eventually the consequences of supply and load imperfections may be assessed (Fig. 1.6b).
  - If the commutations are to be studied (Fig. 1.6c), their duration can no longer be overlooked. But voltage u' increases and decreases in sufficiently short periods for variations in current i' during these periods to be overlooked.

There are clear differences in the time-scales to be adopted for the three graphs in Fig. 1.6: the time intervals to be taken in consideration are of the order of

a second, for the first graph,

a millisecond, for the second,

a microsecond, for the third.

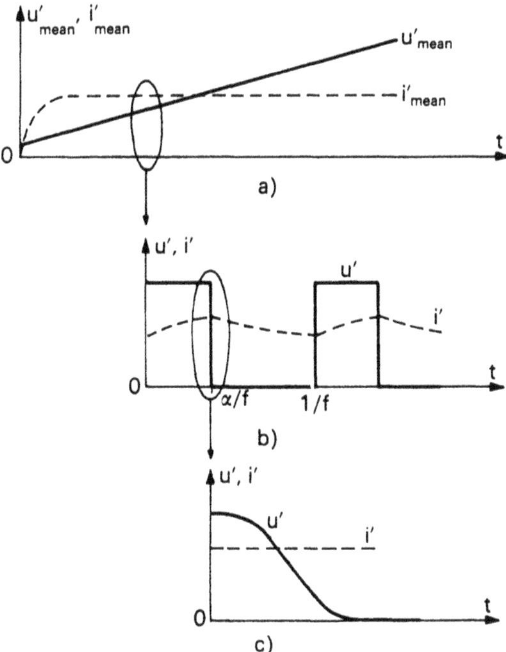

Fig. 1.6

1.3.2 Differences in Time Intervals: Consequences

The study of converters is made greatly easier by the considerable difference normally observed between different time intervals.

- In studying the *operation*, it is assumed that the converter is in steady state, i.e. at the end of a switching period some voltages or currents have the same value as at the beginning. Furthermore, the switching times of the semiconductor devices may be ignored.
  - Firstly, the generator and the load are considered as perfect.
  - Secondly, the influence of imperfections of the generator and/or load is examined.

- In studying *commutations*, some voltages and currents are assumed to remain constant during these intervals. For example, this is the case of a voltage across a voltage source, of a current flowing through a current source, even if the sources are AC or imperfect.

• *Remarks*

- These simplifying assumptions – used for a general study of choppers and inverters – are obviously not always valid. In some cases, the switching period is too short or the switching times too long to allow for the second to be ignored in comparison with the first. Moreover, if the sources cannot be idealized, computer simulation of the system must be used.

- In thyristor converters, the decay to zero of the current flowing in the semiconductor devices followed by the voltage rise across them is imposed by the circuit which carries out the commutations (supply or load in natural commutation, auxiliary commutation circuit in forced commutation). Therefore, even in studying commutations, the thyristor can still be considered perfect except that a minimum reverse bias interval is needed.

In Volume 1, this method was used to study rectifier commutations and will be adopted for the study of forced commutation of thyristors in Chap. 5 of the present work.

Chapter 2

Switching Power Semiconductor Devices

Power electronics use semiconductor devices operating in switching mode, i.e. in an on–off control mode. There are mainly three types of devices:

diodes,
bipolar and field effect transistors,
thyristors or GTO (gate turn-off) thyristors.

For converters in which commutations play an important role, the semiconductor switches behaviour must be known during transitions from ON state to OFF state, or vice versa.

We do not propose to go into the details of semiconductor physics or their manufacturing technology, but to limit our explanation to the minimum necessary for an understanding of the characteristics of devices.

2.1 Diodes

2.1.1 Remarks on the P–N Semiconductor Junction

- In order to exploit the controlled conducting properties of semiconductors, highly elaborate and constantly improved manufacturing processes have been developed.

  Silicon – used for virtually all power semiconductor devices – must be very pure, monocrystalline and oriented in a certain direction. It must be dislocation-free and carefully brased on compatible electrodes.

- In such a pure silicon monocrystal, each atom is next to four others with which it shares its four peripheral electrons, thus forming a highly stable structure.

  At normal temperatures, thermal ionization sets free a number of electrons, leaving holes in the crystalline lattice. These charge carriers have an extremely low density: the resistivity of pure silicon is very high at normal temperature.

2.1.1.1 Doping, Junction, Space Charge

- The basic crystal is doped in order to increase the number of free carriers and thus to reduce silicon resistivity.

  • *In* N-*type* regions or layers, "donor" atoms (phosphorous, arsenic, antimony, . . .) are introduced. These have 5 peripheral electrons. This injection creates fixed positive ions and extra free *electrons* in the lattice.

- *In* P-*type* regions or layers, the "acceptor" atoms or impurities (boron, aluminium, gallium, . . .) have only 3 peripheral electrons. An absence of electron or hole corresponds to each atom of impurity. We thus have an equal number of fixed negative ions and extra *holes* which move about freely.

Several doping processes exist: alloying, diffusion from surfaces, introduction during crystallisation, epitaxy, . . .

Carriers due to doping and those of the same polarity due to thermal ionization are called "*majority*" carriers in their own region (electrons in N, holes in P). Carriers of the opposite polarity, generated by thermal ionization or due to the diffusion of majority carriers in a neighbouring layer, are called "*minority*" carriers.

At any moment, electrons are generated by thermal ionization and recombine with holes. Probability calculation enables a mean electron lifetime to be defined, i.e. a mean time during which an electron remains free and which lasts several microseconds. The electron–hole recombinations are all the more numerous as the concentration of charge carriers is high.

- The transition between two layers – P and N – of the same crystal is called a *junction*. The region in which conductibility passes from the P-type to the N-type is very thin (a few microns)

Immediately next to the junction, electrons and holes recombine with each other, so that, in equilibrium state, only negative ions remain in the P side of the junction and positive ions in the N side. An electric field going from the positive charges toward the negative charges appears, and prevents electrons of the N region from going to the P region and holes from the P region from going towards the N region.

The zone where the internal electric field appears is called the *space charge layer*, or transition layer or depletion layer.

2.1.1.2 Forward-Biased Junction

If an external electric field going from the P region to the N region is applied to a P–N junction, it opposes the internal field.

- As long as the external field remains smaller than the internal one, the resulting field at the junction level remains directed from *N* towards P. The space charge region also remains, but of reduced width. The external field provides some *majority carriers* of each region with enough energy to cross the junction despite the resulting field. A small current flowing from P towards N appears, its value increasing with the value of the external field.
- When the latter becomes equal to the internal field, the space charge region disappears: the majority carriers easily cross the junction. If the external field

is increased once more, there is a sharp increase in the current flowing from P to N.

2.1.1.3 Reverse-Biased Junction

If an external electric field of the same direction as the internal field is applied, the space charge region is widened.

To *minority carriers* of each layer, the electric field applies a force which make them cross the junction. They create a very small current going from the N-side to the P-side. This reverse current depends to a large extent on the temperature since most minority carriers are generated by thermal ionization.

Two phenomena may occur when the reverse bias is increased:

- The *Zener breakdown* appears when the external field is sufficiently strong to free the peripheral electrons of silicon atoms. As these atoms are all identical, the phenomenon appears suddenly, creating a very large number of electron–hole pairs. This sudden generation of numerous minority carriers brings a rapid increase in the current crossing the junction in the direction N–P.
- The *avalanche breakdown* appears when the external field provides a free electron with enough energy to liberate a peripheral electron from a silicon atom with which it collides (impact ionization). This produces two free electrons which, in turn, liberate two others. The phenomenon spreads rapidly throughout the whole crystal and produces a rapid increase in the reverse current. The avalanche breakdown appears for a decreasing field when doping increases.

2.1.2 Diode Steady-State Characteristics

The P–N junction diode is made up of a two-layer silicon wafer. The P-type is connected to anode A, and the N-type to cathode K. Figure 2.1 shows its circuit symbol and indicates the conventional signs for current i and voltage v.

2.1.2.1 ON State

- When voltage v is positive, current i – called forward current – increases rapidly with v. The values of i and v are theoretically linked by:

$$i = I_S(e^{qv/kT} - 1) \tag{2.1}$$

with: I_S, saturation current,
q, electron charge,

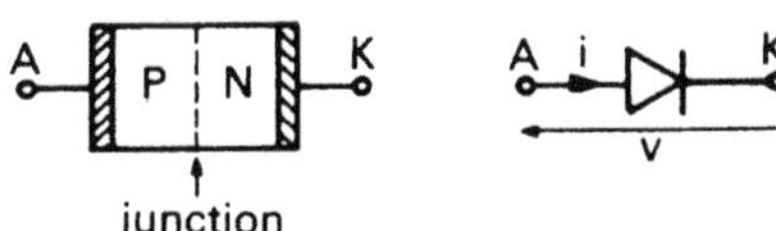

Fig. 2.1

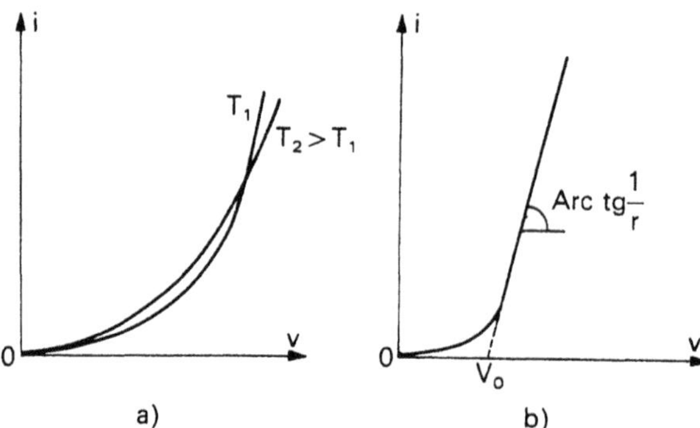

Fig. 2.2

k, Boltzman constant,
T, absolute temperature on Kelvin scale.

In fact, the current deviates from this theoretical value when v increases; but, at the beginning of the characteristics, it has effectively an exponential form (Fig. 2.2).

If the temperature rises, thermal ionization produces more carriers and the saturation current I_s increases, but the coefficient q/kT is smaller. Figure 2.2a shows the beginning of the curve $i(v)$ for two values of temperature T.

- In the case of the current, a diode can be characterised by the *mean value of the forward current* it can sustain. This value is expressed as

$$I_0 \quad \text{or} \quad I_F \quad \text{or } I_{FAV} \text{ (averaged forward current).}$$

The following are also indicated:

- maximum admissible value for a repetitive current peak

  I_{FM} or I_{FPM} (peak forward current),
- maximum admissible value for a non-repetitive current peak

  I_{FSM} (surge non-repetitive forward current).

For power diodes the direct voltage drop for the nominal value of the current I_F is normally between 1.3 and 1.5 V.

These characteristics – and all the others – are defined for a given junction temperature T_j and case temperature T_{case}.

2.1.2.2 On-State Losses

For the normal values of the forward current, the diode voltage drop can be shown by a straight line segment characterized by a threshold voltage V_0 and a resistance r (Fig. 2.2b):

$$v = V_0 + ri\,.$$

For given values of u and i, the instantaneous power in the diode is

$$p = vi = V_0 i + Ri^2.$$

Taking the mean value over a cycle T, the ON-state losses are obtained:

$$P = \frac{1}{T}\int_0^T p\,\mathrm{d}t = \frac{1}{T}\left[V_0 \int_0^T i\,\mathrm{d}t + r\int_0^T i^2\,\mathrm{d}t\right]$$

$$P = V_0 I_{moy} + rI_{rms}^2. \tag{2.2}$$

When the temperature rises, V_0 decreases but r increases.

2.1.2.3 The Surge Factor: Remarks

In order to calculate protections, a surge factor $\mathbf{I}^2 t$ is defined. This integral quantity is defined in the case of a sinusoidal half-wave of the forward current, of peak value I_{FSM} and of duration T, normally taken equal to 10 ms (Fig. 2.3).

In these conditions

$$I^2 t = \int_0^T \left(I_{FSM} \sin\frac{\pi}{T} t\right)^2 \mathrm{d}t = I_{FSM}^2 \int_0^T \frac{1}{2}\left(1 - \cos\frac{2\pi}{T} t\right)\mathrm{d}t$$

$$I^2 t = \frac{1}{2} I_{FSM}^2 T.$$

When $I^2 t$ or I_{FSM} and the corresponding value of T are given, the maximum surge non-repetitive forward current can be deduced for another form of an accidental current spike.

2.1.2.4 OFF State

- When voltage v is negative, the junction is reverse biased: the low value of current i – now negative – is once more given theoretically by Eq. (2.1):

$$i = I_S(e^{qv/kT} - 1), \quad \text{with } v < 0.$$

As soon as v deviates noticeably from zero, the exponential term becomes negligible and the reverse current becomes equal to the reverse saturation current I_S due to minority carriers. In fact, it rises slightly with the voltage (Fig. 2.4).

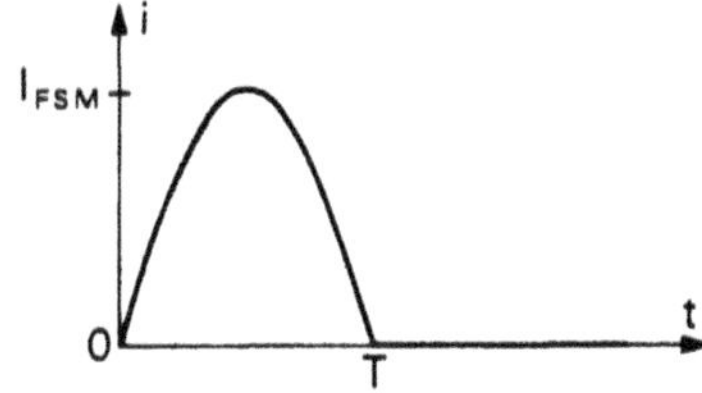

Fig. 2.3

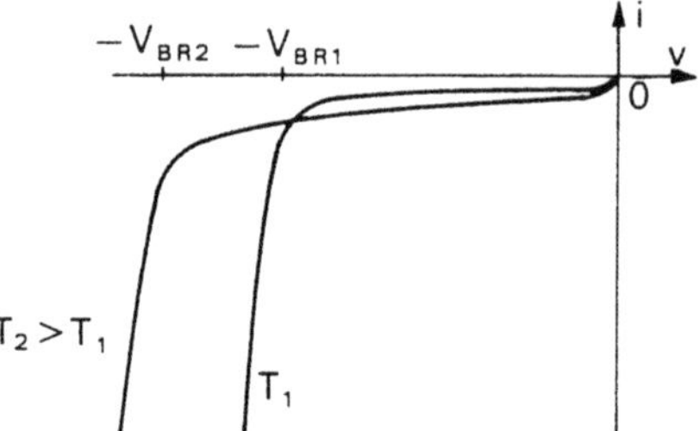

Fig. 2.4

Only if the reverse voltage becomes too high does it reach the avalanche voltage value V_{BR} (breakdown voltage), which would produce an abnormally high reverse current.

In the avalanche region, the saturation current is multiplied by a factor M, such that

$$M = \frac{1}{1 - |v/V_{BR}|^n} \tag{2.3}$$

where n is a coefficient depending on doping and temperature.

When the temperature rises, the reverse current increases since the number of minority carriers generated by thermal ionization is also increased. On the other hand, the mean lifetime of the carriers decreases and the avalanche phenomenon is produced for a higher reverse voltage V_{BR}, as shown in Fig. 2.4.

- In the case of the voltage, a diode is characterized by the instantaneous value of the *repetitive peak reverse voltage* V_{RRM} that it is capable of withstanding.

The reverse current, whose value is normally given for V_{RRM}, is very low. The losses in the OFF state are thus often negligible compared to those in the ON state.

For example,

$$\text{for } i = 100\,\text{A}, \quad v = 1.5\,\text{V}; \quad \text{vi} = 150\,\text{W},$$

$$\text{for } v = -500\,\text{V}, \quad i = -10\,\text{mA}; \quad \text{thus} \quad vi = 5\,\text{W}.$$

2.1.3 Switching Characteristics

The turn-on switching corresponds to the rapid transition from the blocking state to the ON state, and the turn-off switching to the reverse transition. Such rapid changes depend on the circuit in which the diode is embedded.

2.1.3.1 Turn-on Transient

• "*Voltage generator*" *supply*

The voltage e of a voltage generator changes sharply from a negative value $-E'$ to a positive value E which is higher than the diode threshold voltage. Voltage e is applied to the diode across resistance R (Fig. 2.5a).

For the negative value of e, the diode is reverse-biased; current i is negative and very small (equal to the saturation current); voltage v, equal to $e - Ri$, is virtually equal to $-E'$.

When e becomes positive, the majority carriers of both layers gradually penetrate the space charge layer. They first reduce and then eliminate its width. The voltage and current are finally stabilised at the intersection of the load line, $v = E - Ri$, with the static characteristic of the diode.

When voltage v rises, the diode behaves in the same way as the parallel combination of a resistance r and a capacitor C (Fig. 2.5b).

- Resistance r corresponds to that of the semiconductor; it varies with i, as the increase in the number of carriers on either side of the junction reduces the resistivity of both layers.
- Capacitor C corresponds to the charge due to the majority carriers which the external field has brought on either side of the junction. These make up the diffusion charge.

Figure 2.5c shows how voltage v and current i vary during turn-on.

• "*Current generator*" *supply*

In most cases, the current rise speed is imposed by the external circuit. Current i goes from zero to I with a constant rate of rise $\mathrm{d}i/\mathrm{d}t$ (Fig. 2.6).

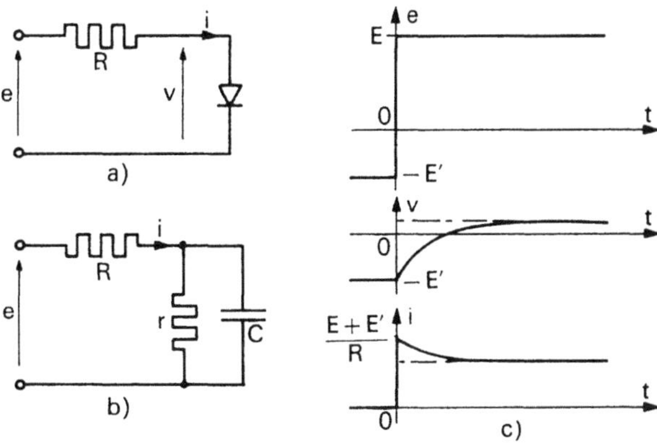

Fig. 2.5

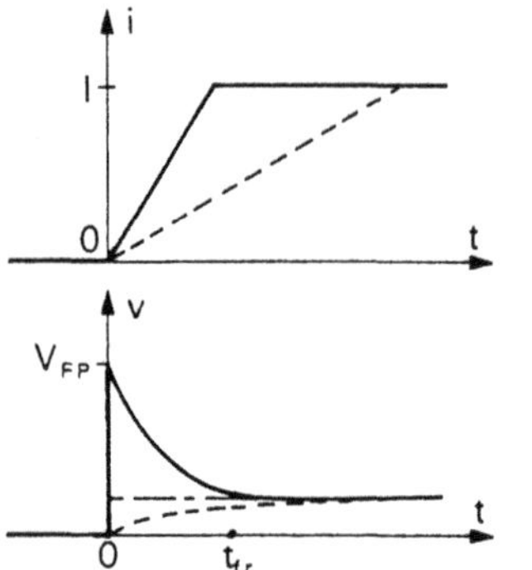

Fig. 2.6

An equivalent circuit of the diode, similar to that presented in the previous paragraph, can be used:

- If di/dt is high, the number of diffusion charges increases rapidly and brings about a rapid rise in the forward voltage, before the latter falls gradually back to the value which corresponds to the steady state (full lines on the diagram).
- If di/dt is low, the capacitive effect of the junction diminishes (broken lines on the diagram).

For the turn-on transient, a diode behaviour can be characterized by:

- the peak forward voltage V_{FP} or V_{FM}, which denotes the maximum value of v during turn-on;
- the forward recovery time t_{fr}, which is the time required for the turn-on voltage surge to disappear.

These parameters depend on applied current I, on its time rate of change di/dt and on the temperature.

Despite the peak forward voltage, the turn-on switching losses are generally negligible compared to the turn-off switching losses.

2.1.3.2 Turn-off Transient

- "*Voltage generator*" *supply*

Voltage e, applied to the diode across resistance R, changes sharply from a positive or zero value to a negative value $-E'$ (Fig. 2.7).

In a first period of time, called storage time t_s, the resulting field extracts the carriers located in the space charge layer, the result is a reverse current; its value is limited by R. But numerous stored charges remain on either side of the junction, imposing on voltage v a value close to that which it had during ON state.

When most of the carriers have been removed from the space charge layer, the diode recovers its blocking capability. Carriers still remaining in excess in this region are gradually eliminated; the current falls progressively back to zero and voltage v tends towards $-E'$.

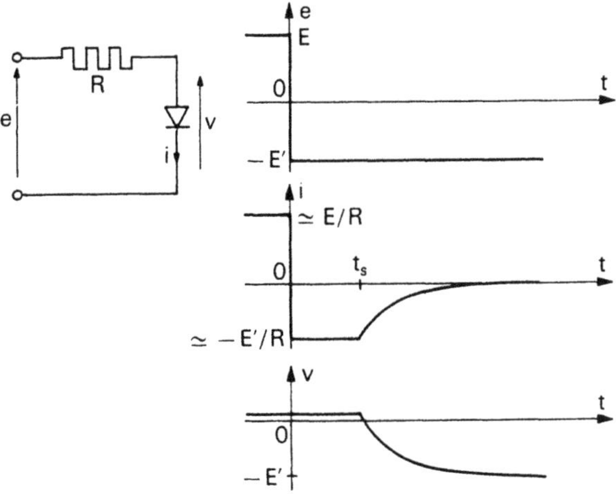

Fig. 2.7

● *"Current generator" supply*

In most converters, there is a limit to the time rate of change of current decrease. Turn-off switching is similar to that which occurs in the circuit of Fig. 2.8a:

current i being set at a value close to $+E/R$, for $t = 0$, the switch K applies voltage $-E'$ to the diode across inductance l.

– We find again the two types of interval seen in the previous case:

- The diode must first recover its blocking capability by sweeping free carriers out of space charge layer. Some of these carriers disappear naturally by recombining. However if there is a rapid current decrease, such recombinations are few in number and the *peak reverse current* I_{RM} reaches a value close to that of the forward current during ON state. During this interval, voltage v remains slightly positive:

$$\frac{di}{dt} = (-E' - v)/l \simeq -E'/l .$$

- Next, when most of the carriers have been extracted on either side of the junction, current i becomes close to zero and voltage v to $-E'$. But the fast decrease in reverse current leads to a *peak reverse voltage*, since

$$v' = -E' - l\frac{di}{dt}, \quad \text{with } \frac{di}{dt} \text{ positive} .$$

This peak reverse voltage is all the higher as the circuit is more inductive and the return to zero of i takes place more quickly.

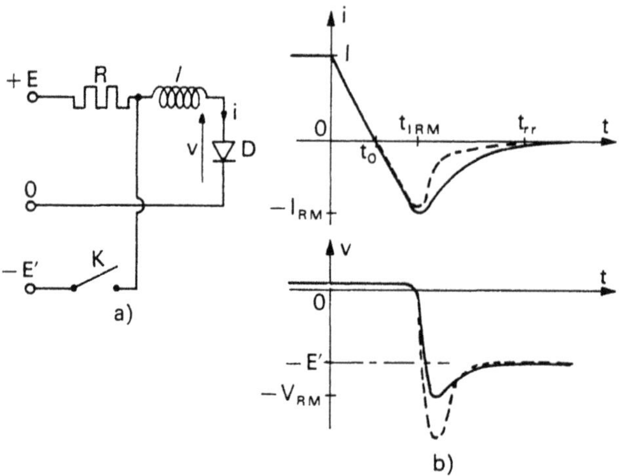

Fig. 2.8

The full lines in Fig. 2.8b show the variations of i and v, in the case of a soft-recovery diode. The broken lines correspond to a snap-off diode.

The turn-off switching can be characterized by:

- the *reverse recovery time* t_{rr}; this is the time needed for the current to turn back to zero; it is timed from t_0;
- the *reverse recovery charge* Q_R; this is the charge corresponding to the existence of a reverse current during the period (t_0, t_{rr}).

In order to give the precise value of t_{rr} and Q_R, the following data must be indicated:

- the forward current intensity I before commutation,
- the time rate of change of the current decrease di/dt,
- the reverse voltage applied during turn-off,
- the junction temperature.

The waveforms in Fig. 2.9 show the influence of time rate of change di/dt and value I of the switched current on the variations of t_{rr} and Q_R.

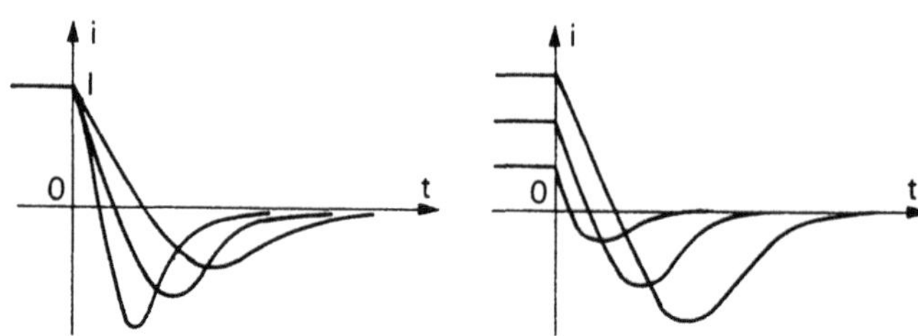

Fig. 2.9

2.1.3.3 Switching Losses

Losses in the diode are usually calculated using idealized waveforms as in Fig. 2.10, i.e. with linear variations of the current.

- From t_0 to t_{IRM}, i goes from zero to $-I_{RM}$:

$$v = 0 = -E' - \frac{l\,di}{dt} \quad \text{gives:} \quad \frac{di}{dt} = \frac{-I_{RM}}{t_{I_{RM}} - t_0} = -\frac{E'}{l}.$$

It can be deduced that

$$t_{I_{RM}} - t_0 = \frac{lI_{RM}}{E'}$$

and the charge Q_1 stored in the diode during this period is given by

$$Q_1 = \frac{I_{RM}}{2}(t_{I_{RM}} - t_0) = \frac{lI_{RM}^2}{2E'}.$$

- From $t_{I_{RM}}$ to t_{rr}, i goes from $-I_{RM}$ to zero:

$$v = -V_{RM} = -E' - \frac{l\,di}{dt} \quad \text{gives:} \quad -V_{RM} = -E' - l\frac{I_{RM}}{t_{rr} - t_{I_{RM}}}$$

and the charge swept out during this period is

$$Q_2 = \frac{I_{RM}}{2}(t_{rr} - t_{I_{RM}}).$$

- Energy dissipated in the diode at each commutation is

$$\begin{aligned} W &= \int_{t_0}^{t_{rr}} vi\,dt = -V_{RM}\int_{t_{I_{RM}}}^{t_{rr}} \left(-I_{RM} + I_{RM}\frac{t - t_{I_{RM}}}{t_{rr} - t_{I_{RM}}}\right)dt \\ &= \left(-E' - l\frac{I_{RM}}{t_{rr} - t_{I_{RM}}}\right)\frac{1}{2}[-I_{RM}(t_{rr} - t_{I_{RM}})] \\ &= E'\left|\frac{I_{RM}}{2}(t_{rr} - t_{I_{RM}})\right| + E'\left|\frac{lI_{RM}^2}{2E'}\right| = E'(Q_1 + Q_2). \end{aligned}$$

Since $Q_1 + Q_2$ equals the recovery charge Q_R,

$$W = E'Q_R. \tag{2.4}$$

It can be seen that it is the recovery charge which enables the losses to be calculated.

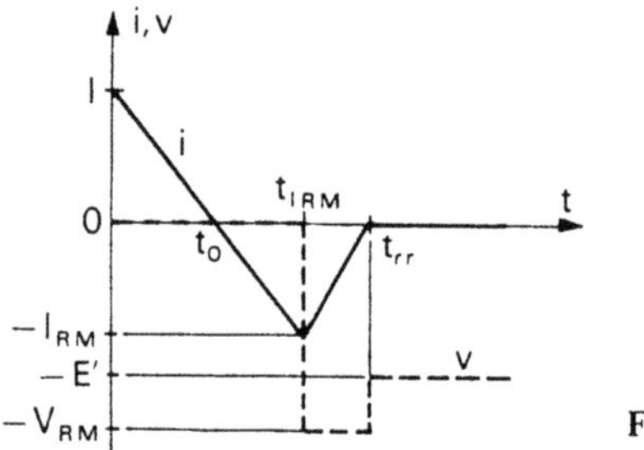

Fig. 2.10

2.1.4 Notes on Special Diodes

A diode's performances can be judged in three main ways:
- if it can block a higher reverse voltage;
- if its forward-voltage drop in ON state is lower;
- if its switching time is shorter.

As all three conditions cannot be attained at one and same time, the choice of a diode must depend on its intended use.

2.1.4.1 High-Voltage Diodes

In order to increase the reverse voltage which can be sustained, the P and N layers should be thick and slightly doped. This is to reduce the probability of electron collision. However, a decrease in doping increases the resistivity; for a given section of wafer, a decrease in admissible forward current results.

Double diffused diodes are used. A thick and weakly doped central layer (P^- or N^-) enables high voltage to be sustained; the two more strongly doped end-regions (P^+ or N^+) enable a strong current to flow. The central layer can be of P- or N-type. In this way, $P^+P^-N^+$ or $P^+N^-N^+$ diodes are produced (Fig. 2.11).

2.1.4.2 Controlled Avalanche Diodes

Some diodes are manufactured to be able to withstand the avalanche effect without damage, for short but repeated periods. In characterizing them, not only the breakdown repetitive voltage V_{BRR} but also the reverse current corresponding to the maximum power dissipation must be indicated.

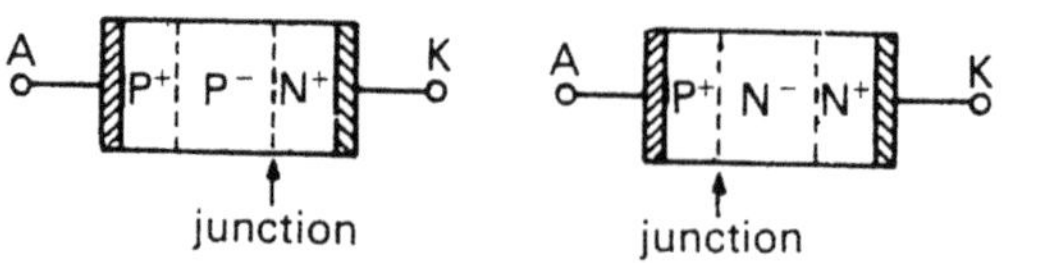

Fig. 2.11

2.1.4.3 Fast Diodes

Diodes which have to operate at high frequency must have a low recovery charge Q_R since commutation losses are proportional to the latter.

To reduce Q_R, the lifetime of minority carriers must also be reduced, in order to increase the number of recombinations during the forward-current decrease. For this purpose, gold doping is used, which has the disadvantage of increasing resistivity and, as a result, the forward-voltage drop in the ON state.

Q_R can also be reduced by creating a thinner central layer which the carriers can leave more quickly. But this reduces the maximum reverse voltage which the diode can sustain. Note that the thinness of the central layer reduces the peak forward voltage V_{FP} and the forward-recovery time t_{fr}.

2.1.4.4 Low-ON-State-Voltage-Drop Diodes

For the *P–N* junction diodes, the ON-state voltage drop is linked to the resistivity of the semiconductor, which decreases by using epitaxy to make the junction and by using a precise amount of doping inside each layer.

A voltage drop of between 0.8 and 1 V can be obtained for the nominal forward current in this way.

Schottky diodes use the properties of metal–semiconductor contact where there appears a potential barrier as in a P–N junction. But the conduction threshold is substantially lower than in a P–N junction. Moreover, the current is only due to the majority carriers; this leads to extremely short switching times, since there are no minority carriers. However, the use of these diodes remains limited to medium or small currents and to low voltages (75 V).

2.2 Bipolar Transistors

Power transistors are transistors which are specifically adapted for use in static converters, i.e. to a switching operating mode.

2.2.1 Principle: OFF State, ON State

Power transistors are normally of the NPN type. Fig. 2.12 gives a diagrammatic representation of them as well as the symbols used.

Collector C and emitter E form the terminals of the controlled switch. Base B is the control electrode which enables

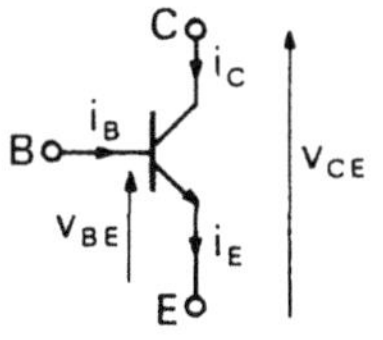

Fig. 2.12

- the transistor to be made saturated or conducting,
- the transistor to be blocked,

with low power consumption

Power transistors are often obtained by triple diffusion. The N layer, which forms the collector, is the thickest, and comprises an N^- region, the size of which increases with the value of the forward voltage which the transistor has to block (see Fig. 2.13).

2.2.1.1 The Transistor Effect

The effect used in transistors appears when the emitter–base junction is forward-biased ($v_{BE} > 0$) and the collector–base junction reverse-biased. These biases can be obtained using the simple circuit shown in Fig. 2.13.

- When voltage v_{BE} is positive the electrons in the emitter, where they are majority carriers, are attracted towards the base, where they become minority carriers. Some electrons which have come into the base recombine with the majority holes there. The electrons which fail to recombine, easily cross the collector–base junction which is reverse-biased and reach the collector. This flow of electrons from emitter to collector, across the base which attracted them, is known as the transistor effect.
- Each time that a base hole recombines with an electron coming from the emitter, the voltage supply E_B provides a new positive charge to keep the voltage v_{BE} constant. This injection of charges forms the base current i_B: its strength is linked to the flow of electrons going from the emitter to the collector and which form current i_C.

  If the fraction of electrons coming from the emitter and reaching the collector is denoted by α, we can write:

$$i_C = \alpha i_E; \quad i_B = (1 - \alpha) i_E .$$

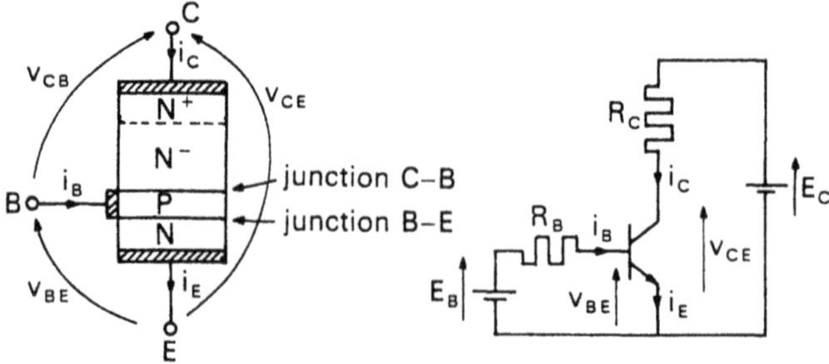

Fig. 2.13

Thus:

$$i_C = \frac{\alpha}{1-\alpha} i_B = \beta i_B \tag{2.5}$$

where β denotes the transistor current gain: it is often written as h_{FE} (F for forward, *E* for the common emitter configuration and *h* for hybrid, on account on the hybrid parameters used in the study of small signal amplification mode of operation).

• *Remarks*

– When v_{BE} and i_E are zero, the reverse-biased collector–base junction lets the saturation current I_{CBO} flow from C towards B. When v_{BE} and i_E are positive, this current must be added to i_C and to $-i_B$, yielding

$$i_C = \alpha i_E + I_{CBO}; \quad i_B = (1-\alpha) i_E - I_{CBO}$$

The equation linking i_C to i_B becomes

$$i_C = \frac{\alpha}{1-\alpha} i_B + \frac{1}{1-\alpha} I_{CBO}$$

$$i_C = \beta i_B + (1+\beta) I_{CBO}. \tag{2.6}$$

The saturation current I_{CBO} is produced by thermal ionization and varies widely with the temperature.

– In order to reduce the losses of the drive circuit, the base current must be reduced, by limiting the number of recombinations in this layer. This is achieved by reducing the thickness of layer P and by a moderating of its doping level.

2.2.1.2 OFF State

At positive voltage v_{CE}, the transistor is in OFF state when *the base current is zero or negative*. Three cases may be considered:

• *Base in open circuit* (Fig. 2.14a)

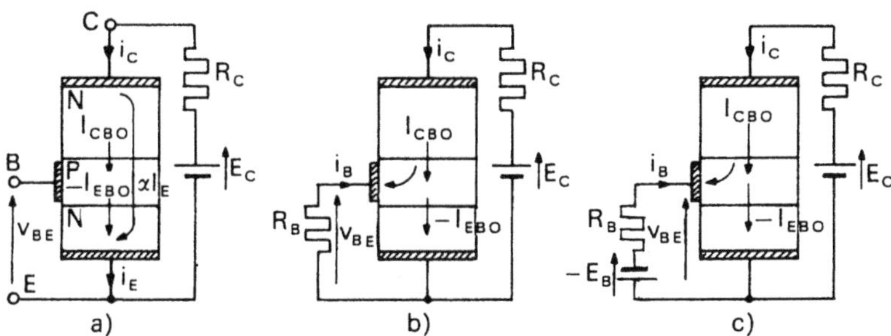

Fig. 2.14

Since i_B is zero, the current entering by the collector leaves by the emitter. The B–E junction is lightly forward-biased and the minority-carrier current across this junction allows for compensation of recombinations between holes and electrons in the base. Equation (2.6) gives:

$$i_C = i_E = (1 + \beta) I_{CBO} .$$

- *Resistance between base and emitter* (Fig. 2.14b).

Voltage v_{BE} and current i_B are then linked by:

$$v_{BE} + R_B i_B = 0 ;$$

they cannot be of the same polarity. If resistance R_B is not too high, v_{BE} is positive and lower than the threshold voltage of junction B–E; as current i_B is negative, it corresponds to the injection of electrons into the base by voltage supply E_C across resistance R_B. This gives

$$i_C = i_E - i_B > i_E , \quad \text{since } i_B < 0 .$$

- *Reverse bias of junction B–E* (Fig. 2.14c).

If junction B–E is reverse-biased, there is no longer any transistor effect. Junction C–B is crossed by the reverse current I_{CBO}, junction E–B by the reverse current I_{EBO}. The negative base current corresponds to electrons sent into the base by voltage $-E_B$ to keep v_{BE} constant:

$$i_C = I_{CBO} ; \quad i_E = -I_{EBO} < I_{CBO} ; \quad i_B = i_E - i_C < 0 .$$

In all three cases, the forward cut-off current i_C is very small; the difference between the three blocking modes come into play especially during the turn-off (see § 2.2.5).

2.2.1.3 Saturation

To enable the transistor to act as a closed switch, it is brought to saturation by means of a *positive base current*. This must be higher than the precise level of current required to compensate for the recombinations between holes and electrons in the base. Three saturation states can be recognised.

- When an excessive number of positive charges are injected into the base, these are diffused in the lightly N^- doped layer of the collector. This is equivalent to a P-type doping and to an *extension of the base thickness*, with a corresponding reduction of the N^- region. To the reduction of this lightly doped – and thus highly resistive – layer, there is a corresponding decrease in the voltage drop across the collector: for a given i_C, v_{CE} decreases.

  The extension of the base region increases the transit time of the minority carriers in this region and, as result, the number of recombinations; current gain β decreases: the transistor is in *quasi-saturation*.

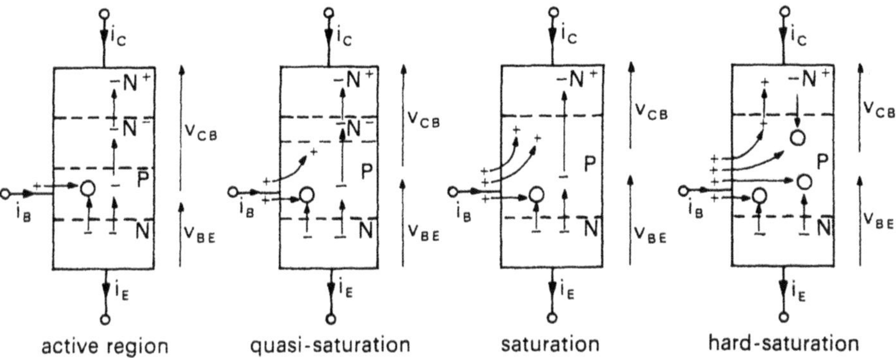

Fig. 2.15

- For a given collector current i_C, the N^- region is further reduced if i_B is further increased. The transistor is *in saturation* when this region has completely changed to the P-type. The highly doped N^+ layer of the collector prevents the base region from extending further.

  The increase in the number of recombinations, caused by the increased thickness of the base, further decreases the current gain. Voltage v_{CB} becomes slightly negative.

- If voltage v_{CB} becomes so negative that it reaches the threshold level of the base–collector junction, the polarity of the collector current is reversed and the transistor is in *hard-saturation*. This operating mode is to be avoided.

Figure 2.15 shows the different types of charge flow inside the crystal and the recombinations (shown by small circles) for a transistor operating

- in the active region,
- in quasi-saturation,
- in saturation,
- in hard-saturation.

2.2.2 Steady-State Characteristics

2.2.2.1 Output Characteristics

The output characteristics give the current i_C flowing through the "switch" as a function of voltage v_{CE} across it, for different values of drive current i_B.

In the active region (Fig. 2.16), segments of straight lines correspond to Eq. (2.5):

$$i_C = \beta i_B + (1 + \beta) I_{CBO} \,.$$

These lines are not quite horizontal as β and I_{CBO} increase slightly with v_{CE}.

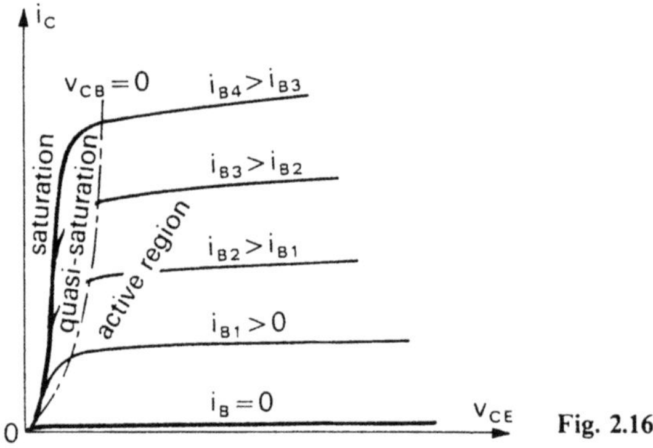

Fig. 2.16

When v_{CB} becomes negative, i.e. when voltage v_{CE} becomes lower than voltage v_{BE} across the forward-biased base–emitter junction, a quasi-saturation zone is reached: the decrease in current gain is shown by a decrease in i_C for a given value of i_B. When saturation is reached, i_C no longer depends on i_B: all the characteristics are expressed by a curve going through the point ($v_{CE} = 0$, $i_C = 0$).

In power electronics, transistors operate in an on–off control mode:

- during the OFF-state intervals, i_B is zero (and i_C negligible);
- during the ON-state intervals, it is used in saturation or quasi-saturation (with very low v_{CE}).

The operating point will only move outside the thicker part of the curves in Fig. 2.16 during commutations.

2.2.2.2 Drive Characteristics, Current Limitation

- As Fig. 2.17 shows, the *drive characteristics* $v_{CE} = f(i_B)$, for given values of i_C, can be deduced from the output characteristics. For a given value of i_C (Fig. 2.18) v_{CE} decreases as i_B increases, and moves towards a minimal value corresponding to the intersection of the straight line at constant i_C with the point of the output characteristics for which the transistor is in saturation. This minimal value of v_{CE} increases with i_C and is denoted $v_{CE\,sat}$.

By imposing i_C and v_{CE}, the value of current i_B to be supplied to the base is obtained. It can be seen that, when v_{CE} reaches $v_{CE\,sat}$, a further increase of i_B is without interest. On the contrary, the amount of excess carriers in the base will be greater, thus increasing the storage time. Normally, the characteristics are limited to i_B equal to i_C.

- Concerning the current, the following values must be indicated in order to characterize a power transistor:
  - the value of the collector current I_C which it allows to flow in *continuous steady state*. The suffix sat is occasionally added, indicating that this

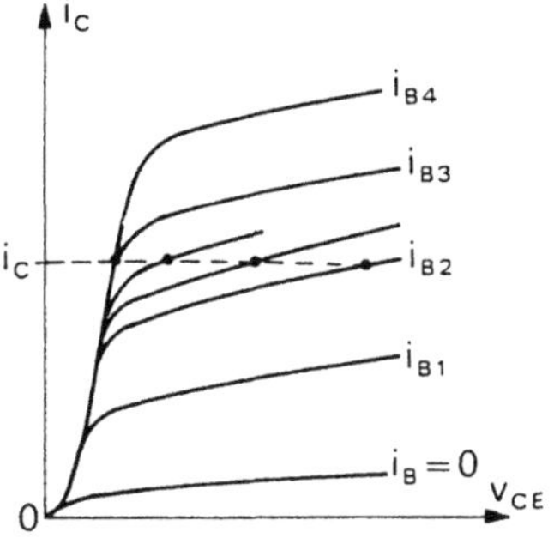

Fig. 2.17

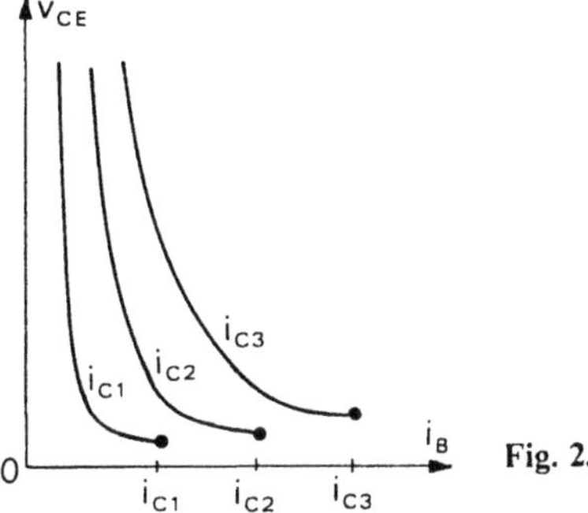

Fig. 2.18

transistor is intended to operate in saturation during its conduction periods;

- the maximum value I_{CM} of the collector current which can be accepted in repetitive pulses.

(The meaning of I_{CSM} will be seen when considering pulse operation in the Forward-Bias Safe-Operating Area).

Voltage drop v_{CE} is given for one or several values of i_C, including $I_{C(sat)}$.

- *Remarks*

– The *input* characteristics $v_{BE} = f(i_B)$ plotted for different values of i_C (Fig. 2.19) have the same forms as those for the P–N junction. For a given current i_B, v_{BE} increases slightly with i_C.

Despite the low current gain corresponding to saturation, there is considerable power gain, since v_{BE} is very low compared with the voltages used in the power part of the circuit.

– In order to speed up the turn-off switching, a negative voltage v_{BE} may be applied, i.e. the base–emitter junction can be reverse-biased. This reverse voltage must not reach the avalanche breakdown voltage, which is normally indicated for a zero value of i_C and denoted as BV_{EBO} or V_{EBO}.

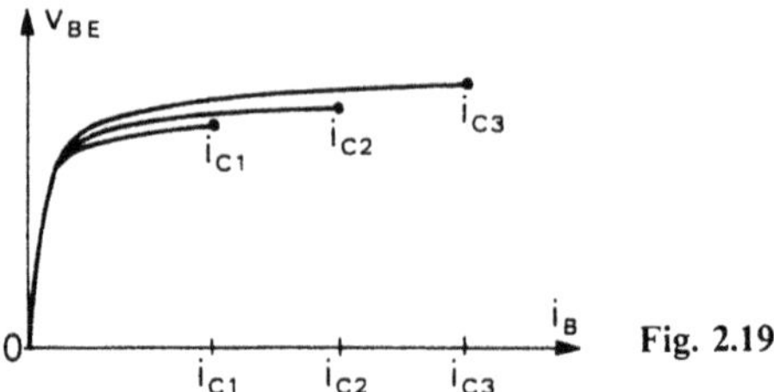

Fig. 2.19

2.2.2.3 Avalanche Breakdown at the C–B Junction, Voltage Limitation

The maximum voltage v_{CE} that the transistor can sustain is limited by the avalanche breakdown of the reverse biased C–B junction.

- *In the active region* ($i_B > 0$)

If BV_{CBO} denotes the avalanche breakdown voltage of the C–B junction (for a zero emitter current), the current flowing through this junction, for high voltage values across it, is multiplied (see Eq. (2.3)) by:

$$M = \frac{1}{1 - (v_{CB}/BV_{CBO})^n}.$$

Since v_{BE} always remains small, the following may be used:

$$M = \frac{1}{1 - (v_{CE}/BV_{CBO})^n}. \tag{2.7}$$

For small values of v_{CE}, M approaches 1 and the equation

$$i_C = \alpha i_E + I_{CBO}$$

remains valid.

But for high values of v_{CE}, current i_C is given by

$$i_C = M\alpha i_E + MI_{CBO} = M\alpha(i_C + i_B) + MI_{CBO}.$$

This gives

$$i_C = M\frac{\alpha i_B + I_{CBO}}{1 - M\alpha}.$$

This equation shows that i_C tends towards infinity as M tends towards $1/\alpha$.

Taking Eq. (2.7) into account, this corresponds to a value V_{CEO} of v_{CE}, such that

$$1 - \left(\frac{V_{CEO}}{BV_{CBO}}\right)^2 = \alpha$$

$$V_{CEO} = BV_{CBO}\sqrt[n]{1-\alpha} = BV_{CBO}\sqrt[n]{\frac{1}{1+\beta}} \simeq \frac{BV_{CBO}}{\sqrt[n]{\beta}} < BV_{CBO}.$$

This maximum value of v_{CE} is independent of i_B (see Fig. 2.20).

- *In the OFF state* ($i_B \leqslant 0$)

- When the base is in open circuit ($i_B = 0$), the transistor effect is very weak. As the value of α is considerably less than 1, the value of v_{CE} which brings about the avalanche is higher than that found previously. When breakdown occurs, the appearance of a strong current i_C places the transistor in conditions similar to those encountered with forward bias and there is an asymptotic tendency of v_{CE} towards V_{CEO} (Fig. 2.20).

- When the base is connected to the emitter by a resistance R_B (see Sect. 2.2.2.2), v_{BE} is positive and very small. There is no transistor effect and the avalanche should occur when v_{CE} reaches BV_{CBO}. In fact, current I_{CBO} across the junction is multiplied by M, and fraction MI_{CBO} which supplies i_B is sufficient to make junction B–E conducting and to make the transistor effect reappear at high v_{CE}. The avalanche occurs for a value V_{CER} lower than BV_{CBO} and dependent on R_B.

- When a generator imposes a negative voltage across junction B–E, the avalanche occurs for value V_{CEX} of v_{CE} making v_{CB} equal to BV_{CBO}. Since

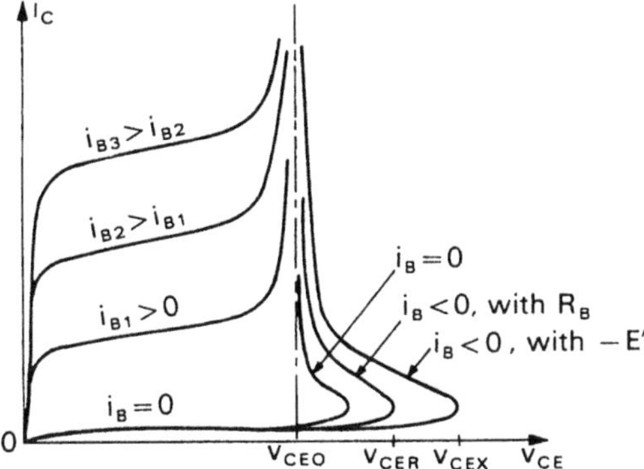

Fig. 2.20

v_{BE} is negative, voltage V_{CEX}, given by

$$V_{CEX} = v_{BE} + BV_{CBO}$$

is still lower than BV_{CBO}.

In all cases, as soon as current i_C increases, there is an asymptotic tendency of voltage v_{CE} towards V_{CEO}.

- In describing a power transistor, the following must be indicated, in the case of the voltages:
  - the *maximum instantaneous value of the collector–emitter voltage* that the transistor can sustain, $V_{CEO(sus)}$ or V_{CEO}, whatever the base bias conditions;
  - the maximum instantaneous value, V_{CEX} or V_{CEV}, of voltage v_{CE} that the transistor can sustain with a negative bias, by specifying the value of v_{BE}.

The maximum value V_{CER} of voltage v_{CE} with resistance between base and emitter is sometimes indicated, by specifying the R_B value.

2.2.3 Safe Operating Areas

In the output characteristics plane (i_C, v_{CE}), a Safe Operating Area (SOA) represents limits beyond which the operating point must not go.

2.2.3.1 ON-State Safe Operating Area

Figure 2.21 recalls the setting-up of the SOA used for operation in the active region, i.e. when B–E junction is always forward-biased, and its extension to pulsed operation. The axes are normally graduated in a logarithmic scale.

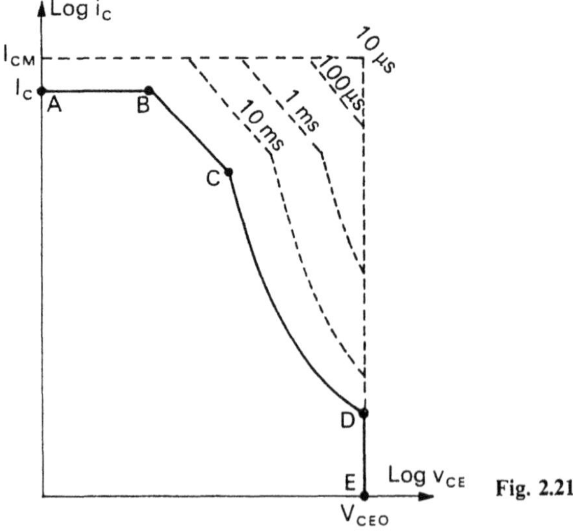

Fig. 2.21

- For a given case temperature, the SOA is delimited by:
  - segment AB corresponding to the admissible *current* I_C in steady state;
  - segment BC corresponding to the maximum *power* the transistor can dissipate. In equation form

$$v_{CE} i_C = P_{max} = (T_{j\,max} - T_c)/R_{th}$$

    $T_{j\,max}$ being the maximum temperature of junction C–B (this can extend from 125 to 200°C depending on the case and on the transistor technology),
    T_c being the temperature of case,
    R_{th} being the thermal resistance junction C–B/case;
  - segment CD corresponding to the second breakdown phenomenon. The P_{max} limitation assumes a homogeneous current density in the various transistor regions. This is difficult to achieve, notably at junction B–E. For high values of $v_{CE} i_C$ – though these may be lower than P_{max} – the difference in current line concentration leads to localized hot spots which accentuate this difference and may cause the destruction of the transistor by thermal runaway. (This "direct" second breakdown must not be confused with the "reverse" second breakdown. The latter arises when the avalanche – appearing in the OFF state if the v_{CE} value becomes too high (see Sect. 2.2.3.3) – brings about the thermal runaway of the transistor);
  - vertical segment DE, corresponding to *maximum voltage* V_{CEO}.

- For *pulsed operation* (shown in broken lines on Fig. 2.21) the v_{CE} maximum remains equal to V_{CEO}; but:
  - the instantaneous value of i_C may reach I_{CM};
  - the acceptable instantaneous power $v_{CE} i_C$ increases as the pulses become shorter.

2.2.3.2 Switch-Mode Operating Areas

As power transistors are switching transistors, SOAs during switching are often simply referred to as "operating areas".

The commutation process closely depends on the bias of the emitter–base junction. Two areas are to be distinguished:

- the forward bias safe operating area (FBSOA),
- the reverse bias safe operating area (RBSOA).

- In the reverse-biased state, the turn-off process depends on the previous state and on the way in which turn-off is achieved:
  - The deeper the saturation of the transistor during conduction, the more difficult it becomes to eliminate the carriers present in the collector in a rapid and uniform manner.

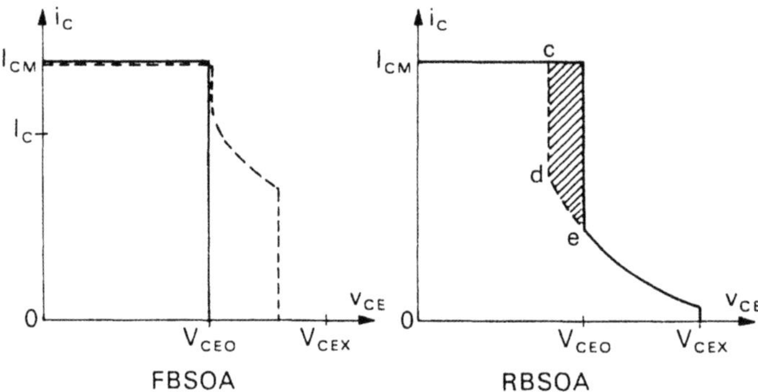

Fig. 2.22

If turn-off is carried out starting from quasi-saturation, the transistor can tolerate a higher voltage than during turn-off starting from saturation.

- The turn-off process is more uniform if the control is softer: the base current time rate of change must be limited when it is changed from its positive to its negative value.

These two remarks show that a certain number of specifications must be added to any description of operating areas.

- Figure 2.22 gives the form of these areas normally drawn using linear scales for v_{CE} and i_C.
  - The FBSOA (in full lines) can be used during turn-off, without reverse bias. At turn-on, the area can be enlarged (curves in broken lines) provided that the extra zone is crossed very quickly (e.g. in less than a microsecond) during the current increase i_C;
  - The RBSOA can be used for turn-off with a negative base drive. This area is slightly reduced (deletion of the hatched area) if the turn-off starts from the saturation region rather than from the quasi-saturation region.

2.2.3.3 Accidental Overload Areas

Current may accidentally go beyond the limits given by the operation areas; this may happen on a limited number of occasions in a transistor life, without the latter being destroyed.

Accidental overload areas are defined

- in forward bias (FBAOA);
- in reverse bias (RBAOA).

Figure 2.23 gives the form of these areas

In the FBAOA, curves are plotted for given values of i_B to show that i_C increase assumes the same in i_B or V_{CEO} will be exceeded. The curves noted in times indicate that the tolerable non-repetitive instantaneous power $v_{CE}i_C$ de-

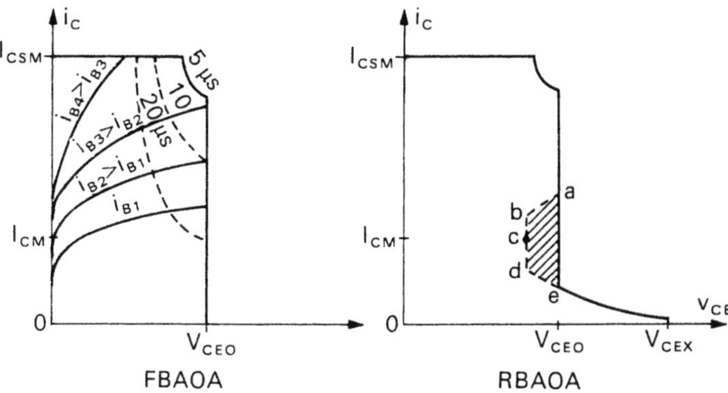

Fig. 2.23

creases when overload duration increases. Beyond a certain value of the surge non-repetitive collector current I_{CSM}, the crystal undergoes irreversible damage.

In the RBAOA, which corresponds to the turn-off of an overload current by reversing the base current, the hatched area can only be crossed (i.e. going from ab towards de) if the departure point is in the quasi-saturation region.

These overload areas allow to know

- whether the transistor can tolerate a short-circuit of the load:
- whether it can tolerate the discharge of the input filter capacitor;
- whether a limiting inductance must be added.

2.2.4 Turn-on Commutation

In static converters, the transistor operates as a switch, either in ON state ($v_{CE} \simeq 0$: switch closed) or OFF state ($i_C \simeq 0$: switch open). If operation at a high frequency is required, the switching from one state to another must take as little time as possible and lead to minimum losses.

On the other hand, commutations are at the origin of current and voltage peaks; these depend on the circuit in which the transistor is embedded.

Commutations thus require special analysis, and parameters describing the behaviour of a switching semiconductor device must be given detailed specifications.

In the case of the transistor and all other controlled switches, we denote as

- *turn-on time* t_{on}, the time between applying the turn-on signal to the control electrode and the end of the current rise in the main circuit.
- *commutation turn-on time* t_{con}, the time between applying the turn-on control signal and the end of the switch turn-on (the current having increased and the voltage decreased to its ON-state value).

2.2.4.1 Establishing the Current

When a current step I_{B1} is applied to the base of a non-conducting transistor with a positive voltage v_{CE}, the collector current increases gradually from zero to a value I, imposed by the external circuit when conduction is established (Fig. 2.24).

- The current total establishment time is the sum of
  - the *delay time* t_d, taken by i_C to go from 0 to $I/10$,
  - the *rise time* t_r, taken by i_C to go from $I/10$ to $9I/10$;

$$t_{on} = t_d + t_r .$$

- In physical terms, the delay corresponds to the time needed by the majority carriers to cover the space charge of the transition zone of junction B–E. The "transition charge", e.g. the charge which corresponds to the disappearance of this zone, is carried by i_B. The emitter electrons have not yet reached the base, the transistor effect has not appeared and the collector current remains negligible.

  The current rise corresponds to the establishment of the electron flux coming from the emitter in layer P. As this flux reaches the collector, it gives current i_C. The charge contained in the base layer when i_C equals I is called "diffusion charge".

- Within the axes v_{CE}–i_C, the operating point must go from point B (blocked) to point S (saturation). If the current gain in the active region is β, the base current I_{B1} must be higher than I/β for the transistor to be brought into saturation.

  Certain constructors indicate the *forced gain* of the transistor in saturation β_F, defined by the equation

$$\beta_F = I/I_{B1} .$$

This is specified for a certain number of points S, given by I and $v_{CE\,sat}$.

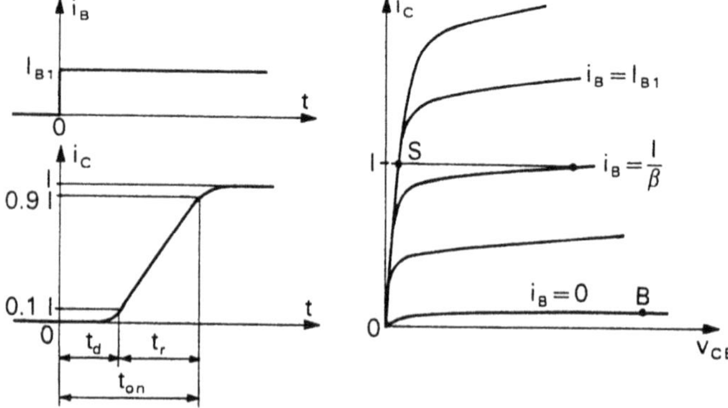

Fig. 2.24

The trajectory followed by the operating point from B to S, the speed at which this trajectory is completed, and the commutation losses depend on the way in which voltage v_{CE} varies, and thus on the circuit into which the transistor is embedded.

2.2.4.2 Description of a Commutation (Fig. 2.25)

The commutation under study usually consists of transferring the current flowing in an inductive path from a diode to the transistor. This process can be followed in the simple diagram of the step-down chopper shown in Fig. 2.25a.

The current I flowing in load R, L is assumed to be constant, at least during commutation. Furthermore, if E is the voltage of the supply to the converter,

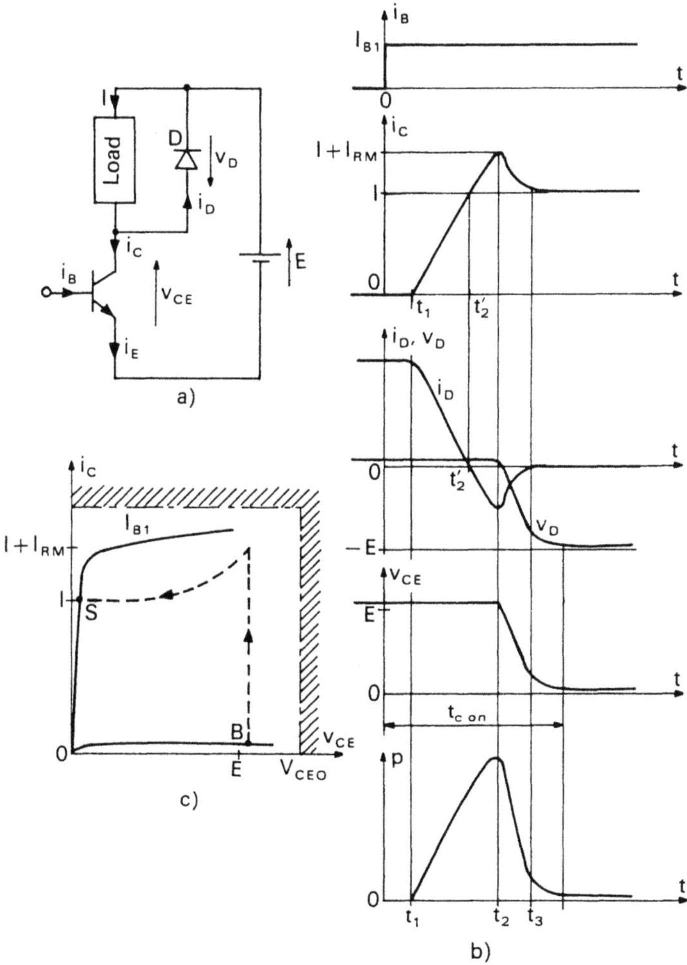

Fig. 2.25

this gives

$$v_{CE} - V_D = E\,.$$

When the diode is conducting, and denoting its ON-state voltage drop by ΔV, we have

$$i_D = I; \quad v_D = \Delta V; \quad i_C = I_{CBO}; \quad v_{CE} = E + \Delta V\,.$$

For $t = 0$, the turn-on control signal is applied. Current i_C begins to rise for $t = t_1$; since $i_C + i_D$, equal to I, is constant, the i_C rise is accompanied by a fall of i_D.

For $t = t_2'$, current i_D becomes negative and remains so during the reverse recovery time of the diode (see Sect. 2.1.3.2). The maximum value I_{RM} of the reverse current of the diode brings about a collector current peak; for $t = t_2$, i_C reaches $I + I_{RM}$.

When the diode reverse current no longer increases, v_D becomes negative and v_{CE} falls towards $V_{CE\,sat}$. Two periods can be distinguished. In a first stage ($t_2 < t < t_3$), voltage v_{CE} decreases rapidly from its initial value, $E + \Delta V$, to a value somewhat higher than $V_{CE\,sat}$. In a second stage ($t_3 < t < t_4$), the variation of v_{CE} are slower because of the gradual extension of the base P region into the collector N^- region (see Sect. 2.2.1.3).

Figure 2.25b shows the waveforms of i_B, i_C, i_D and v_D, v_{CE} and the instantaneous losses p in the transistor. Figure 2.25c shows the operating point trajectory from B to S; it must remain within the FBSOA.

2.2.4.3 Turn-on Losses

A study of Fig. 2.25b indicates the origin of turn-on switching losses:

- current rise i_C takes place when full voltage v_{CE} is present;
- voltage fall then takes place when full current i_C is present.

– In cases where the diode has a fast reverse recovery time t_{rr}, these losses can be calculated using the simplified waveforms of Fig. 2.26:
  - current i_C is assumed to increase linearly from zero to $i_{C\,max}$, equal to $I + I_{RM}$, from $t = t_1$ to $t = t_2$, and then immediately to take on the value I.

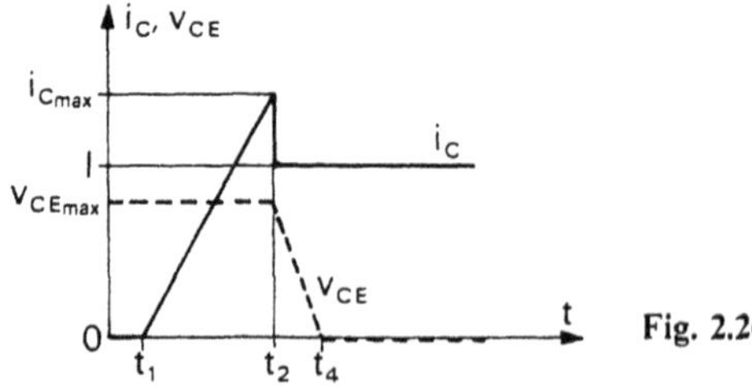

Fig. 2.26

- voltage v_{CE} is assumed to decrease linearly from $v_{CE\,max}$, equal to $E + \Delta V$, to zero, between $t = t_2$ and $t = t_4$.

Under these conditions, the energy dissipated in the transistor, at each turn-on, is

$$W = \frac{1}{2} v_{CE\,max} i_{C\,max}(t_2 - t_1) + \frac{1}{2} v_{CE\,max} I(t_4 - t_2).$$

If t_r is the current i_C rise time, corresponding to its change from $0.1I$ to $0.9I$,

$$t_2 - t_1 = 1.25 t_r \frac{I + I_{RM}}{I}.$$

When the voltage fall time is unknown, a rough approximation can be made:

$$W \simeq v_{CE\,max} i_{C\,max} t_r. \tag{2.8}$$

- In order to *reduce* the turn-on switching losses, a small inductance can be connected in series with the transistor (see Appendix).[1]

2.2.4.4 Reducing the Turn-on Time

Turn-on time of the transistor can be reduced by supplying the latter, during the corresponding period, with a base current i_B higher than I_{B1} which is needed to ensure saturation. It then returns to I_{B1} to avoid unnecessary increase in the stored charge. Figure 2.27 represents two drive signals and shows the corresponding waveforms of current i_C.

The initial overshoot of the base current

- reduces the rise time of i_C,
- may reduce commutation losses,

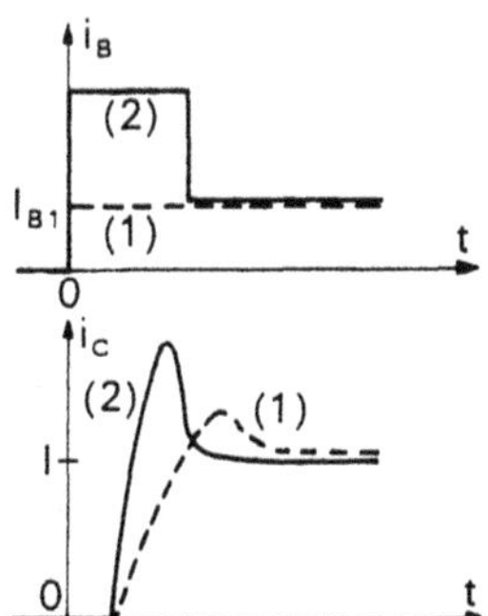

Fig. 2.27

[1] To avoid making this chapter unnecessarily long, we have placed in Appendix the study of turn-on/turn-off snubbers associated with fully controlled semiconductor switches.

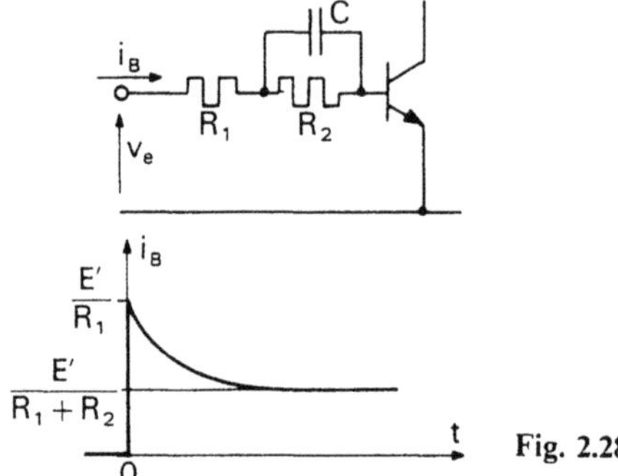

Fig. 2.28

- but increases the i_C current peak by raising the diode recovery current peak (see Fig. 2.9), if i_B is not reduced quick enough.

Figure 2.28 gives a simple circuit in order to obtain this i_B overcurrent. Voltage source v_E, which goes from zero to $+E'$ for $t = 0$, is linked to the base by two resistances R_1 and R_2. The latter is shunted by a capacitor C. Because of C, the voltage across R_2 cannot vary instantaneously: at instant $t = 0$, i_B goes from zero to E'/R_1; then i_B moves towards $E'/(R_1 + R_2)$ with the time constant $R_1 R_2 C/(R_1 + R_2)$.

2.2.5 Turn-off Commutation

For the transistor and any controlled turn-off semiconductor device, we will denote as:

- *total turn-off time* t_{off}, the time between the application of the turn-off signal to the drive electrode and the end of current turn-off in the transistor.
- *total commutation turn-off time* $t_{c\,off}$, the time between the application of the turn-off signal and the end of the switch turn-off (zero current, voltage restored at its terminals).

2.2.5.1 Current Turn-off

In order to interrupt a strong current I in the transistor, the base current is reversed by making it change from $+I_{B1}$ to $-I_{B2}$. It must remain negative until current i_C becomes negligible (Fig. 2.29).

The current turn-off time t_{off} is the sum of

- the *storage time* t_s, taken by i_C to go from I to $0.9I$,
- the *fall time* t_f, taken by i_C to go from $0.9I$ to $0.1I$;

$$t_{off} = t_s + t_f \,.$$

The storage time corresponds to the removal of excess positive carriers in the P zone by the base current. This time increases as the collector current is higher,

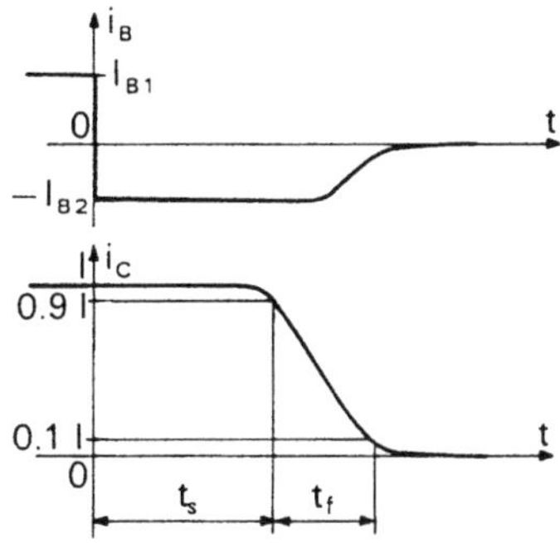

Fig. 2.29

as the saturation is harder and as the removal speed (and thus the $-I_{B2}$ intensity) decreases.

The fall time is the time required for the minority base carriers to be extracted by the emitter or to recombine with holes. Such a disappearance of the diffusion charge depends on I, on $-I_{B2}$, on the junction temperature and on the speed of the voltage rise v_{CE}.

2.2.5.2 Description of a Commutation (Fig. 2.30)

The turn-off commutation is described for the circuit in Fig. 2.30a, the same as that used to describe turn-on. Current I is once more assumed to be constant:

$$i_C + i_D = I; \quad v_{CE} - v_D = E .$$

Period 0, t_1 corresponds to the transistor storage time. Once the excess base carriers have been removed, current i_C cannot yet decrease. As the diode, with a negative voltage across it, cannot start to conduct, current i_B continues to sweep positive charges out of the base, leading to an increase in the resistivity of the latter and of the internal resistance of the transistor (period t_1, t_2).

When the diode starts to conduct (for $t = t_2$), this leads to a voltage peak (see Sect. 2.1.3.1) to be found in v_{CE}. The i_D rise is linked to the i_C decrease. When i_D reaches value I, the transistor cuts off.

Figure 2.30b shows the waveforms of i_B, i_D and v_D, of i_C and v_{CE}, and of the instantaneous power dissipated in the transistor. Figure 2.30c shows the operating point trajectory in the RBSOA.

2.2.5.3 Turn-off Losses

Turn-off switching losses are mostly caused by the fact that current i_C only starts to decrease after voltage v_{CE} is restored.

In order to simplify the calculation of the energy dissipated in the transistor at each turn-off, i_C is assumed to decrease linearly from I to zero, during a time interval equal to $1.25t_f$, and voltage v_{CE} is assumed to have the value which it

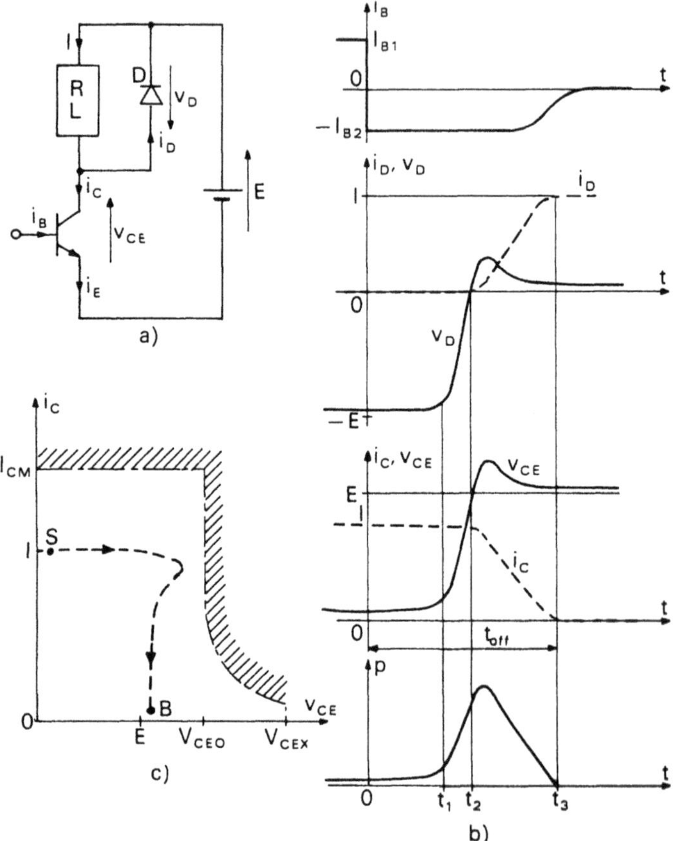

Fig. 2.30

will reach at the end of commutation (see Fig. 2.31) during the whole of this period. Denoting the voltage drop in the diode by ΔV, this value V_{CE} is equal to $E + \Delta V$.

Under these conditions,

$$W = \frac{1}{2} V_{CE}\, I \,\frac{t_f}{0.8}\,.$$

However, we have not taken into account the losses corresponding to voltage rise v_{CE} (before the current starts to decrease) and those due to voltage peak at the end of the voltage rise (hatched areas on Fig. 2.31). Thus, as a first approximation, we have

$$W \simeq V_{CE} I t_f\,. \tag{2.9}$$

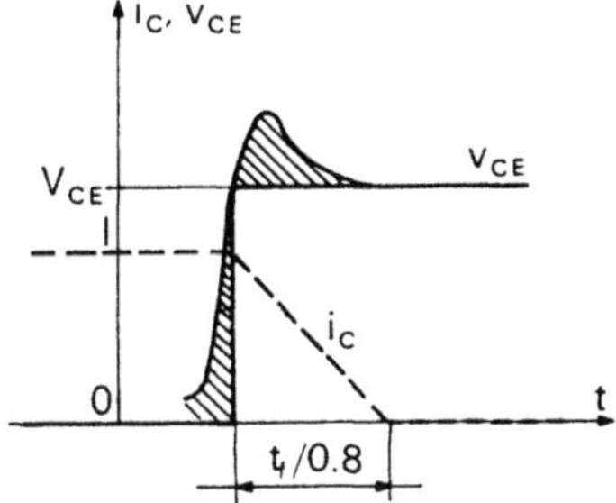

Fig. 2.31

The losses are reduced by use of a turn-off snubber, basically made up of a capacitor in parallel with the transistor (see Appendix). This capacitor links voltage rise v_{CE} to current decrease i_C.

2.2.5.4 Reducing Turn-off Time

The turn-off time (and especially the storage time) can have a limiting effect on high-frequency operation. It can be reduced by

- reducing the amount of charges to be removed,
- using a negative current pulse which is strong but well-dosed.

a) Anti-saturation circuit

The storage time t_s depends on the carriers in excess in the base during conduction: the more they are, the longer t_s lasts. If the saturation of the conducting transistor is limited to quasi-saturation, voltage v_{CE} is slightly increased but t_s is greatly reduced.

To avoid saturation of the transistor, the circuit shown in Fig. 2.32 can be used. If the voltage drop ΔV is assumed to be the same in all conducting diodes, voltage v_{CE} cannot fall below

$$v_{CE\,min} = v_{BE} + \Delta V\,.$$

Diode DAS, which is connected to the collector, is called the antisaturation diode. Diode D′ enables the negative base current to flow when turn-off occurs.

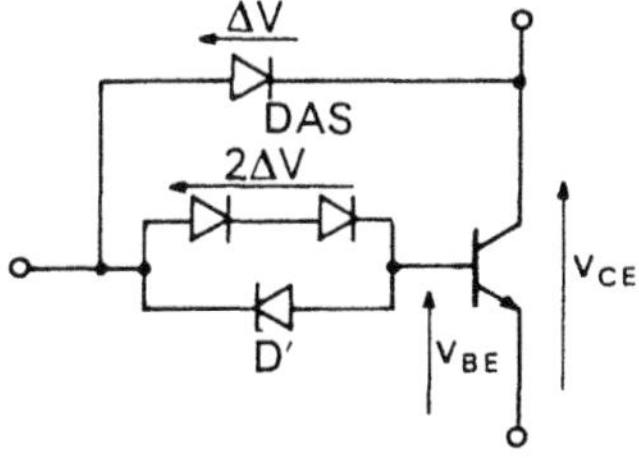

Fig. 2.32

b) Base negative current peak

The storage time also depends on the speed at which carriers in excess are removed, and thus on the negative current I_{B2}. However, if the value of I_{B2} is too high, this may lead to substantial losses when turn-off occurs.

- A moderate base current I_{B2} (in broken lines in Fig. 2.33a and 2.33b) removes excess carriers in the base during t_s, and then ensures that i_C decreases gradually as the transistor effect disappears.
- A high base current I_{B3}, applied sharply (Fig. 2.33a), makes the transition zone of junction $B-E$ reappear very quickly, by extracting the holes from the base. It thus cancels out current i_E. Time t_s is reduced as the decrease in i_C appears more rapidly. But all the positive charges stored do not have time to be removed: those of the base leave by means of i_B, and those of the collector slowly recombine since the electron flux coming from the emitter via the base has disappeared. A long tail appears in current i_C while voltage v_{CE} has been restored. This leads to an increase in commutation losses.
- However, after the storage time, the increase of the base current from I_{B2} to I_{B3} enables i_C to decrease more rapidly. Virtually the same result can be

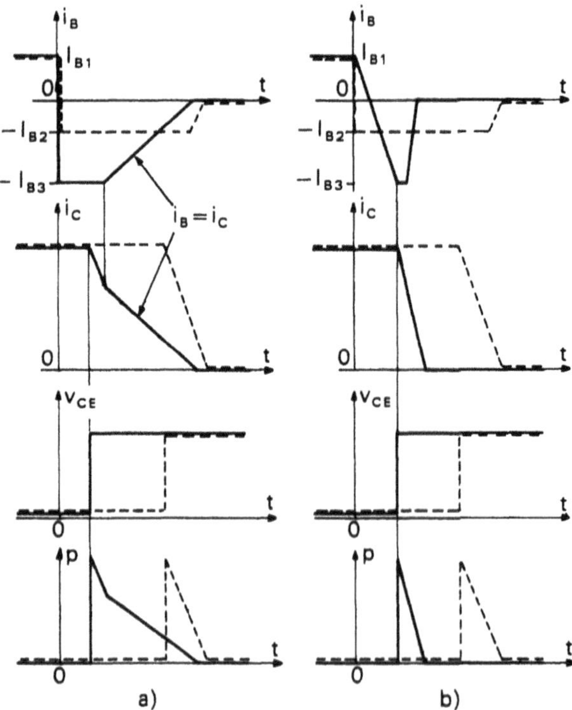

Fig. 2.33

obtained by applying a strong negative base current with a suitable $-di_B/dt$ (Fig. 2.33b). This di_B/dt can be obtained by placing, in the circuit, a small inductance which applies a negative voltage at the base. This has the added advantage of avoiding a concentration of current lines which would create hot spots.

2.2.5.5 Reverse Current: Remarks

The transistor is intended to work as an unidirectional base-controlled switch. But it has a symmetrical structure, and a negative voltage v_{CE} can make it conduct in the reverse direction. If the reverse current is high, the dissipated power at the emitter–base junction can rapidly exceed the maximum allowed, and lead to the destruction of the transistor.

- In many cases, *a diode is connected in antiparallel* across the transistor. When the diode is conducting, it protects the transistor from excessive negative values of v_{CE}. However, the ON-state voltage drop of this diode can be enough to make the base–collector junction conduct. If the base is not in open circuit but connected to the emitter by a resistance, a current can flow from E to C across this resistance R_B and the junction B–E without damaging the transistor. If the base was forward biased, this reverse current would be greater, since the base bias voltage would be added to $-v_{CE}$.
- However, this *reverse transistor current creates problems* in most circuits. Let us show this in Fig. 2.34, by following the diode D_1 to transistor T_2 commutation.

When D_1 was conducting, v_{CE_1} was negative and low. Turning on T_2 ensures that the load current could switch from D_1 to T_2, but v_{CE_1} does not vary as long as D_1 remains conducting. When the current in D_1 falls to zero, the rapid decrease in v_{CE_2} brings about a rapid increase in v_{CE_1}. The reverse current charge carriers of transistor T_1 quickly leave the latter via the base–emitter junction, which is forward-biased by R_B. This current is added to the recovery current of diode D_1, leading to an overcurrent in T_2.

It is therefore advisable to keep the voltage across the emitter–base junction negative, before turning on T_2, in order to limit the current peak in the latter.

Moreover, the danger of simultaneous conduction of T_1 and T_2 is avoided, Indeed if both transistors are simultaneously turned on, this could lead to voltage supply E being short-circuited and both transistors being destroyed.

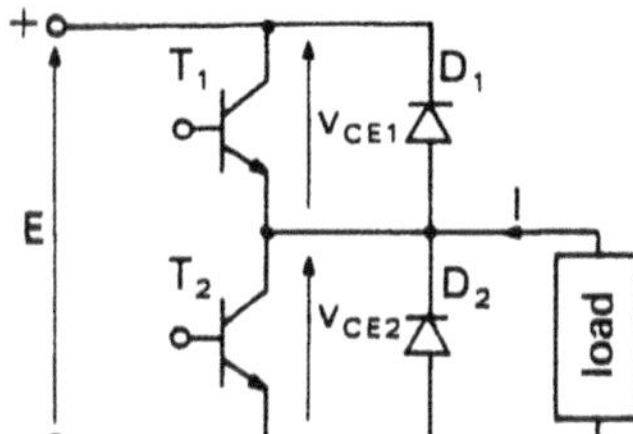

Fig. 2.34

2.2.6 The Darlington Circuit

The current gain is very low in a bipolar transistor operating a saturation especially for high-voltage transistors. For a given collector current, the Darlington circuit enables the transistor to be in saturation with a lower base current: drive can thus be achieved more easily.

A Darlington circuit (Fig. 2.35) comprises two bipolar NPN-type transistors, T_1 and T_2, with the base current of T_2 equal to the emitter current of T_1.

2.2.6.1 Current Gain

- Let us recall the standard calculation of current gain concerning the Darlington circuit in the active region.

If β_1 and β_2 denote the current gain of both transistors, the equivalent transistor current i_C is given by

$$i_C = i_{C_1} + i_{C_2}$$
$$i_C = \beta_1 i_{B_1} + \beta_2 i_{B_2} = \beta_1 i_{B_1} + \beta_2 i_{E_1} = \beta_1 i_{B_1} + \beta_2(1 + \beta_1) i_{B_1}$$
$$i_C = (\beta_1 + \beta_2 + \beta_1\beta_2) i_B \,. \tag{2.10}$$

The current gain of the equivalent transistor is thus equal to

$$\beta = \beta_1 + \beta_2 + \beta_1\beta_2 \,.$$

- *In saturation*, Eq. (2.10) can be used, provided that the values of β_1 and β_2 corresponding to i_C are chosen. However these values are heavily dependent on i_C. It is more convenient to construct the characteristic $i_C = f(i_B)$, using the corresponding characteristics of T_1 and T_2 (Fig. 2.36).

In quadrant 1, $i_{C_1} = f(i_{B_1})$ is plotted; in quadrant 3, $i_{C_2} = f(i_{B_2})$ is plotted. Since i_{E_1} equals $i_{C_1} + i_{B_1}$, the characteristic $i_{E_1} = f(i_{C_1})$ plotted in quadrant 2 can be deduced from the first curve.

For a value OA of current i_B, i_{C_1} is read in OB. Current i_{E_1} is given by OC, equal to OA + OB. Since i_{B_2} equals i_{E_1}, current i_{C_2} is read in CD. Current i_C, equal to $i_{C_1} + i_{C_2}$, is given by CD + OB; it is carried over in OE. The point of abscissa OA and ordinate OE is a point on the curve $i_C = f(i_B)$, shown in quadrant 4.

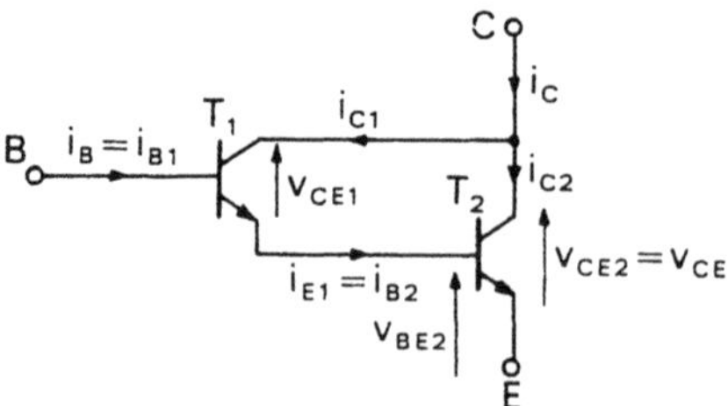

Fig. 2.35

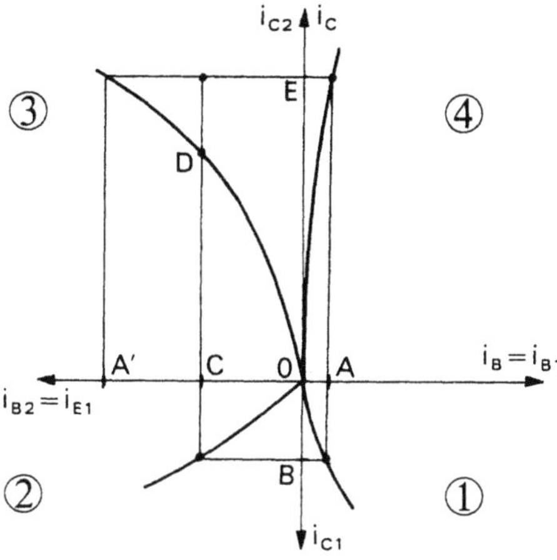

Fig. 2.36

In order to have the same collector current with transistor T_2 alone, a base current shown by OA' would have been necessary.

- *Remarks*

– Figure 2.36 shows that the Darlington circuit requires a much lower drive current than if transistor T_2 had been used on its own. It also indicates that, at a given i_C, this transistor operates with lower collector and base currents (CD instead of OE, OC instead of OA'). For a given controlled power, a smaller (hence cheaper) transistor can be used for T_2.

– Figure 2.35 shows that voltage v_{CE_2}, being equal to $v_{CE_1} + v_{BE_2}$, remains higher than v_{BE_2}. Transistor T_1 acts as anti-saturation device for transistor T_2. The latter never reaches a state of full saturation, but remains in quasi-saturation.

The voltage drop of a Darlington circuit in the ON state is thus greater, for a given collector current, than that of a bipolar transistor in a saturation. It is between 1.5 and 2.5 V.

2.2.6.2 Stabilization Resistances

- In setting up Eq. (2.10), the saturation currents of the transistors were not taken into account. If I_{CEO_1} and I_{CEO_2} are included, this gives

$$\begin{aligned}
i_C = i_{C_1} + i_{C_2} &= \beta_1 i_{B_1} + I_{CEO_1} + \beta_2 i_{B_2} + I_{CEO_2} \\
&= \beta_1 i_{B_1} + I_{CEO_1} + \beta_2 i_{E_1} + I_{CEO_2} \\
&= \beta_1 i_{B_1} + I_{CEO_1} + \beta_2 (i_{C_1} + i_{B_1}) + I_{CEO_2} \\
&= \beta_1 i_{B_1} + I_{CEO_1} + \beta_2 (\beta_1 i_{B_1} + I_{CEO_1}) + \beta_2 i_{B_1} + I_{CEO_2} \\
&= (\beta_1 + \beta_2 + \beta_1 \beta_2) i_{B_1} + (1 + \beta_2) I_{CEO_1} + I_{CEO_2} \,. \qquad (2.11)
\end{aligned}$$

The current gain β of the equivalent transistor is expressed in the same way, but the saturation collector current of T_1 is multiplied by $(1 + \beta_2)$ and added to that of T_2. Such saturation currents show considerable variation with temperature and give the Darlington circuit a bad *thermal stability*.

- This problem can be avoided by connecting low-value resistances between the base and the emitter of each transistor (Fig. 2.37).

  We thus have:

$$i_{B_1} = i_B - \frac{v_{BE_1}}{R_1}; \quad i_{B_2} = i_{E_1} + \frac{v_{BE_1}}{R_1} - \frac{v_{BE_2}}{R_2}$$

and a new expression of i_C:

$$\begin{aligned} i_C &= i_{C_1} + i_{C_2} = \beta_1 i_{B_1} + I_{CEO_1} + \beta_2 i_{B_2} + I_{CEO_2} \\ &= \beta_1\left(i_B - \frac{v_{BE_1}}{R_1}\right) + I_{CEO_1} + \beta_2\left(i_{E_1} + \frac{v_{BE_1}}{R_1} - \frac{v_{BE_2}}{R_2}\right) + I_{CEO_2} \\ &= \beta_1\left(i_B - \frac{v_{BE_1}}{R_1}\right) + I_{CEO_1} + \beta_2\left(\beta_1 i_{B_1} + I_{CEO_1} + i_{B_1} + \frac{v_{BE_1}}{R_1} - \frac{v_{BE_2}}{R_2}\right) \\ &\quad + I_{CEO_2} \\ &= \beta_1\left(i_B - \frac{v_{BE_1}}{R_1}\right) + I_{CEO_1} + \beta_2\left[(1 + \beta_1)\left(i_B - \frac{v_{BE_1}}{R_1}\right) + I_{CEO_1}\right. \\ &\quad \left. + \frac{v_{BE_1}}{R_1} - \frac{v_{BE_2}}{R_2}\right] + I_{CEO_2} \\ &= (\beta_1 + \beta_2 + \beta_1\beta_2) i_B + (1 + \beta_2)\left(I_{CEO_1} - \beta_1\frac{v_{BE_1}}{R_1}\right) \\ &\quad + \left(I_{CEO_2} - \beta_2\frac{v_{BE_2}}{R_2}\right). \end{aligned} \tag{2.12}$$

It can thus be seen that, by appropriate choices of R_1 and R_2, the saturation-current effect can be considerably reduced and thermal stability improved.

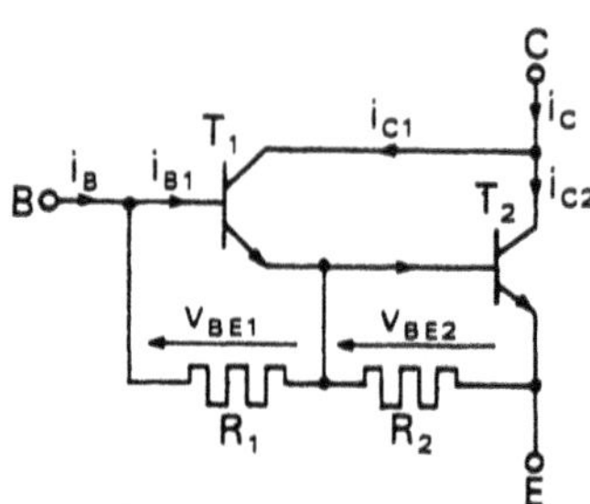

Fig. 2.37

Note that resistances R_1 and R_2 have a low power consumption since only voltages v_{BE_1} and v_{BE_2} are applied across them.

2.2.6.3 Turn-on Switching

When the drive signal is applied on the base of T_1, the latter is naturally the first to react since the collector current of T_2 can only rise after T_1 – which supplies its base – begins conducting.

Figure 2.38 gives the current and voltage waveforms of the Darlington circuit when it replaces the transistor of Fig. 2.25a. Current i_{C_1} only appears after the delay time t_{d1} of transistor T_1. Current i_{C_2} then appears after delay time t_{d2}. In the next stage, i_{C_1} and i_{C_2} increase to their final values. Voltage v_{CE}, equal to v_{CE_2}, can then fall to its saturation value.

Current i_{C_1} rises more slowly than i_{C_2} and reaches its final value at the same time since, with voltage v_{CE_1} equal to $v_{CE_1} - v_{BE_2}$ and v_{BE_2} unable to exceed the ON-state voltage drop of a conducting P–N junction, the decrease of v_{CE_1} is linked to that of v_{CE}. When T_1 begins to conduct, the losses are thus relatively high and this transistor must have the appropriate size.

By using an adapted base drive, the setting time can be reduced, as with an ordinary transistor (see Sect. 2.2.4.4).

2.2.6.4 Turn-off Switching

- Figure 2.39 shows waveforms obtained when a Darlington circuit replaces the single transistor of Fig. 2.30a.

  When current i_B falls to zero or is reversed, T_1 is once more the first to react. After its storage time t_{s_1}, i_{C_1} decreases gradually (t_{f_1}). If the load maintains i_C at a constant value, the increase in i_{C_2} makes up for the reduction in i_{C_1}.

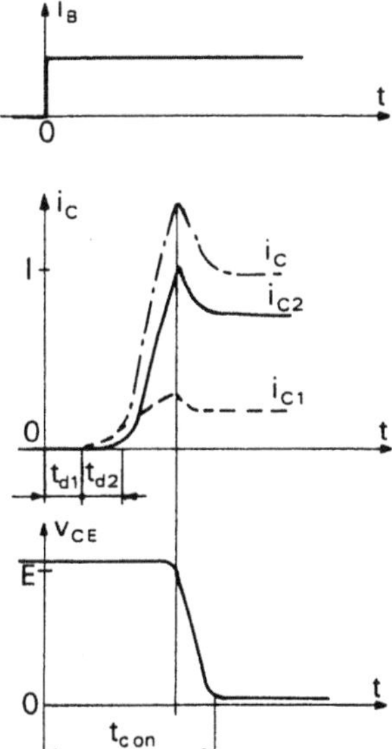

Fig. 2.38

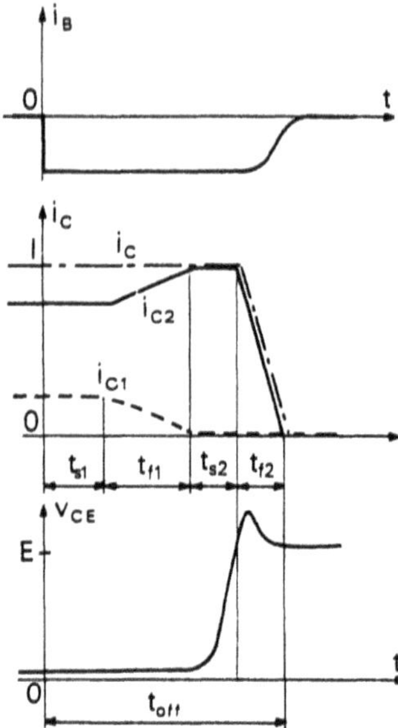

Fig. 2.39

Current i_{C_2} does not decrease until after the storage time t_{s_2} of transistor T_2. This is a relatively rapid decrease (t_{f_2}) as T_2 was in quasi-saturation, and voltage v_{CE_2} is high.

- Stabilizing resistances R_1 and R_2 can play a role in reducing the storage times since they enable negative base current to flow and stored charges to be removed.

  - A *negative* bias of the T_1 base enables storage time t_{s_1} to be reduced by increasing the reverse base current value i_{B_1}. When T_1 is in the OFF state, the bias can also reduce the storage time t_{s_2}, if a diode is placed between the emitter and the base of T_1 (Fig. 2.40a) to enable the reverse base current of T_2 to flow.

    This diode enables T_1 to remain in the OFF state during the turn-off of T_2. Two diodes are often connected in series to make the blocking more effective. Indeed, the Darlington circuit is highly sensitive to dv/dt and, when it is in the OFF state, the voltage between T_1 emitter and base must be sufficiently negative to prevent conduction from starting again if there is a high dv/dt. When using a single diode, T_1 could once again become conducting after T_2 is turned off, leading to substantial losses.

  - To avoid excessive saturation of T_1 before switching, and to reduce storage time t_{s_1}, an *anti-saturation device* similar to that described in Sect. 2.2.5.4 can be used (Fig. 2.40b). This gives

$$v_{CE_1} = v_{BE_1} + v_{D_1} - v_{DAS}.$$

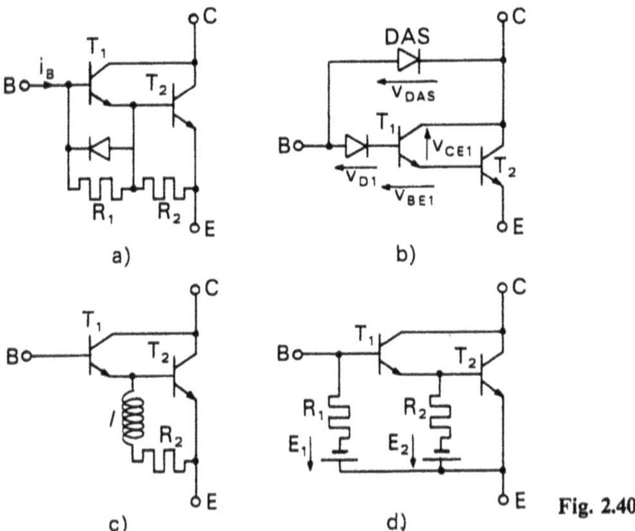

Fig. 2.40

- The i_C fall time depends essentially on T_2. It can be reduced by adding a *small inductance l* in series with R_2 (Fig. 2.40c). When turn-off begins, current decrease in this inductance generates a voltage across it which tends to give a reverse bias to the E–B junction of transistor T_2 and speeds up this transistor turn-off.
- Finally, since the relatively long turn-off time is caused by successive blocking of T_1 and T_2, this can be reduced by *simultaneous turn-off* of both transistors. This can be achieved (Fig. 2.40d) by a negative bias, at turn-off, of both bases, using two appropriate sources E_1 and E_2.

- When used individually or together, the four procedures shown in Fig. 2.40 aim at reducing the main disadvantage arising from the basic principle of the Darlington circuit:

  To increase the current gain, one transistor is driven by another connected between its collector and its base. But this leads to an increase in the commutation times.

The use of a *triple Darlington circuit* (Fig. 2.41) in the drive of some high-voltage transistors increases both advantages and drawbacks of this procedure.

2.2.6.5 The Monolithic Darlington Circuit

For low and medium power (up to about 500 V and 100 A) Darlington circuits, all the components can be diffused in the same silicon wafer. Such integration takes less room, makes it easier to operate, and improves thermal stability. But it has some drawbacks, as must be shown.

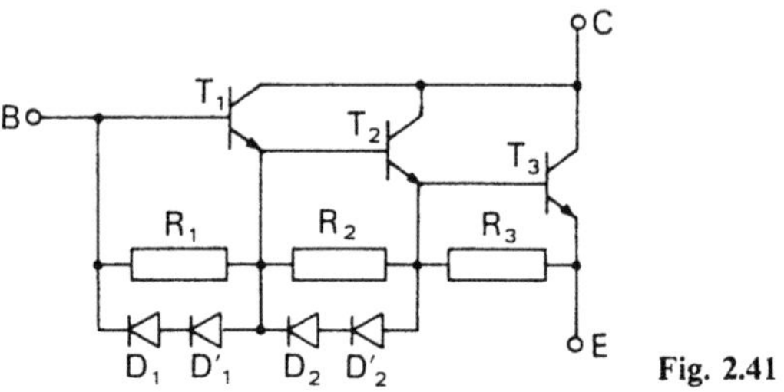

Fig. 2.41

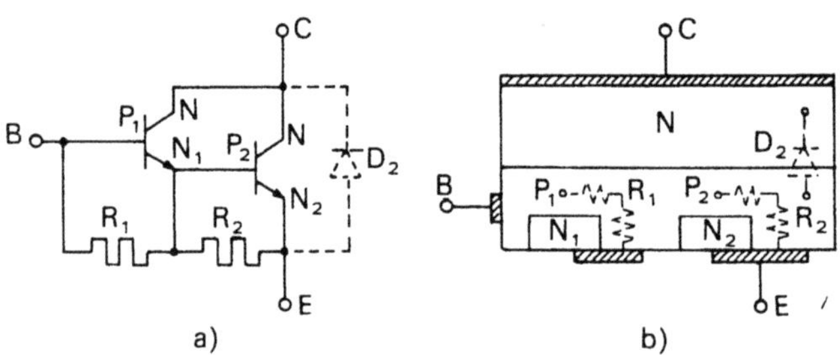

Fig. 2.42

Figure 2.42a shows the various components needed for the circuit; the latter is shown in diagrammatic form in Fig. 2.42b. Stabilization resistances R_1 and R_2 are integrated and appear in the P zone.

Linking R_2 to E requires a metallization on the P zone surface. This brings out a *diode* D_2 in antiparallel on T_2. As a large part of the N zone of the collector is only slightly doped, this diode is not fast and cannot be used in circuits where there is a diode in antiparallel across the transistor, since its long recovery time would lead to excessive commutation losses. A separate fast recovery diode must be used.

If the terminal common to N_1 and P_2 is inaccessible, the behaviour of transistor T_2 during commutations cannot be modified by any action concerning the base current or bias.

It should be noted that the forward breakdown voltage V_{CEO} of the transistor equivalent to the monolithic Darlington circuit is voltage V_{CER} of transistor T_2, which has resistance R_2 between its base and emitter. Thus it is higher than the forward breakdown voltage of T_2 with unconnected base. Despite the higher apparent value of V_{CEO}, the width of the T_2 safe operating area is not modified.

2.3 Field Effect Transistors

Like the bipolar transistor, the FET (Field Effect Transistor) can act as a switch. As its operating principle, is very different, this component has both advantages and disadvantages compared to the bipolar transistor in the design of low-power converters.

There are two types of field effect transistors:

- junction transistors or JFET,
- insulated gate transistors or MOSFET.

We shall limit the discussion to this second type, as it alone is used in making "power" components.

2.3.1 Description and Operation

2.3.1.1 Low-Power MOSFET

In a doped P type semiconductor (Fig. 2.43a), two N type zones are diffused, on which the terminals of both source S and drain D are soldered. A layer of oxide covers the P zone situated between the source and the drain; the gate terminal G is soldered on this layer. These successive layers – Metal–Oxide–Semiconductor – provide the prefix MOS which is used to describe this type of transistor.

Figure 2.43b shows the symbol normally used. The connection which goes from the terminal linked to bulk ends in an arrow showing the conducting direction of the bulk–source and the bulk–drain junctions.

The bulk and the drain are usually connected. Figure 2.43c shows the notations.

2.3.1.2 Operating Principle

If a positive voltage v_{GS} is applied between gate and source, the electric field which comes through the oxide layer and appears on the surface of the P layer between drain and source, expels the majority carriers (positive charges) from this zone and attracts the minority carriers to it.

Above a given v_{GS} value – the *threshold voltage* v_T – there are more negative than positive charges in the immediate vicinity of the oxide layer. The N type zone formed in this way acts as a *channel* linking source and drain.

If a positive voltage v_{DS} is then applied between drain and source, drain current i_D can flow via the channel. The section of the channel – and thus its resistance – depends on the difference between v_{GS} and the threshold voltage.

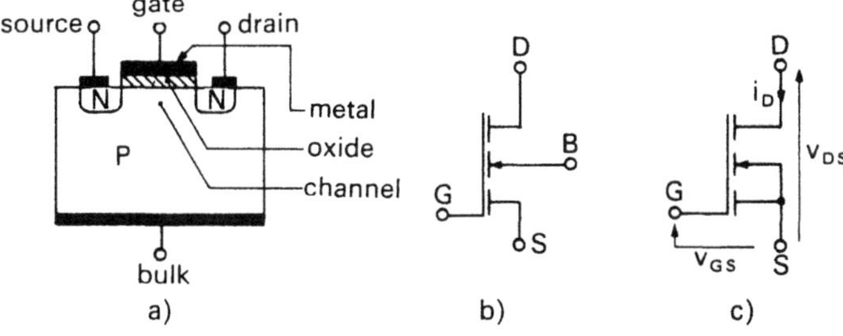

Fig. 2.43

2.3.1.3 "Power" MOSFETS

In the preceding structure, the only path available to the drain current was "horizontal", i.e. parallel to the wafer surface. Doping reversal only occurs over a thickness of a few microns and thus a prohibitive increase in the wafer surface would be needed to obtain a relatively high current.

Power MOSFETS have resulted from new technologies which provide the current with a vertical path, i.e. perpendicular to the surface, as in the bipolar transistor. At first, VMOS-type transistors were produced with the V standing for both the vertical path of the current and the form of the notch in the wafer surface. At present, the most frequently used power components are of the D MOS-type, with D standing for the double diffusion process used in their fabrication.

As in the case of the power bipolar transistor collector, the bulk is made up of two N type layers (Fig. 2.44). The N^- layer gives the component its voltage characteristics. In this layer P type cells are introduced during the first diffusion; during the second, N type cells are introduced into the P type cells.

A positive v_{GS} voltage attracts the minority charge carriers of the P zones near to the oxide layer, while repulsing the positive charges. When v_{GS} exceeds v_T, N type channels appear via which the current can flow from drain to source.

A little part of the current filaments remains horizontal but a substantial drain current can be obtained since the transistor is composed of a very large number of elementary cells, similar to that shown in Fig. 2.44 and electrically connected in parallel.

2.3.2 Steady-State Characteristics

2.3.2.1 Output Characteristics

Figure 2.45b shows the form of the characteristics which give drain current i_D, as a function of the drain–source voltage v_{DS}, for various values of the gate–source voltage v_{GS}. Figure 2.45a gives an enlarged view of the initial part of the curves.

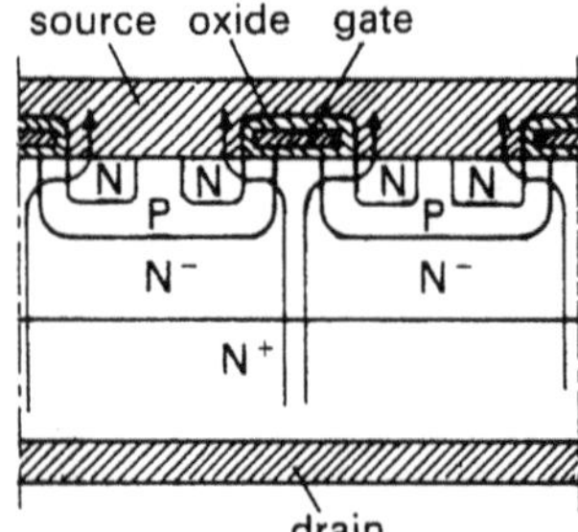

Fig. 2.44

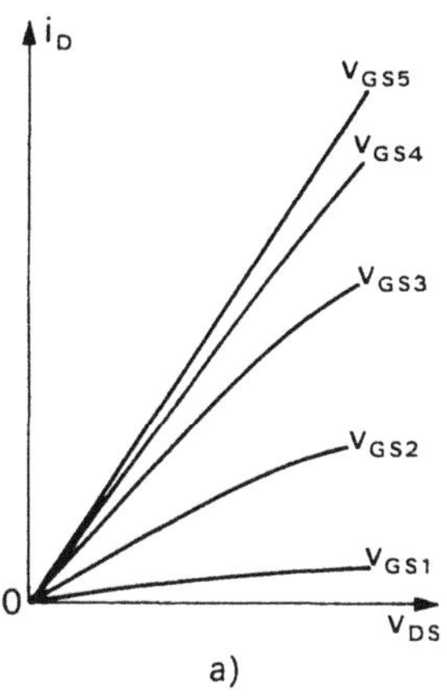

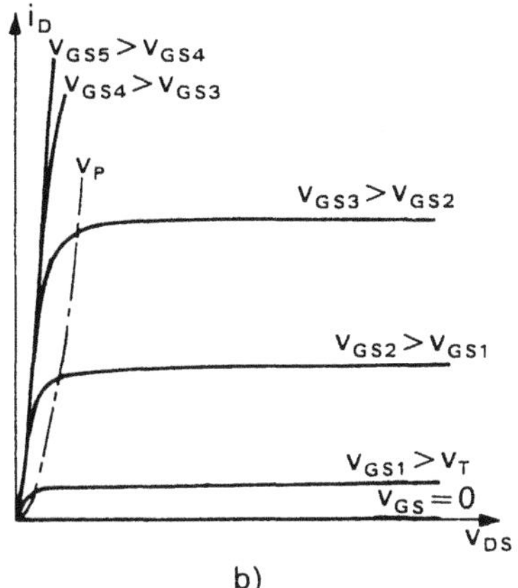

Fig. 2.45

Current i_D can only flow if N-type channels are created by a voltage v_{GS} greater than the threshold voltage v_T. The thickness of these channels therefore depends on the difference $v_{GS} - v_T$.

When v_{DS} has a low value – less than a few volts – current i_D is weak; its flow through the channel has no effect on resistivity and current i_D increases linearly with v_{DS}.

When i_D is so high that the electron fluxes saturate the channels, i_D characteristics are horizontal. The v_{DS} value corresponding to the channel saturation is called the "pinching" voltage v_P.

Operation in linear electronics concentrates on the zone in which the characteristics $i_D = f(v_{DS})$ are horizontal. In power electronics, the transistor is required to operate as a switch:

- the switch is OFF for v_{GS} lower than v_T; in fact, OFF state occurs for $v_{GS} = 0$. Resistance R_{DSOFF} between drain and source is then almost infinite.
- the switch is ON when v_{GS} is sufficiently high for voltage drop v_{DS} to be very low, for a given i_D. As Fig. 2.45a indicates, an increase in v_{GS} reduces v_{DS}; however, above a voltage of about 10 V, there is no further reduction discernable in v_{DS} for a given i_D.

2.3.2.2 Input Characteristics

As the gate is insulated, there is, in theory, *no current* between gate and source. In fact, there is a low leakage current of well below a microamp. The main *advantage* of the MOSFET transistor compared to the bipolar transistor is the very high input impedance. As will be seen (Sect. 2.3.3.1), the input impedance is mainly capacitive.

The transfer characteristics giving i_D as a function of v_{GS} at constant v_{DS} can be plotted. They show how current i_D rises rapidly, as soon as v_{GS} goes beyond the threshold voltage v_T (typically a few volts). The transfer characteristics slope is called forward transconductance g_{FS}.[1]

The rise in temperature increases the number of minority carriers in the P zones but makes the majority carriers less mobile. As the temperature rises, the threshold voltage decreases, but i_D rises less rapidly as a function of v_{GS}.

2.3.2.3 Apparent Resistance in the ON State

Instead of plotting the characteristics which give the voltage drop v_{DS} as a function of i_D in the ON state, data sheets usually give the variations of the apparent resistance $R_{DS\,ON}$, as a function of this current, for various values of v_{GS}. Since

$$R_{DS\,ON} = v_{DS}/i_D,$$

the corresponding curves can be deduced directly from those in Fig. 2.45a.

Let us compare (Fig. 2.46) the characteristics of a MOSFET and a bipolar transistor, equally able to let the same current flow and to sustain the same voltage. It can be seen that, when conducting, the bipolar device has a substantially lower voltage drop than the MOSFET: this is one of the main *drawbacks* of the latter.

In order to be able to sustain high voltages, both types of transistor require a lightly doped N layer. The higher the voltage to be sustained, the thicker this latter must be. In the conduction state, the N^- zone of the bipolar device is invaded by the minority carriers of the base and its resistivity decreases, while the carrier concentration in the N^- zone of the MOSFET shows hardly any variation.

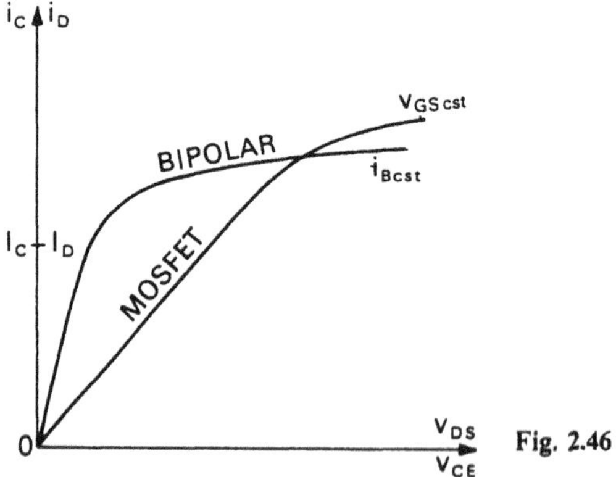

Fig. 2.46

[1] Both bipolar and field effect transistors were originally used in active region. This explains why parameters of little use for switching operation are still given for power transistors.

The multicellular structure of power MOSFETS enables the $R_{DS\,ON}$ resistances of all the cells to be connected in parallel and to avoid a prohibitive value of v_{DS}. This gives, in the case of $R_{DS\,ON}$, values from a few hundreds of an ohm (large wafers, low voltage) to about 10 ohms (small wafers, several hundred volts).

2.3.2.4 Safe Operating Area

In the output characteristics plane, the axes are normally graduated in logarithmic coordinates and the safe operating area is limited when MOSFET is in steady state (Fig. 2.47) by

- the maximum drain current in steady state I_D (segment BC),
- the maximum drain–source voltage $V_{(BR)DSS}$, corresponding to the avalanche breakdown of the P–N junction (segment DE).
- the maximum power dissipation (segment CD),
- the limit imposed by $R_{DS\,ON}$; for low values of v_{DS}, higher values of i_D cannot be obtained (segment AB).

Unlike the bipolar transistor, the MOSFET is not affected by thermal runaway. Indeed, the positive temperature coefficient of $R_{DS\,ON}$ reduces the current in paths where resistivity increases due to a temperature rise.

For the pulses of around a microsecond, the safe operating area is extended to the area in dotted lines AB′C′E. The rated maximum current becomes I_{DM}; the limit corresponding to the maximum power dissipation has disappeared. For longer pulses, the limits of the safe operating area lie between the two lines plotted in Fig. 2.47; these limits will indicate the maximum pulse duration.

2.3.2.5 The Parasitic Diode: Remarks

An unwanted bipolar transistor appears in the MOSFET structure as a result of the succession of diffused N and P zones and of the N^- zone (See Fig. 2.44). The effects of this transistor are reduced by the source metallization covering both the diffused P and N zones, thus short-circuiting the emitter–base junction.

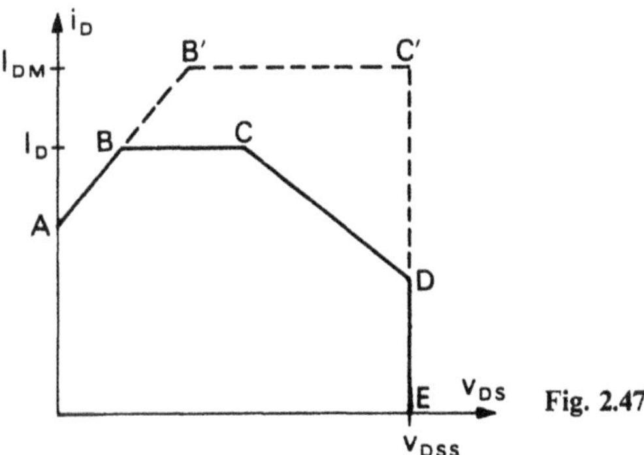

Fig. 2.47

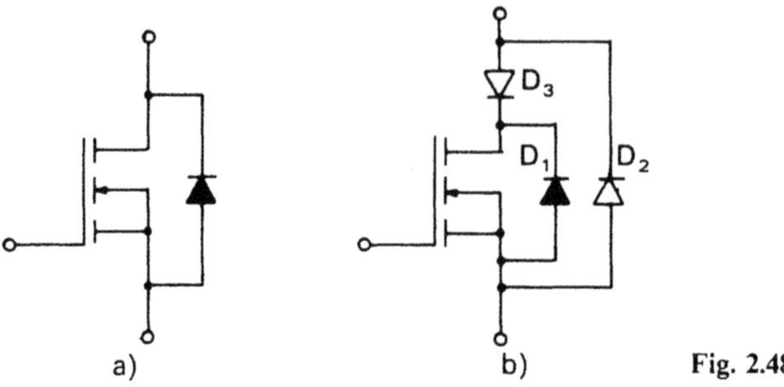

Fig. 2.48

However, the presence of the collector–base junction produces a parasitic diode of which the source is the anode and the drain, the cathode. In some equivalent circuits, this diode is shown in antiparallel across the MOSFET (Fig. 2.48a).

This body diode has a forward-voltage drop of the same amount as that of a rectifying diode. But this diode cannot be used in converters in which the transistors must be shunted by a fast-recovery diode, as its recovery time is too long.

In some transistors, doping with gold or platinium makes this diode fast, but this causes a substantial increase in $R_{DS\,ON}$.

If a fast-recovery diode D_2 has to be added across the transistor (Fig. 2.48b), the body diode effect has to be cancelled out by connecting a low voltage drop diode D_3 in series with the drain.

2.3.3 Commutations

2.3.3.1 Stray Capacitances and Their Charge

- In a MOSFET transistor, the stray capacitances limit the switching speed by the time required to charge or discharge them.

Three types are to be distinguished:

- the gate-to-source capacitance C_{GS}. Its dielectric is the oxide layer isolating the gate from the source. This capacitance is hardly affected by variations in voltage v_{DS}.
- the gate-to-drain capacitance C_{GD}. This accounts for the space charge region which appears in the P area beneath the gate. It varies considerably with voltage v_{DS}, going from a value similar to C_{GS} when the transistor is conducting (low v_{DS}) to a negligible value when it is nonconducting (high v_{DS}). This is shown in Fig. 2.49b.
- the drain-to-source capacitance C_{DS}. As its effects are hidden by those of C_{GD}, it is less important.

The sum $C_{GD} + C_{GS}$ is called input capacitance C_i.

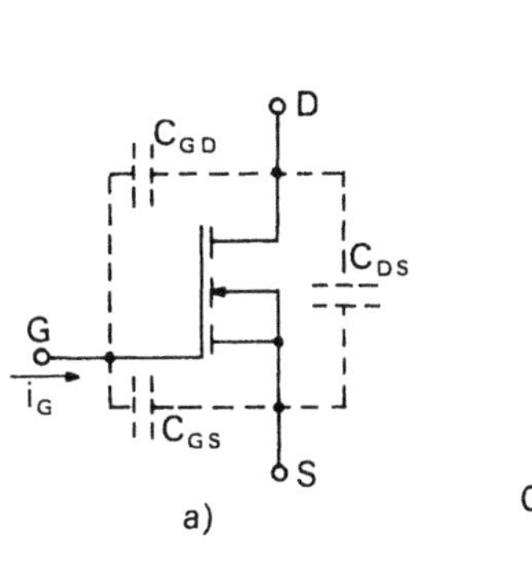

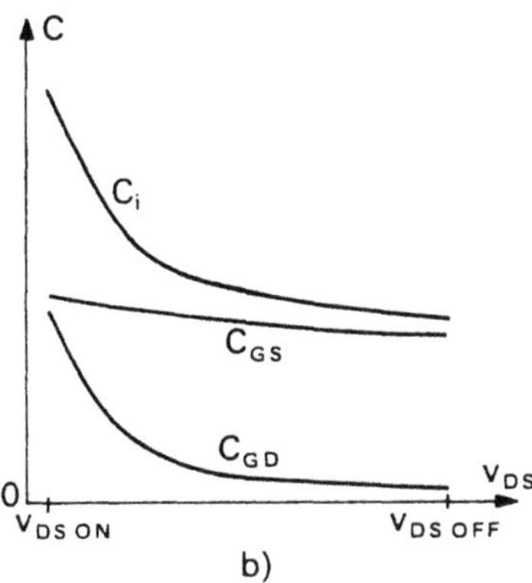

Fig. 2.49

- Figure 2.50 shows the increase in voltage v_{GS} as a function of charge Q_G carried to the gate by current i_G *during turn-on*. The OFF state is the starting point: v_{GS} equal to zero, v_{DS} having value E imposed by the supply of the transistor.

There are three distinct sections on this curve:

- *Section OA* corresponds to the charging of the input capacitance under full voltage v_{DS}. C_i is then virtually equal to C_{GS}. The amount of charge to be supplied depends on the drain current I the transistor must take.
- *Section AB* corresponds to the v_{DS} decrease from E to $v_{DS\,ON}$. Voltage v_{GS} remains unchanged. The amount of charge supplied is used to vary the voltage across C_{GD}; its value varies according to the initial value E of v_{DS}.
- *Section BC* correspond to the input capacitance charge when the transistor is on. The input capacitance is then equal to $C_{GS} + C_{GD\,ON}$ and is independent of I and E.

The amount of charge Q_2 enables the turn-on time to be calculated, using the values of the gate current i_G, the drain current I and initial voltage E. The

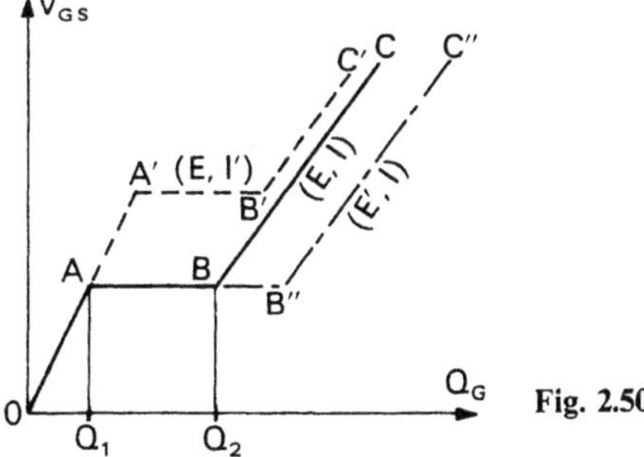

Fig. 2.50

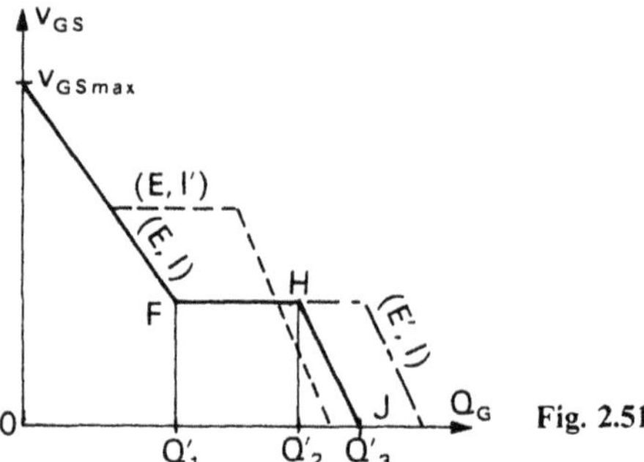

Fig. 2.51

dashed line OA′B′C′ corresponds to a current I' higher than I, the mixed line OAB″C″ corresponds to a voltage E' higher than E.

- *During turn-off* (Fig. 2.51), there will similarly be the removal of the excess charge (section CF), the discharge of C_{GD} during voltage rise (Section FH) and capacitance C_{GS} discharge during the current fall (Section HJ).

  The amount of charge Q'_3 must be removed via the gate during turn-off time.

- *Remark.* By reducing the thickness of the oxide layer surrounding the gate, transistors can be obtained with a threshold voltage which is sufficiently low for them to be directly driven by logic circuits with a 5 V supply. These are L2 FET (Logic Level Gate FET). Their gate-to-drain capacitance is higher.

2.3.3.2 Turn-on Switching

We study the normal case where the transistor is in series with an inductive load, shunted by a free-wheeling diode D, under a voltage E (Fig. 2.52a). Current I is assumed to be constant during commutations.

For $t = 0$, the control voltage v_i passes from 0 to E' (Fig. 2.52b); the input capacitance of the non-conducting transistor is charged via resistance R_i. Voltage v_{GS} reaches the threshold voltage, for $t = t_1$.

From $t = t_1$, current i_D increases. Until the latter reaches value I, the diode remains conducting and voltage v_{DS} equal to E. The diode current falls to zero for $t = t_2$.

From $t = t_2$, the diode recovery current is added to I and leads to a current peak i_D. At the same time, voltage v_{DS} decreases until it reaches the value $R_{DSON}\ I$ in $t = t_3$; turn-on is then finished.

Value E' of the control voltage must at least be at the level reached by v_{GS} between t_2 and t_3. For safety reasons, however, it is assumed to be higher and, after t_3, the input capacitance of the conducting transistor goes on charging and v_{GS} increases to E'.

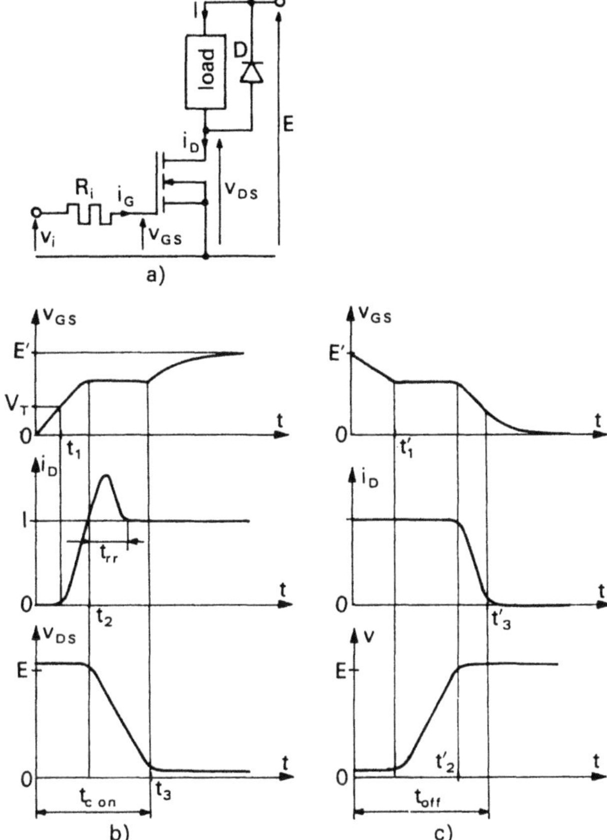

Fig. 2.52

The turn-on time is thus the sum of the delay time caused by the threshold voltage, the current rise time and the voltage fall time. Its reduction, with a given transistor, necessitates a decrease in the control circuit resistance R_i and the use of a fast diode with a very short recovery time t_{rr}.

2.3.3.3 Turn-off Switching

The transistor is assumed to be embedded in the same circuit (Fig. 2.52a) as before. Figure 2.52c gives the waveforms of voltage v_{GS} and v_{DS} and of current i_D, from the instant $t = 0$ at which the control voltage v_i goes from E' to zero. The input capacitance is discharged via R_i; current i_G is negative.

From $t = 0$ to $t = t'_1$, the gate removes the excess charges of the conducting transistor. From $t = t'_1$ to $t = t'_2$, voltage v_{DS} rises, but current i_D remains equal to I since the voltage across the diode is still negative. From $t = t'_2$ to $t = t'_3$, current I is transferred from the transistor to the diode. The transistor reaches the OFF state for $t = t'_3$; input capacitance discharge then continues to bring v_{GS} to zero.

When resistance R_i is reduced, the time constants of the input capacitance discharge circuit are reduced as well as the turn-off time t_{off}.

- *Remarks*

- Stray capacitances have a value of a few hundred picofarads; it is thus possible to obtain very short switching times and operating frequencies up to several hundred kilohertz.
- Current i_D varies so quickly during commutations that, if there is a stray inductance in this current circuit, it produces substantial variations of voltage v_{DS}. This rapid variation leads to a decrease of v_{DS} during the interval (t_1, t_2); it increases this voltage during the interval (t'_2, t'_3). The transistor may have to be protected against this voltage surge at turn-off.

2.3.4 MOS Transistor – Bipolar Transistor Associations

Several circuits enable a switching device combining the advantages of both MOS and bipolar transistors to be produced.

2.3.4.1 BIPMOS Circuit

This association (Fig. 2.53a) can be deduced from the Darlington circuit, if the input bipolar transistor is replaced by a MOSFET.

The high input impedance of the MOSFET, allowing for simplified control, is combined with the lower forward-voltage drop of the bipolar transistor.

The circuit can be integrated or made from separate elements. Figure 2.53b shows elements which may be introduced into this to take full advantage of certain parameters:

- Resistance R_{DC} enables the quasi-saturation of the bipolar device to be adjusted. As it increases, the voltage drop in the ON state is higher but the turn-off time is lower.

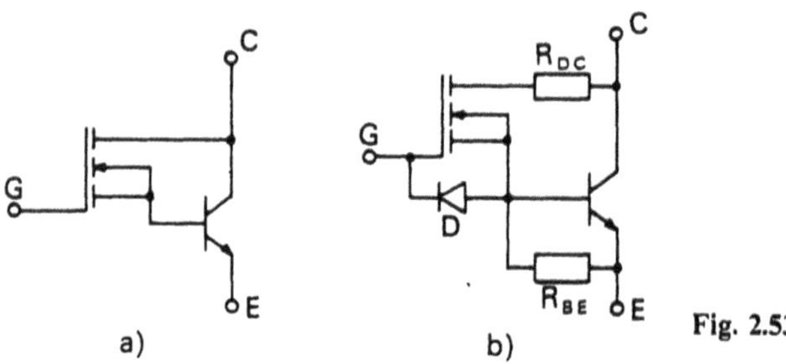

Fig. 2.53

- Resistance R_{BE} provides a path for the negative base current during the turn-off of the bipolar transistor. The weaker it is, the faster the turn-off will be. However, by reducing the base current during conduction, it increases the forward-voltage drop.
- Diode D also makes the negative base current disappear more quickly during turn-off of the bipolar device, provided that the gate control circuit can let such a current flow.

2.3.4.2 Cascode Circuit

By placing two components in series (Fig. 2.54), we can take advantage of the fast turn-off of the MOSFET to accelerate that of the bipolar transistor.

Such an arrangement does not improve turn-on switching. On the contrary, each transistor slows the current rise of the other. However, this is a minor disadvantage as the turn-on losses are much lower than the turn-off losses.

The other drawback is that the forward-voltage drop v_{CS} is higher than that of the bipolar device v_{CE}, since the MOSFET drop v_{DS} is added to it. But, in the OFF state, the voltage across the MOSFET is low: a low-voltage component with highly reduced R_{DSON} can be used.

The advantage of such a circuit lies in its turn-off time when the voltage at the MOSFET is reduced to zero, the latter blocks rapidly and forces the entire collector current of the bipolar device to leave by the negative base current. The excess charges in the bipolar transistor are thus removed more quickly.

2.3.4.3 Parallel Circuit

In the parallel grouping of a MOSFET and a bipolar transistor (Fig. 2.55), the former takes care of the commutations and the latter of the reduction in voltage drop during most of the conducting period.

At turn-on, the MOSFET gate is the first to be supplied, and the MOSFET switches on rapidly. The transistor base is supplied next, the bipolar transistor reaches the saturation state and v_{CE} falls to $v_{CE\,sat}$.

To prepare for turn-off, the bipolar transistor base is no longer supplied and this transistor reverts to the OFF state. The turn-off corresponds to that of the MOSFET which rapidly switches off when the gate voltage falls to zero.

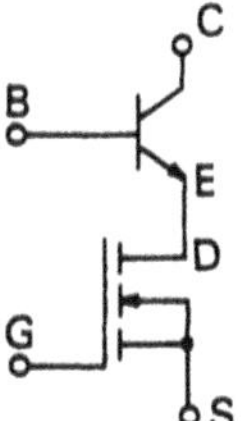

Fig. 2.54

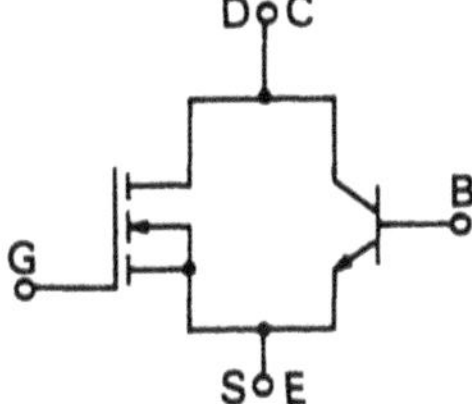

Fig. 2.55

2.3.5 Notes on the Insulated Gate Bipolar Transistor

The name of this component varies according to the manufacturers: IGBT (Insulated Gate Bipolar Transistor); GEMFET (Gain Enhanced Mosfet) or COMFET (Conductivity Modulated FET). In the integrated structure, two bipolar transistors and one MOSFET can be distinguished (Fig. 2.56).

The MOSFET appears in the layout of the P and N zones, located above the P^+ layer, which reproduces the layout seen previously. The NPN transistor appears in the three successive layers found between collector C and emitter E. Resistance R corresponds to the P zone surrounded by the two N-type cells, which are in contact with the same emitter metallization. All elementary emitters are connected to the E terminal of the IGBT.

Insulated gate control allows the input impedance to be similar to that of the power MOSFET. Adding the P^+ layer brings about two series of results.

- *Turn-on. ON state.* When voltage v_{CE} is positive and voltage v_{GE} becomes greater than threshold voltage v_T, channels appear; electrons are injected, via these channels, into the N^- layer. The latter's potential decreases. The P^+N^- junction becomes conducting as does the device.

The N^- layer receives electrons from the emitter *and* holes from the collector. Its resistivity decreases and its apparent resistance becomes lower than that of a MOSFET where no injection of holes into the N^- zone is found.

The steady-state characteristics $i_C = f(v_{CE})$ indicated in Fig. 2.57a show two differences compared to those of the MOSFET (see Fig. 2.45):
- current i_C only moves away from zero when v_{CE} has gone over the threshold voltage of the P^+N^- junction;
- the apparent resistance in the saturation state is lower.

- *Turn-off.* Conduction by minority carriers has the advantage of reducing the voltage drop in the ON state, but the drawback of increasing turn-off time, and thus limiting the frequency range of operation of IGBT.

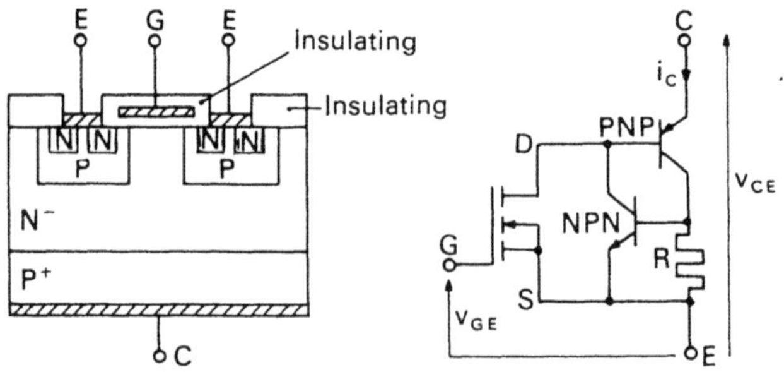

Fig. 2.56

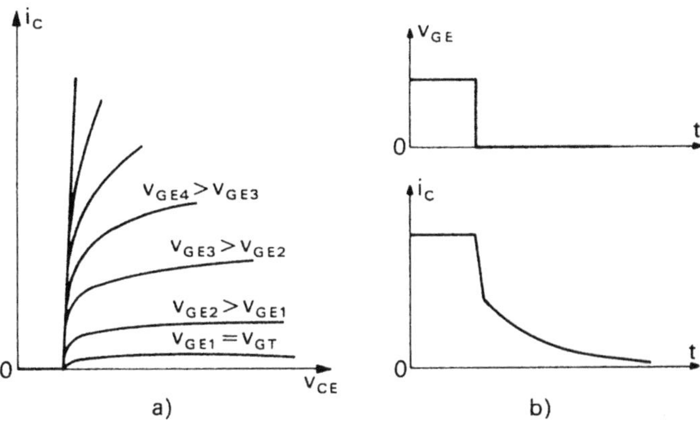

Fig. 2.57

When the gate control signal is suppressed, the turn-off takes place in two stages (Fig. 2.57b):

- the channels disappear and the MOSFET blocks quickly, leading to an initial decrease of i_C;
- the excess carriers of the N^- layer then gradually recombine and the i_C decrease becomes relatively slow.

- *Remark.* In normal operation mode, the current across the low resistance R is not strong enough for the emitter–base junction of transistor NPN to becomes conducting. The collector–base junction of this transistor has a stray capacitance across which there is a voltage close to the drain–source voltage of the input MOSFET.

If the drain–source voltage falls too quickly, at turn-off, the current which the discharging stray capacitance sends into R may be strong enough to make the transistor NPN conduct: the two bipolar transistors behave, in this case, like a thyristor in the avalanche state and current i_C is no longer controlled by voltage v_{GE}. The value of i_C, above which this avalanche appears, is called the *latch-up current.*

If the IGBT is to be controlled by the gate again, i_C must fall below the avalanche holding current.

2.4 Thyristors

The thyristor or SCR (Silicon Controlled Rectifier) is the semiconductor device which lies at the origin of the rapid development in power electronics. It acts as a switch unidirectional in current, of which only the turn-on can be controlled. It is thus the basic device for converters operating in natural commutation.

In order to use it in forced-commutation converters, an extra turn-off circuit must be added (see Chap. 5) or a controlled turn-off thyristor used in its place. This is the GTO thyristor which will be studied in the last part of this chapter.

This difficulty in turn-off has led to the thyristor being replaced by transistors in low- and medium-power equipments.

2.4.1 Description and Operation

2.4.1.1 Description

As shown in Fig. 2.58a, the thyristor is a semiconductor device with four alternately P and N doped layers. Metallization is used to obtain the three terminals – anode A, cathode K and gate G.

Theoretically, there are two types of thyristor:

- N type or anode gate thyristors, in which the gate is connected to the N layer next to the anode:
- P type or cathode gate thyristors, in which the gate is connected to the P layer next to the cathode.

In practice, only the second type, corresponding to Fig. 2.58a, is found. Figure 2.58b gives the circuit symbol for the thyristor and indicates the voltage and current conventions.

The thickness and doping of each layer are different and affect the thyristor properties:

- the N_2 layer (cathode layer) is very thin and highly doped;
- the P_2 control layer is thicker and less highly doped. Together with N_2, it forms the cathode junction J_K;
- the N_1 layer (blocking layer) is the thickest and the less doped. It gives to thyristor its blocking possibility; together with P_2 it forms the control junction J_C;
- the P_1 anode layer has similar characteristics to those of P_2. Together with N_1, it forms a third junction – the anode junction J_A.

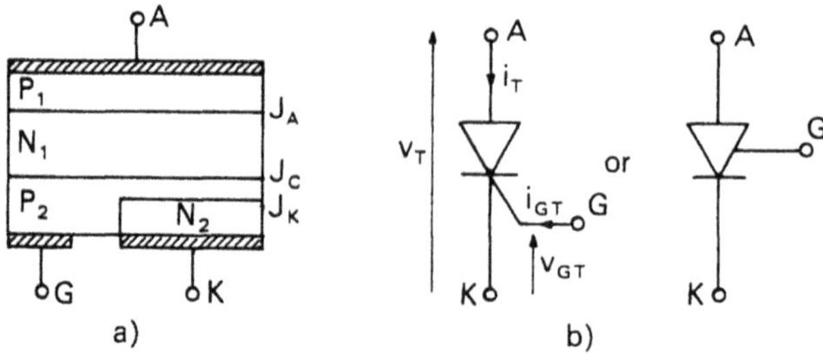

Fig. 2.58

2.4.1.2 Forward-Biased Thyristor Triggering

- The thyristor is forward-biased when voltage v_T between anode and cathode is positive. The potential continuously decreases from A towards K and thus junctions J_A and J_K are forward-biased. Junction J_C is, however, reverse-biased. Almost the whole of voltage v_T is therefore applied between the N_1 and P_2 layers which form this junction.

If the number and energy of the minority free electrons in the P_2 layer become so high as to produce an avalanche breakdown in junction J_C, a strong reverse current can cross the latter: a substantial forward current can flow from anode to cathode.

The resistivity of the silicon in the N_1 layer is reduced by the electron flux which comes into this zone after the avalanche: voltage v_T becomes very low, the thyristor is in the ON state, and current i_T is only limited by the circuit into which the thyristor is embedded.

Several phenomena can produce an avalanche breakdown of junction J_C.

- *Controlled avalanche.* To make the avalanche appear at a chosen moment, a positive voltage pulse is applied between the gate and the cathode. This pulse attracts a large number of electrons from the highly doped N_2 layer into the P_2 layer which is very thin and partly invaded by the depletion zone of the reverse-biased junction J_C. The saturation current of this junction then increases substantially and, if voltage v_T is sufficiently positive, the avalanche breakdown appears and the thyristor turns on.

Positive voltage v_{GT} only initiates the effect which is then kept up by the large number of electrons moving from K to A. Voltage v_{GT} can thus be brought down to zero after triggering.

(There are light-triggered thyristors in which the rise in the reverse current of junction J_C – which sets off the avalanche – is obtained by use of a photoelectric effect. The photons sent on to the P_2 layer bring about a rise of number of minority carriers.)

- *Uncontrolled avalanche.* Other phenomena can bring about the breakdown of junction J_C by avalanche. As they are more difficult to control, their appearance must be avoided when the thyristor is used as a controlled turn-on switch.
  - If voltage v_T is increased, without acting on the gate, the depletion zone of junction J_C is enlarged and its reverse current increases. When v_T reaches the forward-breakdown voltage V_{BO}, the avalanche effect appears and the thyristor turns on.
  - Without the voltage v_T reaching V_{BO}, firing can occur if voltage v_T is increased too rapidly, without any action on the gate. When the reverse voltage across junction J_C increases sharply, the depletion zone of this junction is rapidly enlarged and the minority carriers of P_2 are repulsed

into N_1. The number and speed of these electrons, corresponding to the charge of the space charge capacitance of junction J_C, depend on the rate of rise dv/dt of voltage v_T. If the latter is too strong, the thyristor is triggered.

- When there is a temperature rise, the reverse current of J_C may become sufficiently high for the avalanche effect to appear at a value of v_T lower than V_{BO}.

2.4.1.3 Thyristor Turn-off

A conducting thyristor can only be turned off by making the avalanche effect in it disappear. This is done by reducing the current below the minimum needed to maintain this effect.

This minimum is called the *holding current* and it is denoted I_H. It decreases with the temperature and increases if a negative voltage v_G is applied. This enables the turn-off commutation in some devices to be speeded up (see Sect. 2.4.4.2).

In AC supplied circuits, turn-off can be achieved either by current i_T falling to zero (free natural commutation) or by transferring i_T to another path by means of the supply (line commutation).

In DC-supplied circuits, an external turn-off (forced commutation) circuit should normally be added to the thyristor, thus creating a considerable drawback.

2.4.1.4 Two-Transistor Analogy

The thyristor triggering can also be explained by considering the four-layer structure as made up of two interconnected bipolar transistors $P_1N_1P_2$ and $N_1P_2N_2$ (Fig. 2.59).

If i_{CBO_1} denotes the saturation current of the collector–base junction of transistor $P_1N_1P_2$ and i_{CBO_2} that of transistor $N_1P_2N_2$, the equations linking the collector and emitter currents for both transistors can be expressed as

$$i_{C_1} = \alpha_1 i_{E_1} + i_{CBO_1}$$

$$i_{C_2} = \alpha_2 i_{E_2} + i_{CBO_2} .$$

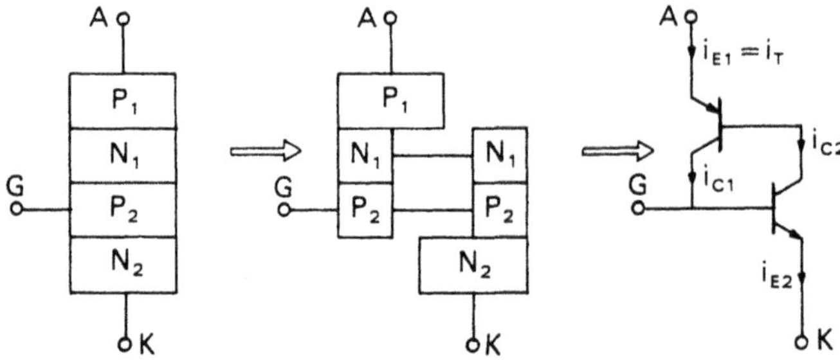

Fig. 2.59

For zero gate current, current i_T is given by

$$i_T = i_{E_1} = i_{E_2} = i_{C_1} + i_{C_2}$$

or, if i_{C_1} and i_{C_2} are replaced by their expressions,

$$i_T = \alpha_1 i_{E_1} + \alpha_2 i_{E_2} + i_{CBO_1} + i_{CBO_2}.$$

The sum of $i_{CBO_1} + i_{CBO_2}$ is equal to the total saturation current i_{CBO} across junction P_2N_1. Current i_T flowing through the thyristor is thus given by

$$i_T = \alpha_1 i_T + \alpha_2 i_T + i_{CBO}$$

$$i_T = \frac{i_{CBO}}{1 - (\alpha_1 + \alpha_2)}.$$

- For low values of i_{C_1} and i_{C_2}, the current gains α_1 and α_2 are low. Current i_T is small but still higher than i_{CBO}.
- If the base current of transistor $N_1P_2N_2$ increases, whether by injection of a gate current in addition to i_{C_1} or by an increase in i_{CBO} leading to a rise in i_{C_1}, current i_{C_2} increases as well. For low values of i_C and i_E, the current gain α increases in the same way as these currents. If such an increase brings the sum of $\alpha_1 + \alpha_2$ to nearly 1, current i_T becomes very high and is only limited by the thyristor load circuit.
- Thyristor turn-off can only be achieved by reducing i_T to a value low enough for $\alpha_1 + \alpha_2$ to become much lower than 1.

2.4.1.5 Reverse Bias

When the anode voltage is less than that of cathode, junction J_C is forward-biased while junctions J_A and J_K are reverse-biased. However, when v_T is negative, these junctions behave differently: due to the heavy doping of the N_2 layer, the avalanche voltage of junction J_K is well below that of junction J_A whose N_1 layer is less heavily doped.

In reverse bias, most of voltage v_T appears across junction J_A. Since the doping of the P_1 and P_2 zones is similar, the avalanche voltage of J_A is virtually equal to that of J_C. The reverse-breakdown voltage of a conventional thyristor is more or less the same as its forward-breakdown voltage V_{BO}: it is thus known as a *symmetrical thyristor*.

When the thyristor has become reverse-conducting, the holes in the P_1 layer invade the N_1 layer: therefore, the resistivity of the latter does not decrease as was the case during forward-bias avalanche. The reverse current increases under a reverse voltage which remains high: the thermal dissipation limits are reached very quickly. A negative voltage higher than its reverse-breakdown voltage V_{BR} must therefore not be applied to the thyristor.

2.4.2 Steady-State Characteristics

2.4.2.1 Output Current–Voltage Characteristics

- Figure 2.60 gives the anode current i_T variations as a function of the anode-to-cathode voltage v_T, for zero gate current.

  This characteristic comprises three distinct branches:

- *With negative voltage* (OA). This characteristic has the same form as that of a PN junction. The thyristor is characterised by the tolerable maximum repetitive reverse voltage V_{RRM}.
- *With positive voltage and blocking state* (OB). Voltage v_T can reach V_{BO} without the thyristor allowing a significant forward current to flow. When v_T reaches v_{BO}, the current becomes sufficiently strong to bring about the avalanche breakdown: the corresponding value is called the latching current.

  The thyristor is characterized by the maximum repetitive forward voltage in the blocking state, V_{DRM}.
- *With positive voltage and in the on-state* (CD). The ON-state voltage drop is low. Often indicated for the peak maximum current I_{TM}, it goes from 1.5 to 2 V for slow thyristors and from 2.5 to 3 V for fast thyristors.

  In order to characterize the forward current, the following values are indicated, for usual waveforms and various case temperatures:

  I_{TAV}, average forward current,
  I_{TRMS}, rms forward current,
  I_{TM}, maximum repetitive peak current,
  I_{TSM}, maximum overload peak current,

  When the temperature rises, the values of V_{BO} and I_H decrease; on the contrary the reverse value of I_T increases.

The current–voltage characteristic of the ON-state thyristor can be assimilated to a straight line segment of the equation

$$v_T = V_0 + ri_T\,.$$

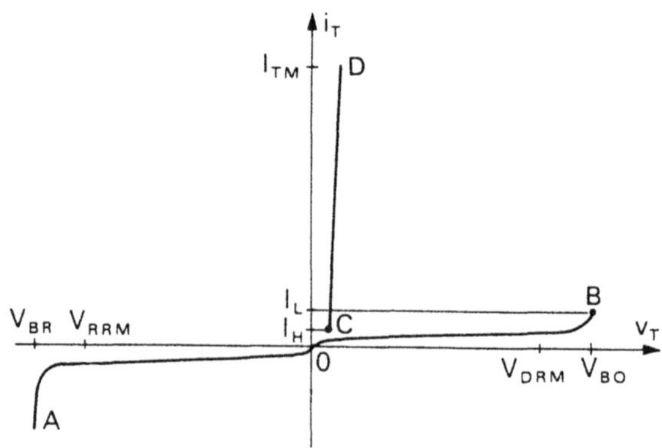

Fig. 2.60

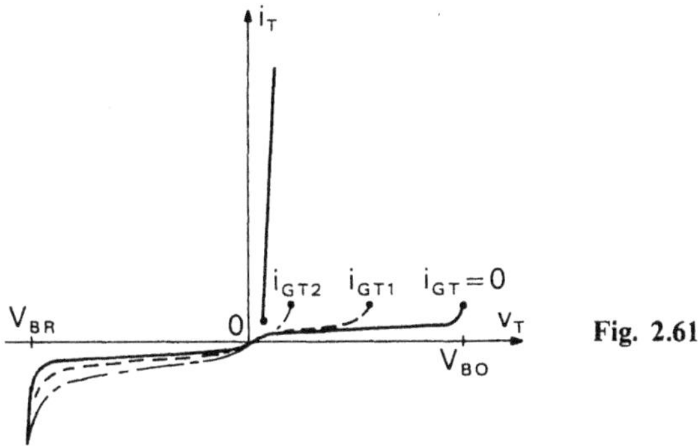

Fig. 2.61

When the temperature rises, r increases but V_0 decreases.

The ON-state losses can be computed in the same way as for a diode (see Sect. 2.1.2.2).

- Figure 2.61 shows the current–voltage characteristics for different values of the gate current; the characteristics at zero i_{GT} are used as a reference.

  For a positive value i_{GT_1} of the gate current, the avalanche breakdown occurs for a value of v_T smaller than V_{BO}.

  With negative voltage, the value of i_{GT} has no effect on voltage V_{BR} which only depends on the P_1–N_1 junction. On the other hand, the gate current increases the base current of transistor $N_1P_2N_2$ and thus its collector current i_{C_2} and, in turn, i_{C_1}, since i_{C_2} is the base current of transistor $P_1N_2P_2$ (see Fig. 2.59): current i_{GT} increases the reverse current of the thyristor and the corresponding losses when v_T is negative. The advantage of controlling the gate by pulses is clear.

  If i_{GT} increases, the previous modifications become more important, as can be seen in Fig. 2.61, where i_{GT_2} is greater than i_{GT_1}.

- *Remark*: The overload factor I^2t is defined, as in the case of a diode (see Sect. 1.2.3), by

$$I^2t = \frac{1}{2} I^2_{TSM} T$$

  with a sinusoidal half-wave of amplitude I_{TSM} and duration T equal to that of a half-cycle of the industrial network.

2.4.2.2 Gate Characteristics

The characteristics giving gate-to-cathode voltage v_{GT} as a function of gate current i_{GT} have the same shape as for a diode (Fig. 2.62).

For the thyristors of a same series, they show a noticeable dispersion on account of the specific layout of zones P_2 and N_2. Therefore, manufacturers will

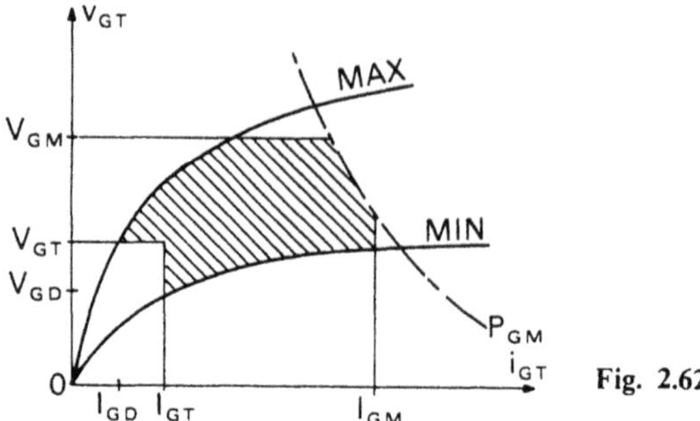

Fig. 2.62

usually indicate the maximum and minimum limits between which all the gate characteristics of the same thyristor type are to be found.

- The minimum values of v_{GT} and i_{GT} which are required to trigger any thyristor of a given series are shown for given junction (J_K) temperature and voltage v_T. They are denoted V_{GT} and I_{GT}.

  However, the voltage and the current must remain below the maximum values V_{GM} and I_{GM}, above which junction J_K may be destroyed. Similarly the product v_{GT} i_{GT} must remain below the maximum power P_{GM} which can be dissipated by the gate–cathode junction.

  The firing pulse must thus bring the operating point within the hatched area in Fig. 2.62.

- The value V_{GT} remains between 1 and 3 V, while value of I_{GT} varies from tens to several hundreds of milliamperes depending on the thyristor size.

  As the temperature increases, the thyristor becomes more sensitive and the values of V_{GT} and I_{GT} decrease.

- The value below which even the most sensitive thyristors in the series cannot be triggered are also specified. They are denoted V_{GD} and I_{GD}.

  The maximum reverse gate voltage and current, above which junction J_K may become conducting by avalanche breakdown, are denoted V_{GRM} and I_{GRM}.

2.4.3 Switching Characteristics

2.4.3.1 Turn-on Transient

If voltage v_T is positive, avalanche breakdown occurs when the gate current reaches a sufficient value.

The *total rise time* t_{on} of current i_T is the sum of

- delay time t_d, corresponding to the difference between the moment when the electrons are attracted from N_2 into P_2 and when they become sufficiently numerous to bring about avalanche breakdown in the control junction

- rise time t_r of the current, corresponding to the time taken by the avalanche to spread across the entire control junction starting from the place where it appeared.

The t_r time which makes up the major part of t_{on} varies with the nature of the load.

Figure 2.63a corresponds to a resistance supplied from a DC voltage source E through the thyristor. Since v_T equals $E - Ri_T$, the fall in v_T is linked to the rise of i_T. Commutation time $t_{c\,on}$ equals t_{on}.

Figure 2.63b corresponds to an RL load supplied from a DC voltage source across the thyristor. The inductance slows down the current rise which goes on after fall in v_T and the switch turn-on. In this case, commutation time $t_{c\,on}$ is shorter than t_{on}.

Figure 2.63c corresponds to the buck converter circuit which we used in studying commutation with transistors (see Sects. 2.2.4.2, 2.2.6.3 and 2.3.3.2); voltage v_T can only fall after the current rise. There is always a peak caused by the diode recovery. In this case, $t_{c\,on}$ is longer than t_{on}.

The turn-on time is generally much shorter than the thyristor turn-off time and does not introduce significant limitation on the operating frequency (in fact

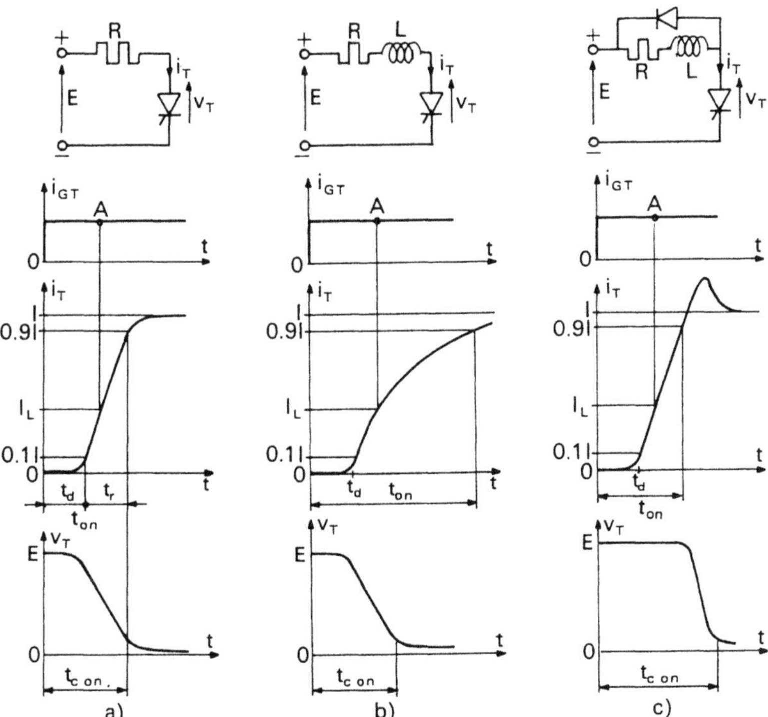

Fig. 2.63

t_r may, in some cases, be deliberately increased, as will be seen in the next paragraph). For relatively high operating frequencies, it may be necessary to reduce the switching losses by using the same method as for the transistors (see Appendix).

When current i_T exceeds the value of the latching current I_L, beyond points A on the graphs of Fig. 2.63, the gate current can be eliminated. Under inductive load, the slower increase in i_T requires longer gate pulses than with resistive load.

2.4.3.2 Limitation of the Current Rate of Rise (di/dt)

An avalanche phenomenon initiates the change of the thyristor from the OFF state to the ON state. The avalanche occurs locally before spreading across the entire control junction at a speed close to a tenth of a millimeter per microsecond.

If the current rises too fast, the current density in the part of the wafer where the avalanche appears may go over the permissible limits and destroy the device. The manufacturers have thus given the gate an interdigited structure, intended to increase its surface, in order to multiply the number of initial points where local avalanches might appear and to quicken their diffusion in the entire junction. It is thus possible to obtain gates which enable a di/dt of 1000 A/μs to be reached.

For each type of thyristor a limiting value of di/dt is indicated. This must not be exceeded. If the load circuit inductance is insufficient to limit the current rate of rise, the device must be protected by connecting an inductance with it in series.

2.4.3.3 Turn-off Transient

The 4-layer structure of the thyristor is a drawback when turn-off is required, since each layer has to return to its equilibrium state for this to occur.

A thyristor can be turned off by the application, during a sufficient period, of a negative voltage v_T, so that current i_T is reversed and, as a result, becomes lower than the holding current. The waveforms of current i_T and voltage v_T are those shown in Fig. 2.64.

- *Turn-off time* t_q is the minimum time period to be respected between the instant when current i_T falls to zero and that when a positive voltage can be reapplied to the thyristor, without the latter conduct again. Also known as recovery time, t_q thus denotes the specified time period required by the thyristor to recover its ability to block a forward voltage after a conducting period. The turn-off time is the sum of two separate times:
  - *the reverse recovery time* t_{rr}. This is the time which junctions J_A and J_K need to recover their blocking ability. The same phenomena as were observed during diode switching are found (Sect. 2.1.3.2).

    Under the reversal of the electric field direction, the load carriers which correspond to forward current I and have not disappeared by recombina-

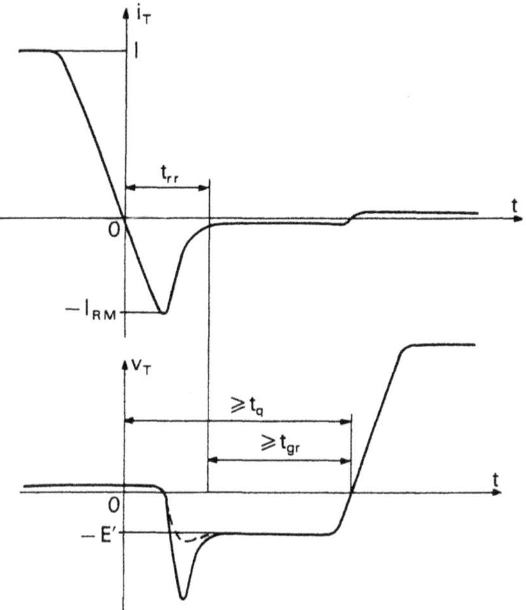

Fig. 2.64

tion during the decrease in i_T, move in opposite direction and set off the peak reverse current i_{RM}.

When the space charge zones of J_A and J_K have emptied themselves of these charge carriers, they recover their blocking ability. J_A blocks just before J_K since the N_1 layer is less heavily doped than N_2. The reverse current is stabilized at a very low value, near to that of the saturation current of J_A and J_K.

- *the gate recovery time* t_{gr}. This is the time needed for the residual charge carriers – which have remained trapped in the P_2 and N_1 layers after J_A and J_K have blocked – to disappear. They can only disappear by recombination or through the saturation current of these junctions. This requires a considerable time which makes up most of t_q.

- Voltage v_T remains close to zero as long as J_A and J_K are reverse-conducting. As soon as these two junctions have recovered their blocking ability, the negative supply voltage is applied across the thyristor; it is increased by the voltage surge caused by the variation of current i_T in the inductance in series with the thyristor; this can lead to a negative voltage peak.

  If this peak is likely to damage the thyristor, it must be reduced (broken line in Fig. 2.64) using an RC snubber across the device.

- The successive phenomena involved in the thyristor turn-off, as well as the definition of this time itself, explain why the *turn-off time value* is dependent to a high degree on numerous factors.

In particular, t_q increases with

- the junction temperature,
- the forward current I before turn-off,
- the rate of fall of this forward current,
- the rate of rise of the reapplied voltage.

It decreases when the applied reverse voltage is increased.

In the case of converters operating in the natural commutation mode at industrial frequencies, long turn-off times are acceptable: slow turn-off thyristors are used, with a t_q which may reach hundreds of microseconds for very high power devices. These thyristors are often called rectifier thyristors.

In forced commutation, fast turn-off thyristors must be used: gold doping enables carrier recombination to be speeded up, to the detriment of the reverse blocking voltage and of the ON-state forward-voltage drop. t_q in tens of microseconds can thus be obtained for high current rating devices.

2.4.3.4 Reapplied Forward Voltage Rate of Rise (dv/dt): Limitation

The avalanche breakdown may appear – at zero i_{GT} and before v_T has reached V_{BO} – if the forward voltage settles too quickly across the blocked thyristor (see Sect. 2.4.1.2). Manufacturers therefore specify the critical rate of rise of v_T beyond which there is likely to be an unintended turn-on.

This rate is specified in the conditions shown in Fig. 2.65. Voltage v_T increases exponentially towards value E, with a time constant τ. The rate of rise is defined by the ratio between the value of v_T, at instant t equal to τ, and this time constant:

$$\frac{dv}{dt} = 0{\cdot}63\,\frac{E}{\tau};$$

$(dv/dt)_{crit}$ is the maximum value of this ratio which does not lead to turn-on.

By varying the doping in order to make the charge carriers recombine rapidly, $(dv/dt)_{crit}$ of 1000 V/μs can be reached. This value decreases when the temperature increases as the charge carriers are then more numerous.

By itself, the turn-on of a thyristor by an excessive (dv/dt) will not destroy the component. But the ensuing modification of the circuit topology can lead to this destruction (e.g. see Vol. 1, Chap. 7, § 4).

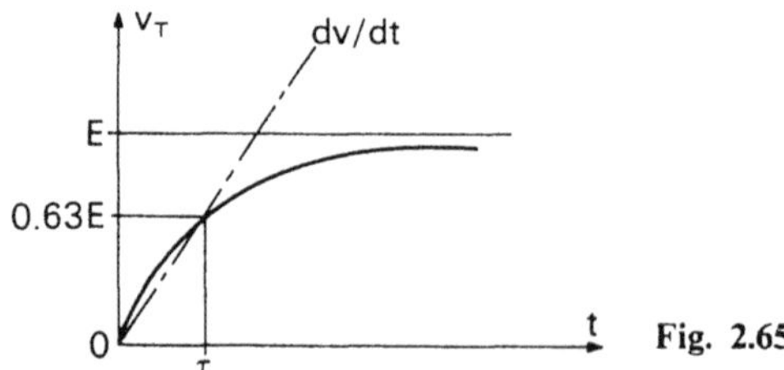

Fig. 2.65

2.4.3.5 Overvoltage Protection

Current variation in the circuit inductances (di/dt limitation or stray inductances) bring about voltage surges which must not go beyond the maximum acceptable values. Moreover, when the thyristor turn-off is achieved, the return of voltage v_T to value E should take place at a rate smaller than the critical value of dv/dt.

The thyristor is therefore protected by an RC circuit (Fig. 2.66a), which reduces the voltage surge amplitude and decreases the rate of rise of v_T. When the free-wheeling diode starts to conduct, voltage E is applied across series-connected elements l, R and C. Resulting waveforms of current i, and voltages v and v_T are shown in Fig. 2.66b.

Resistance R must be rated so that the current peak due to the capacitor discharge at turn-on thyristor does not exceed the maximum acceptable value for i_T. Figure 2.67 shows different topologies which enable the charge and discharge time constants of the capacitance to be varied.
(In Figs. 2.66 and 2.67, the auxiliary circuit, which turns off the thyristor, is not shown.)

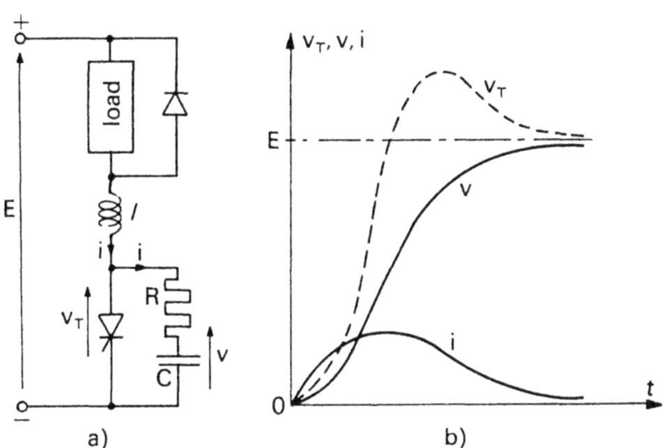

Fig. 2.66

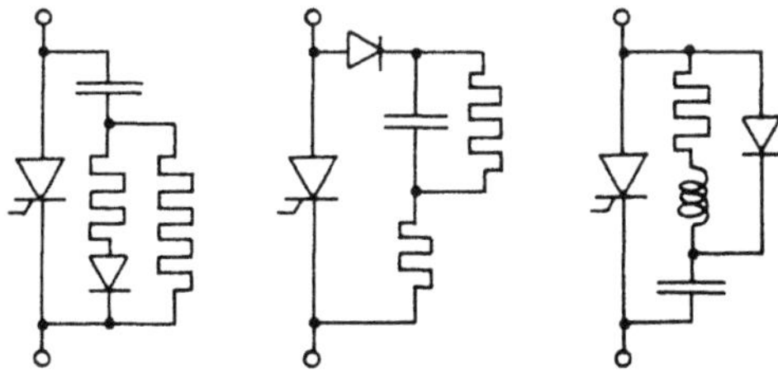

Fig. 2.67

2.4.4 Special Thyristors

Special thyristors operate on the same principles as conventional thyristors but are slightly modified so as to improve one or other characteristic.

2.4.4.1 Amplifying Gate Thyristor

An extra N'_2 zone is diffused into the P_2 layer to form a second thyristor, comprising $N'_2P_2N_1P_1$ (Fig. 2.68). As surface of the $P_2N'_2$ junction is small, only a very weak gate current is needed to turn on the second thyristor, by causing an avalanche into the P_2N_1 junction. The current which then flows across the thyristor is fairly weak but enables the avalanche to spread across the entire control junction more quickly, by increasing the number of electrons injected into P_2.

The second thyristor acts as a driver thyristor. Amplifying the gate signal enables weaker control signals to be used, the turn-on time to be reduced and higher di/dt to be accepted.

2.4.4.2 Gate-Assisted Turn-off Thyristor

In order to reduce the gate recovery time, a reverse voltage can be applied to the gate during turn-off. Carriers trapped in the thyristor are thus enabled to escape via the gate: thyristors obtained by this means are called "ultra-fast" thyristors.

This procedure is used in the Gate-Assisted Turn-off Thyristors (GATT); it cannot be applied to all types since there will be an increase in commutation losses if the diffused zones have not been suitably designed.

2.4.4.3 Asymmetrical Thyristor

The asymmetrical thyristor or ASCR (Asymmetrical Silicon Controlled Rectifier) is made by inserting a heavily doped N'_1 layer between the N_1 and P_1 layers of a conventional thyristor (Fig. 2.69a).

This addition has virtually no effect on the forward blocking voltage which is dependent on the avalanche breakdown voltage of the P_2N_1 junction. On the other hand, the reverse blocking voltage is substantially reduced since it is dependent on the avalanche breakdown voltage of the $P_1N'_1$ junction, which is well below that of the P_1N_1 junction of a conventional thyristor. This is due to

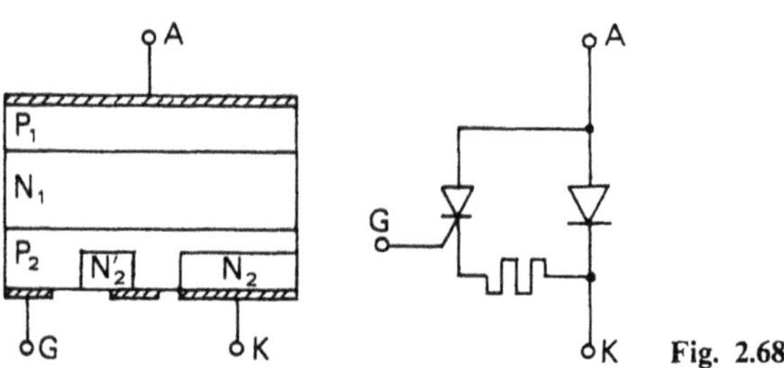

Fig. 2.68

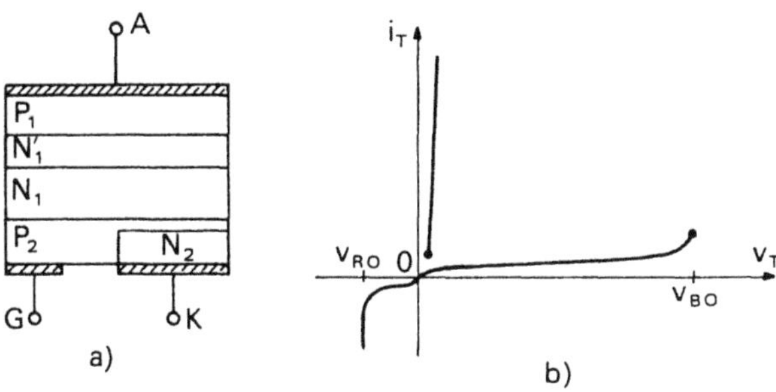

Fig. 2.69

the N'_1 layer being much more heavily doped than the N_1. The reverse blocking voltage V_{RO} is only about 20 V; this explains the name of asymmetrical thyristor (Fig. 2.69b).

The reduction of V_{RO} has no adverse effects in converters where a diode is connected in reverse parallel across the thyristor, the latter not having to sustain a high reverse voltage.

This structure has the advantage of reducing the gate recovery time by decreasing the volume of the N_1 layer; there are fewer carriers trapped when turn-off occurs, and they take less time to recombine. We thus obtain fast thyristors intended for high-frequency applications (tens of kHz) with an anti-parallel diode.

2.4.4.4 Reverse-Conducting Thyristor

Usually denoted by the abbrevation RCT (Reverse Conducting Thyristor) or ITR (Integrated Thyristor Rectifier), this thyristor integrates an asymmetrical thyristor (N_2, P_2, $N_1 + N'_1$, P_1) and a diode ($P_2, N_1 + N'_1$) connected in antiparallel in one and the same structure (Fig. 2.70).

Use of this device is normally limited to low-power applications.

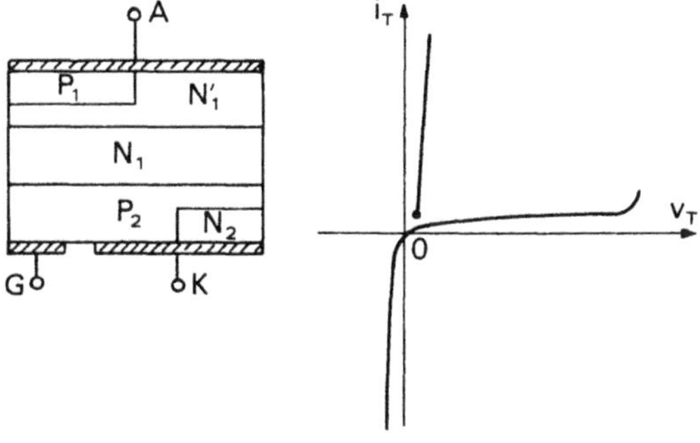

Fig. 2.70

2.4.5 Triacs

The triac is a device used as static switch and in AC regulators (see Vol. 2). It can be considered as the antiparallel arrangement of two thyristors. It is for this reason that its study is included in this chapter. However, it has some specific aspects which must be briefly presented.

2.4.5.1 Description

The triac comprises three alternating layers P_1, N_2, P_2 (Fig. 2.71). An N_1 zone is diffused into the P_1 layer and two zones – N_3 and N_4 – into the P_2 layer. By metallization on N_1, N_3, N_4 and the adjoining P zones, three terminals are produced – the A_1 and A_2 anodes and gate G.

2.4.5.2 Steady-State Characteristics

Figure 2.72 shows characteristics $i_T = f(v_T)$ when no current is applied to the gate.

For a positive v_T, the $N_1P_1N_2P_2$ structure acts as a first thyristor which starts to conduct when v_T reaches the breakdown voltage of junction P_2N_2, in reverse bias.

For a negative v_T, the $N_4P_2N_2P_1$ structure acts as a second thyristor which conducts when v_T reaches the breakdown voltage of junction P_1N_2, in reverse bias.

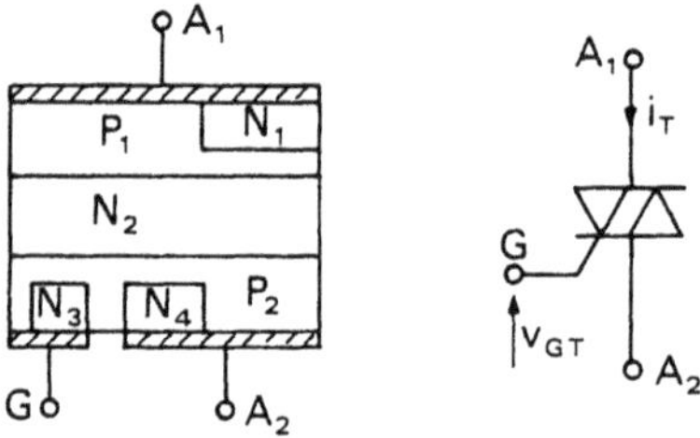

Fig. 2.71

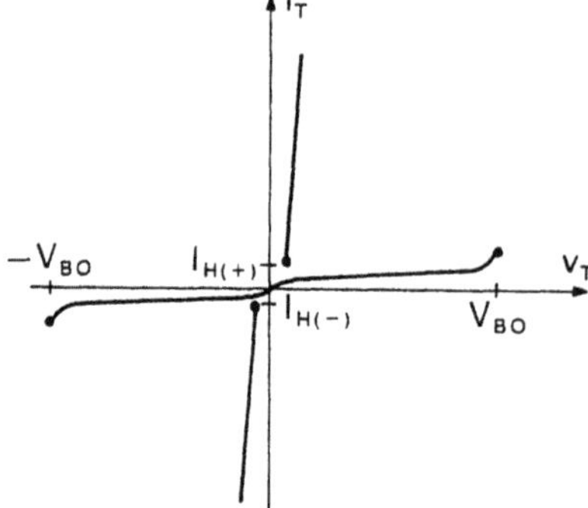

Fig. 2.72

Since the doping of the P_1 and P_2 zones is similar, the two breakdown voltages have nearly the same value. But, because of the lack of symmetry of the structure, the holding current for positive v_T is substantially higher than the holding current for negative v_T.

The mean values, RMS values or surge values, for the current or voltage, are noted as for the thyristors.

2.4.5.3 Gate Pulse Triggering

When applied to the gate, a positive or negative voltage pulse leads to the turn-on of the triac, whether the voltage v_T across it be positive or negative. There are thus four different firing modes, depending on to the biases of A_1 and G with respect to A_2. These correspond to the four quadrants formed by v_T, v_{GT} in Fig. 2.73; in the latter, full lines limit the zones involved in the explanation of the firing.

1st quadrant: v_T is positive, and a positive voltage pulse is applied between G and A_2.

The $P_1N_2P_2N_4$ thyristor is triggered like a conventional thyristor of which A_1 would be the anode and A_2 the cathode.

2nd quadrant: v_T is positive, and a negative voltage pulse is applied between the gate and A_2.

The active thyristor is still $P_1N_2P_2N_4$ but junction P_2N_2 – in reverse bias – breaks down for a different reason. It can be explained in simplified terms: as junction P_2N_3 is forward-biased, the N_3 zone injects electrons into P_2 and

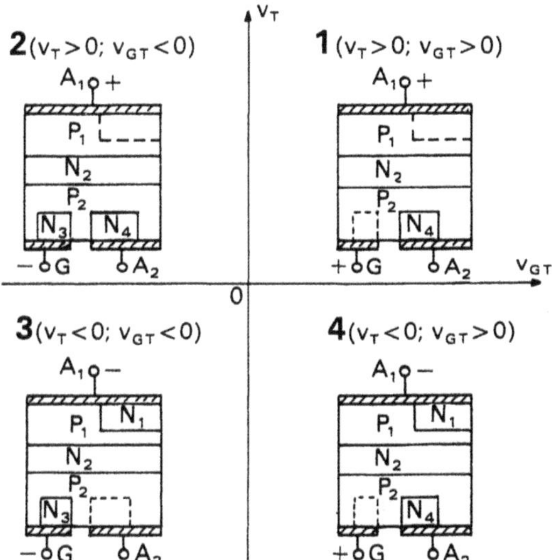

Fig. 2.73

brings about the avalanche breakdown of the P_2N_2 junction adjoining N_3; the silicon resistivity in this area falls sharply and the potential of P_2 around N_3 is close to that of A_1; the left side of junction P_2N_4 is thus forward-biased and the N_4 zone sends electrons into P_2, leading to the avalanche breakdown of the whole P_2N_2 junction.

3rd quadrant: v_T is negative and a negative voltage is applied between the gate and A_2.

The thyristor $P_2N_2P_1N_1$ is considered now. Under the effect of v_{GT}, the N_1 zone injects electrons into P_2. Some of these electrons can be found in the space charge zone of the P_2N_2 junction and are expelled into N_2 by the internal field of this junction. The reverse current of the N_2P_1 junction – which acts as a control junction here – increases and leads to the avalanche.

4th quadrant: v_T is negative and v_{GT} positive.

Firing takes place as in the previous mode, since the triggered thyristor is once more made up of $P_2N_2P_1N_1$. The electrons which provide the increase in reverse current in junction N_2P_1 are now supplied by the left side of the forward-biased N_2P_2 junction.

Because of the different mechanisms used, the minimum gate current required to trigger the triac varies according to the quadrant. It is weak for quadrants 1 and 3. Firing in quadrant 2 requires a slightly stronger current. In quadrant 4, the triac is much less sensitive and this firing mode must thus be avoided if possible.

The latching current also varies according to the triggering mode. It is much higher for quadrant 4 but, in the three others, the values are quite similar.

2.4.5.4 Switchings

The triac switchings raise the same problems as for the thyristor. Moreover, the bidirectional nature of the triac leads to two limits being defined for the voltage rate of change across the blocked device.

- As in the case of the thyristor (see Sect. 2.4.3.4), the *first limit* is the critical rate of rise of the voltage in the OFF-state. If this dv/dt is excessively high, the space charge capacitance of the anode junction charging current acts as a gate current and leads to firing. This limit may reach a few hundreds of volts per microsecond.

- The *second limit* is the critical rate of rise of the voltage at the turn-off switching. This problem was mentioned for the thyristor (see Sect. 2.4.3.3), but is of greater importance in the case of the triac. The latter is used as a bidirectional switch and, as soon as the current falls to zero, the voltage across it becomes very different from zero. It tends to make the current flow in the opposite direction to the previous one (see Vol. 2, the waveforms of v_T in Figs. 1.3 and 2.3). Figure. 2.74 shows the slope examined here.

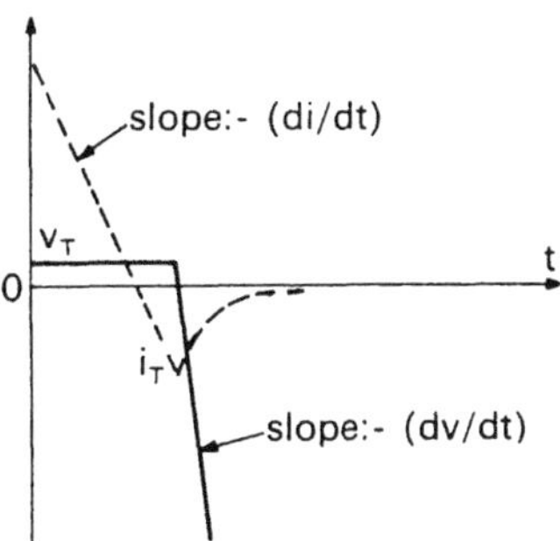

Fig. 2.74

Thus the triac has less time than the thyristor to recover its blocking power. The carriers trapped in the median junction (P_1N_2 or P_2N_2, depending on the previous current direction) greatly limit the dv/dt beyond which the triac is likely to conduct again immediately.

The number of carriers trapped depends on the intensity of current i_T before the turn-off, and also on the rate of fall of this current: the higher the rate of fall, the fewer recombinations there are and the more carriers which remain trapped.

The limiting value $(dv/dt)_C$ indicated by the manufacturer corresponds to a particular value of (di/dt) which must also be indicated. If this is not the case, the value given for $(dv/dt)_C$ corresponds to the negative slope of a sinusoidal current i_T at 50 Hz, of amplitude equal to the rated current. This gives:

$$\frac{di}{dt} = \frac{d}{dt}(I_{TRMS}\sqrt{2}\sin \omega t), \quad \text{for } \omega t = 0$$

$$= 100\,\pi\sqrt{2}\,I_{TRMS}.$$

This second critical rate of change is considerably below the first, and is hardly more than a few volts per microsecond. This is one of the reasons why, for high power, two thyristors connected in antiparallel (i.e. with two gates and complex control circuitry) are preferred to the triac.

To protect the triac against voltage surges, an RC series circuit is connected in parallel across it, as for the thyristor.

2.4.5.5 Increasing the dv/dt

In order to increase the dv/dt sustainable, the triac structure can be modified, as shown in Fig. 2.75:
- The two integrated thyristors $P_1N_2P_2N_3$ and $P_2N_2P_1N_1$ are farther apart.
- An extra triac is installed between these two thyristors by use of zones N_5 and N_6 diffused in P_2 and P_1; this triac is used as a gate amplifier (see Sect. 2.4.4.1).

The device so obtained – also called an *alternistor* – can withstand a stronger rate of rise of reapplied voltage than the conventionally structured triac.

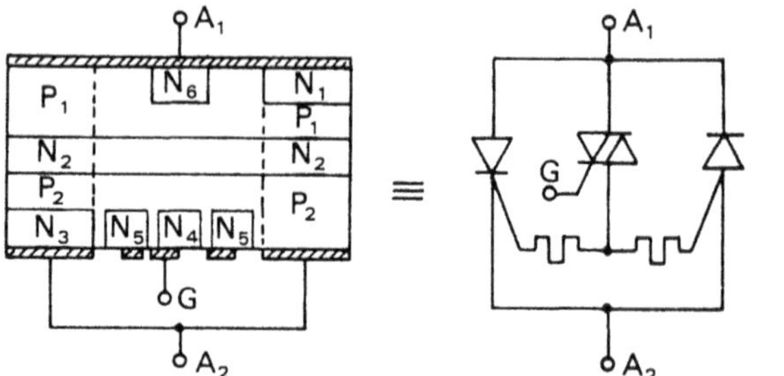

Fig. 2.75

2.5 GTO Thyristors

At its name indicates, the GTO (Gate Turn-Off) thyristor is a device which uses the same principles as the thyristor studied in the previous part of this chapter. It has, however, the specific feature of being turned off by the same gate which enables it to be triggered.

Of all the devices described in this chapter, this is the one which has more recently been developed industrially, and its characteristics may thus change more rapidly.

Its schematic representation is shown in Fig. 2.76a, which indicates the notations used in this section. The line across its gate connection indicates that the gate current may be positive or negative, according to whether turn-on or turn-off is required. In the studies of converter diagrams (as in the following chapter) either of the circuit symbols shown in Fig. 2.76b is frequently used to show the double role of the gate.

2.5.1 Description

- Like the thyristor, the GTO is a four-layer device (Fig. 2.77):
  - the N_2 layer, or cathode layer, is thin and heavily doped;

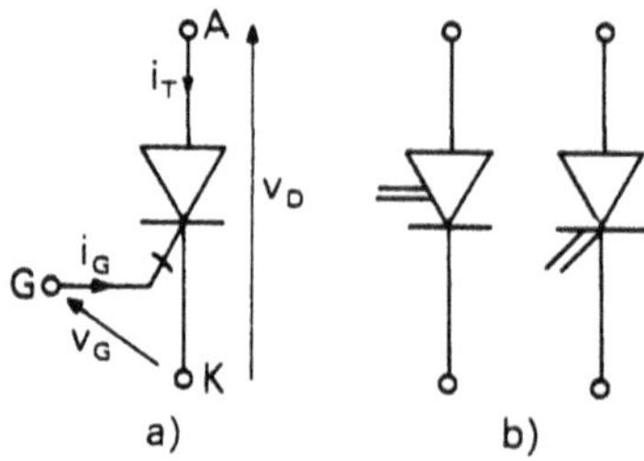

Fig. 2.76

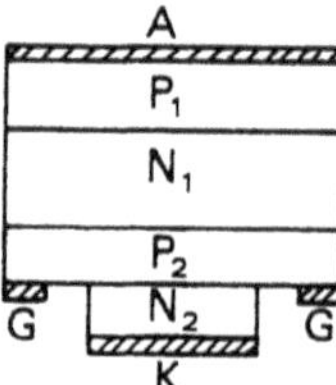

Fig. 2.77

- the P_2 layer, or control layer, is also heavily doped and relatively thin. Together with N_2, it makes up the cathode junction J_K;
- the N_1 layer, or blocking layer, is thick and lightly doped. Its thickness is directly dependent on the forward voltage which can be sustained. Together with P_2, it makes up the control junction J_C;
- the P_1 layer is produced by diffusion in N_1, as for the P_2 layer; it has the same characteristics as the latter. Together with N_1, it makes up the anode junction J_A.

- The N_2 layer is removed by etching in places where the gate contacts are situated. This produces cathode cells in the form of fingers. Cathode fingers of a width between 50 and 500 μm are created on the surface of the silicon wafer; depending on the device current rating their number can reach several hundreds. All of these elementary cathodes are surrounded by gates; they are brought together by a cathode plate clamped on the wafer. Such a clamping must be carried out with care as there must be adequate contact between each cell in the N_2 layer and the cathode plate.

  The GTO must be seen as a large number of small GTOs in parallel; the diagram in Fig. 2.77 shows one of them.

2.5.2 Operation, Characteristics

2.5.2.1 Turn-on

- The GTO is turned on in the same way as a conventional thyristor: when voltage v_D between anode and cathode is positive, a forward gate current i_{GF} of sufficient magnitude brings the blocking junction into avalanche.

  But, on account of the GTO structure, all the elementary GTOs must be triggered almost at the same time. There would otherwise be a current overload in the first GTO starting to conduct, before the voltage drop v_D prevents the others from firing. The firing control must, therefore be sudden: current pulse i_{GF} must rise sharply and show an amplitude of 2 to 5 times the minimal value required for triggering.

- Fig. 2.78 shows the voltage and current waveforms at turn-on.

  We find the delay time t_d and the rise time t_r (defined paradoxically as the fall time of voltage v_D) as previously seen for the firing of the thyristor (Sect 2.4.3.1). The turn-on time of the GTO is denoted t_{gt}:

$$t_{gt} = t_d + t_r.$$

  For components of the same ratings, the turn-on time of a GTO is slightly above that of a thyristor. However, the GTO withstands higher current rates of rise i_T, because the current filaments are distributed among the multiple cathodes. This enables local overheating to be avoided.

When current i_T has exceeded the latching current value the gate current can

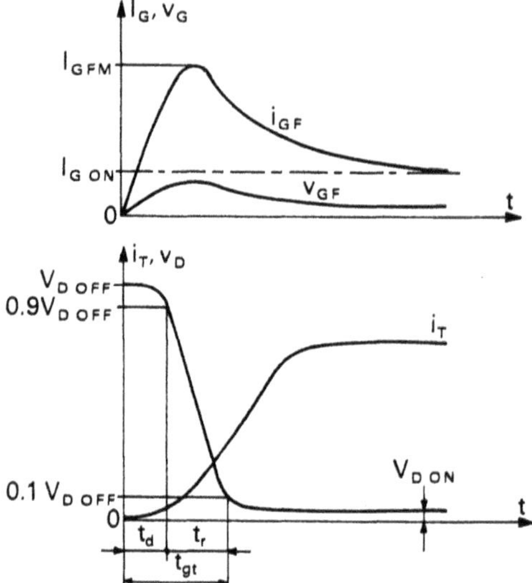

Fig. 2.78

be brought down to a lower value. However, this current must remain sufficiently strong to keep all the elementary GTOs conducting; eliminating the i_{GF} could lead to some being turned off and others being overloaded as a result.

This gate current $I_{G\,ON}$ also enables the ON-state forward voltage drop across the GTO to be reduced.

2.5.2.2 Turn-off

The turn-off switching of the GTO requires a more detailed analysis since it is what distinguishes it from the thyristor.

- *Presentation of the phenomenon*

- During the ON-state phase, the blocking junction P_2N_1 is in avalanche and a strong current flows from anode to cathode: this corresponds to a large number of holes going from P_1 to N_2 and electrons going from N_2 to P_1 (Fig. 2.79a). It should be noted that the current lines are concentrated under the

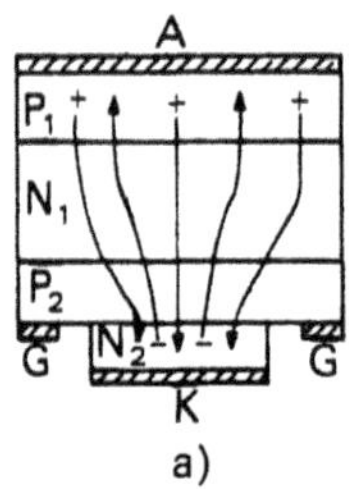

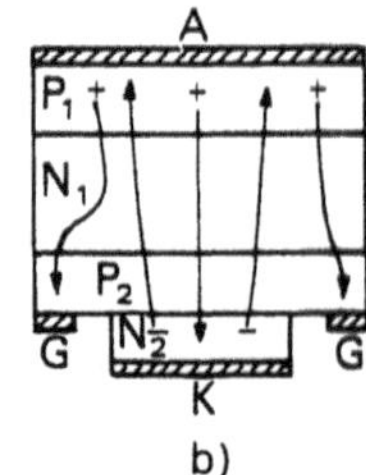

Fig. 2.79

cathode fingers and that only the part of the control junction above the fingers is in avalanche.

- The extraction of charges by the gate brings about the GTO turn-off. Applying a negative voltage between gate and cathode brings out a reverse gate current i_{GR} which deviates some of the current lines towards the anode – gate axes (Fig. 2.79b). With a sufficiently strong reverse gate current, the current density at the control junction level becomes less than the threshold value needed to maintain the avalanche and the GTO blocks.

  While, at turn-on, the gate current is only used to initiate the avalanche in the control junction, turn-off requires that a considerable amount of the main current should be deviated via the gate. This fraction may exceed 20%: the reverse gate current must thus be able to reach very high values for high-rating devices.

- *Two-transistor model*

Progress achieved in semiconductor manufacturing technology has enabled the GTO to develop. It was stated for a considerable time that a thyristor could be turned off by its gate and this has been tested on small-rating devices. The two-transistor model – presented earlier (Sect. 2.4.1.4) – allows the conditions of this turn-off to be explained.

Figure 2.80 shows the two-transistor model of a four-layer structure (see Fig. 2.59).

If α denotes the common base current gain of a transistor and i_{CBO} the reverse saturation current of the collector–base junction, we can express:

$$i_{C_1} = \alpha_1 i_{E_1} + i_{CBO_1}$$

$$i_{C_2} = \alpha_2 i_{E_2} + i_{CBO_2}$$

$$i_{E_2} = i_{E_1} + i_G .$$

From $i_{E_1} = i_{C_1} + i_{C_2}$, the following can be deduced:

$$i_{E_1} = \alpha_1 i_{E_1} + i_{CBO_1} + \alpha_2 (i_{E_1} + i_G) + i_{CBO_2}$$

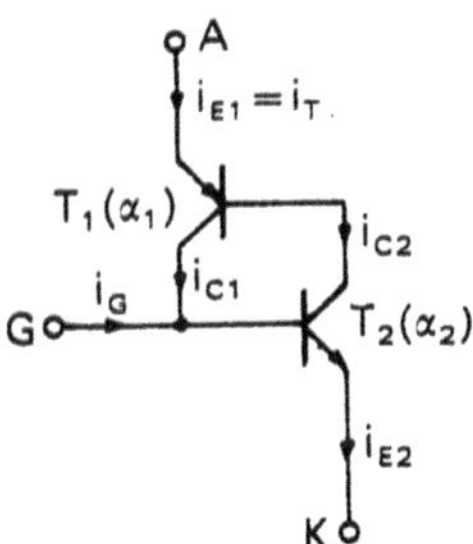

Fig. 2.80

or, by denoting i_{CBO}, equal to $i_{CBO_1} + i_{CBO_2}$, the total saturation current of the GTO:

$$i_{E_1} = \frac{\alpha_2 i_G + i_{CBO}}{1 - (\alpha_1 + \alpha_2)}. \tag{2.13}$$

When the GTO is in the ON state, i_G is negligible and

$$i_{E_1\,(ON)} = \frac{i_{CBO}}{1 - (\alpha_1 + \alpha_2)} \tag{2.14}$$

gives the current to be turned off

By comparing with the expression (2.13) of current i_{E_1}, it can be seen that this current can be reduced to zero, if there is a reverse gate current such as

$$i_G = \frac{-\,i_{CBO}}{\alpha_2}.$$

If i_{CBO} is replaced by its expression as a function of $i_{E_1\,(ON)}$ deduced from Eq. (2.14), the relationship between the gate current needed for turn-off and the anode current to be turned off can be given is

$$i_G = -\frac{i_{E_1\,(ON)}\,[1 - (\alpha_1 + \alpha_2)]}{\alpha_2}. \tag{2.15}$$

This yields the *current gain at turn-off*:

$$G_C = \frac{i_{E_1\,(ON)}}{i_{GR}} = \frac{\alpha_2}{1 - (\alpha_1 + \alpha_2)}. \tag{2.16}$$

Equation (2.16) shows that, in order to have sufficient gain, transistor $N_1P_2N_2$ must be given as high a current gain α_2 as possible; this underlines the importance of the quality of production of the N_2 and P_2 layers.

- *Waveforms, gate turn-off time*

The principle of the GTO turn-off is relatively simple but the phenomena which occur during a turn-off switching are more complex. They are shown in Fig. 2.81.

Concerning the various time intervals, the figure shows

- the *storage time* t_s, which separates the instant when the reverse gate current begins to increase from the instant when the anode current reaches 90% of its initial value. During this period, the gate–cathode junction behaves, in practice, like a short-circuit (see Sect 2.1.3.2) and i_{GR} rapidly reaches its maximum I_{GRM}. This time is of the order of a few microseconds.

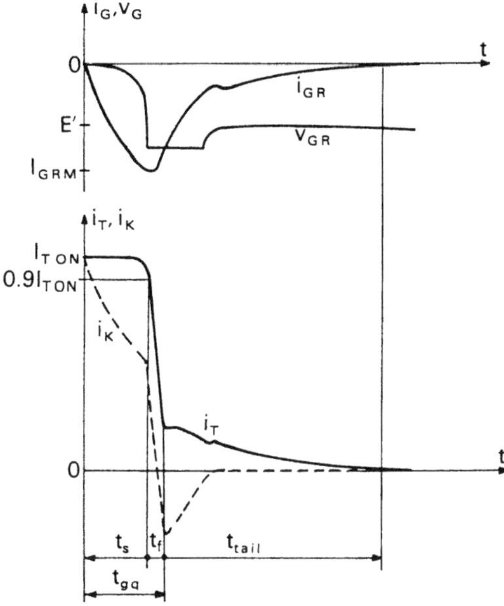

Fig. 2.81

- the *fall time* t_f, which is the time separating the instant when current i_T falls to $0.9\ I_{T\,ON}$ from the instant when it stops falling sharply. The fall time is very short and often less than a microsecond.

  As soon as the avalanche phenomenon disappears from the blocking junction, the reverse gate current decreases and falls gradually to zero.

  Owing to the stray inductances of the gate circuit, the reversal of the sign of $\mathrm{d}i_{GR}/\mathrm{d}t$ leads to a reverse voltage peak v_{GR} which can lead in turn to the breakdown of junction GK, the Zener voltage of which is relatively low (about 25 V). This Zener effect has no adverse effects on switching. When the stray inductance effect has disappeared, voltage v_{GR} stabilises at value E', imposed by the gate control circuit.

- The *gate turn-off time* t_{gq}, is defined by

$$t_{gq} = t_s + t_f .$$

As t_f is very short, t_{gq} does not exceed tens of microseconds for the most powerful devices.

In Fig. 2.81, the cathode current variations i_K, equal to $i_T + i_G$, are also shown. This current passes by a minimum value at the end of t_f, when junction GK becomes reverse-conducting; it then gradually falls towards zero during this reverse conduction.

- *Tail current*

– Once the breakdown transient has disappeared, excess charge carriers still remain in the GTO; it lets through a substantial and slowly decreasing leakage current, which is called the "tail current". During most of the corresponding period, current i_K is zero and i_G equals i_T.

This current corresponds to the free charge carriers which still exist in the blocking layer N_1. This layer is thick and lightly doped, the carriers remaining there are relatively numerous and they recombine slowly. (The higher the forward voltage to be blocked, the thicker the layer and the longer the tail current lasts.)

– Commutation losses are considerably increased by this tail current. There are two means of reducing it:

  - *Gold doping* or *electronic irradiation* of the blocking layer to increase the number of recombination centers and thus reduce the lifetime of minority carriers. The drawback of this process is that the resistance of the blocking layer is increased as well as the forward ON-state voltage drop.
  - *Anode shorts.* As shown in Fig. 2.82, a heavily doped cell N_3 which produces a short-circuit between the anode and the blocking layer, is introduced directly opposite each N_2 finger of the cathode.

    Such N_3 cells avoid a concentration of holes injected by P_1 during the ON-state, in the middle of the N_1 layer. When turn-off occurs, these heavily doped N_3 cells make the minority carriers trapped in N_1 recombine more quickly.

In this way, it is possible to obtain turn-off transitions as rapid as with gold doping, without increasing the forward voltage drop. But the structure is no longer symmetrical and the sustainable reverse voltage is well below the breakover forward voltage (asymmetrical GTO).

- *Voltage spike*

The sudden current decrease in a GTO at turn-off (di_T/dt slope of about 10^9 A/s) would give an unbearably high voltage spike across the device, if care was not taken to limit this. We will come back to the GTO snubbers in the Appendix. However, it should be noted that the forward voltage rise across the GTO cannot be studied without taking the snubber into account.

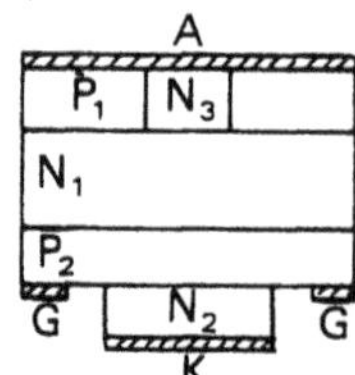

Fig. 2.82

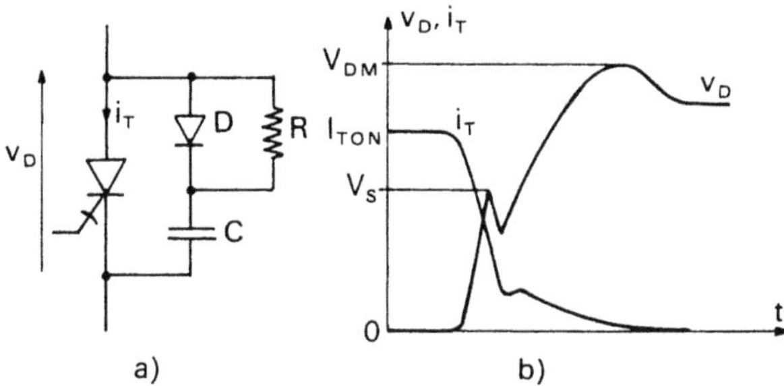

Fig. 2.83

Figure 2.83a provides a diagram of the simplest of these circuits: it is made up of a capacitor C, a diode D and a discharge resistance R.

At turn-off, a fraction of current previously flowing through the GTO is transferred to C via D. This transfer is slowed down by the stray inductances and by the turn-on of the diode. A voltage spike V_S appears across the GTO during the fall time t_f (Fig. 2.83b). The maximum value of this spike V_S is one of the limits of the GTO. When only the tail current remains in the GTO, diode D turn-on is ended and the rate dv_D/dt is limited by capacitor C.

The overvoltage which appears at the end of the turn-off is due to the inductance in series with the GTO (see Appendix, A. 4.2). Its maximum value V_{DM} must remain below the maximum value V_{DRM} of the sustainable repetitive peak forward voltage.

These few remarks on the voltage recovery show that the description of the turn-off transient must include the effects of the snubber components, in the same way as those of the load circuit and the typical parameters of the GTO.

2.5.2.3 Outstanding Parameters

The recent development of the GTO explains the absence of international standards to define its specific parameters. The maximum anode current which can be turned off by the gate is usually denoted I_{TGQ}, but I_{TCM}, I_{TQRM} and I_{TQSM} are also found. Moreover, as this parameter varies according to the test conditions, great care must obviously be taken in the use of manufacturer specifications.

Generally speaking, however, it can be said that

- the anode–cathode *voltage* is denoted V_D in the forward bias and V_R in the reverse bias. Subscripts RM corresponds to the maximum repetitive voltage, SM to the non-repetive peak voltage, DC to the breakover voltage. The values shown for V_D correspond to a negative given value for the gate voltage. The forward ON-state voltage drop is usually shown by $V_{D\,ON}$ or V_{TM};
- the anode *current* in the ON state is denoted I_T. Subscripts GQ correspond to the maximum controllable current (measurement conditions must be pre-

cisely given), RMS to the RMS value, and AV to the average value in the ON state. RM and SM have the same meaning as for the voltage. In the OFF state, the anode current is denoted I_D in the forward direction and I_R in the reverse direction.

The maximum values of the *gate* voltage and current are denoted with subscripts FGM in the forward direction and RGM in the reverse direction. V_{GT} and I_{GT} represent the gate voltage and current values required for turn-on.

2.5.3 Remarks on the Gate Control

Control signals applied to the gate play a large role in the safe operation of the GTO.

- *At turn-on*

In order to reduce the turn-on time t_{gt}, the positive pulse of the gate current must rise very sharply and reach a peak value which is much higher than the minimum value I_{GT} required for turn-on.

The gate circuit must behave like a current source and the series stray inductances must be reduced as much as possible. Use of a pulse transformer cannot be envisaged. Voltage V_{GT}, between gate and cathode, is low – about one volt.

- *During conduction*

Current I_{GT} must flow through the gate permanently during conduction, in order to maintain the avalanche effect throughout the whole GTO. This current also enables the forward ON-state voltage drop to be reduced but must not cause the junction GK to overheat.

- *At turn-off*

When a sufficient amount of charges has been removed by the gate, the avalanche disappears from the control junction. This amount depends on the negative gate current, as well as on the rate of increase of this current. The control circuit must then behave in the same way as a negative *voltage source* which has low internal impedance and is able to withstand a strong current.

- The negative value of the source voltage must not exceed the Zener voltage of the GK junction, i.e. about 25 V. However, by increasing V_{GR}, the admissible value for dv_D/dt is also increased. Negative 15 V voltages are thus commonly used.

- The negative gate current increases with the intensity of the current to be turned off. A turn-off current gain equal to 5 is usually recommended by the manufacturers. In the case of GTOs of high current rating, this can lead to negative gate currents of hundreds of amperes.

- The inductance of the gate circuit L_G must be less than one μH, if the negative gate current has to rise rapidly.

Figure 2.84 illustrates the previous remarks. It shows the influence on the negative current of

- voltage V_{GR}. The three waveforms of Fig. 2.84a correspond to the same controllable current and the same inductance;
- the controllable current. The three waveforms of Fig. 2.84b correspond to the same values of V_{GR} and L_G. The waveforms of current i_T are shown as a reminder that the storage time increases with the controllable current;
- the gate circuit inductance. The lines in Fig. 2.84c correspond to the same values of V_{GR} and $I_{T\,ON}$, but to three values of L_G.

The negative gate voltage must be maintained until the GTO has been completely turned off. However all elementary GTOs do not block simultaneously; if the gate signal is cut off prematurely, this may leave some GTOs in the conducting state and there would be thus the risk of a slow spread of remaining avalanche spots and of conduction reappearing throughout the structure. In most converters such an unwanted turn-on would lead to the device being destroyed.

Figure 2.85 provides the functional diagram of the gate control circuit. In some cases, in order to improve the quality of the control signals, these are made dependent of

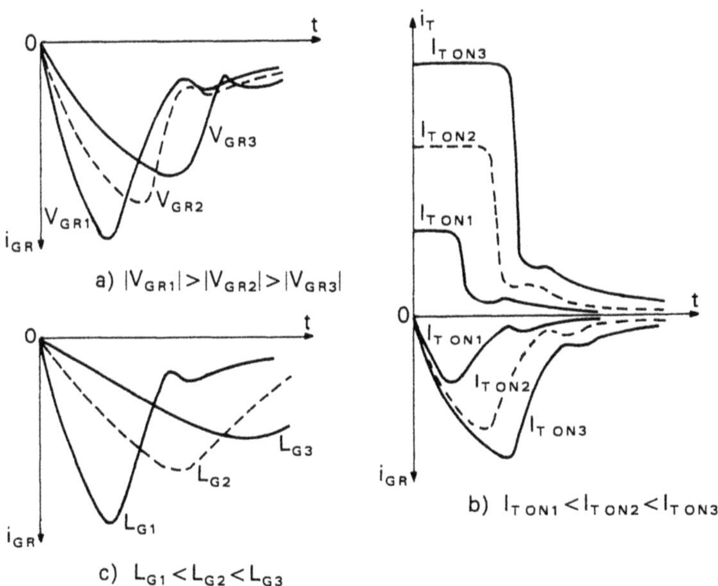

Fig. 2.84

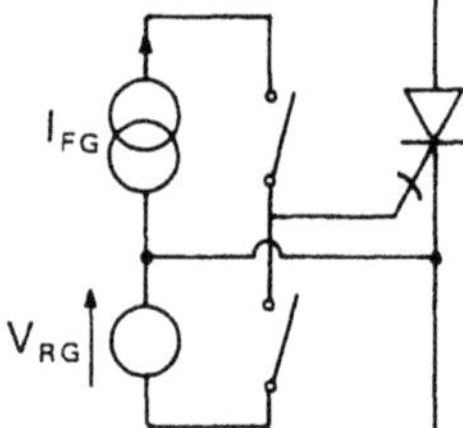

Fig. 2.85

- the controllable current,
- the device temperature,
- the voltage across it.

2.5.4 Remarks Concerning the Losses

- *Switching losses*

– Losses occurring during turn-on are mainly caused by the snubber capacitor discharge, when the GTO becomes conducting. These losses can be reduced with the aid of a snubber which enables the capacitor energy to be recovered and by increasing the peak value I_{FGM} of the forward gate current.

– Losses at turn-off are basically caused by the tail current. To reduce them, the rate of rise of the forward voltage must be reduced by means of a snubber.

The snubber capacitor requires a value which is high enough at turn-off but not too high at turn-on. The choice of the snubber capacitor is therefore the result of a compromise between the controllable anode current, the switching frequency and the corresponding losses.

- *Conduction losses*

Unlike the conventional thyristors, the GTO gate signal has to be maintained during the conduction period. Moreover, the heavy gold doping or electron irradiation of the blocking layer, intended to reduce the tail current, increases the forward ON-state voltage drop.

Conduction losses can be reduced by increasing the value of the positive gate current. But care must be taken to verify whether the ensuing heating can be withstood by junction GK.

Bibliography

There are numerous works which deal with semiconductors and manufacturer manuals in this field abound. We have first indicated [1–10] some *books* which deal exclusively with power semiconductors or which devote a considerable amount of space to their study.

Progress in the manufacturing technology of power semiconductors means that they are developing very rapidly. New techniques make it possible to incorporate a large number of elementary semiconductors on the same silicon wafer and, by grouping them in parallel, to obtain, at high power, possibilities which were previously available only to low-power devices.

This leads to

- the improvement in the performance of conventional power devices: reduction in the forward voltage drop, the reverse current peak, the switching times, …;
- the increase in the power of certain devices, such as fast diodes, MOS transistors;
- the appearance of new devices: high power GTO thyristors, field-controlled thyristor, MOS thyristor, double-injection diode, …;
- the integration of the power semiconductor as well as the whole or a part of its control and protection on the same wafer.

Our aim has been to classify the references and to follow as closely as possible the plan of this chapter. This is often difficult to achieve since many articles concern different themes simultaneously. However, the articles can be found successively under the following headings.

- *Junctions and diodes* [11–29]. The studies deal notably with the reduction of the forward voltage drop, of the reverse current peak.
- *Bipolar transistors* [30–63]. The studies deal especially with the second breakdown, the influence of the control on switching times and the integration techniques.
- *MOS transistors* [64–84]. The articles show the procedures used to reduce the on-resistance and to increase the sustainable forward voltage.
- *FET–bipolar combinations* [85–98]. These combinations were initially produced using discrete components, but are now produced by integration. This

process has lead to the production of new devices: Insulated Gate Bipolar transistor, MOS thyristor, ...

- *Field–controlled thyristor* [99–110]. Also known as a Static Induction Thyristor, this recent component uses a PN junction, when biased in the conducting direction. Its conduction is modulated by a field effect transistor.
- *Thyristors* [111–144]. Most studies on the conventional thyristor deal with the physical mechanisms of turn-on/turn-off or with the ways to reduce the turn-off time t_q and to increase the di/dt and dv/dt.
- *Special thyristors* [145–165]. As well as articles on the Gate-Assisted Turn-off Thyristor, Asymmetrical Thyristor, Reverse-Conducting Thyristor and the Triac, there are studies on the Light-Triggered Thyristor, used in High-Voltage Direct-Current links. Some articles deal with the Double-Injection Diode and its use as a low-power thyristor.
- *Gate turn-off thyristors* [166–179]. Articles on the high-rating GTO thyristor have become increasingly common since the development of this device. They deal with its production and operation, as well as with improvement of its performance and the conditions in which it may be used.
- *Thermal analysis* [180–191]. Articles on localising losses, on their flow and on cooling techniques appear at regular intervals. Developments in integration have given a new impetus to such studies.

Books

1. Maggetto G (1971) *Le Thyristor: Définitions – Protection – Commande.* Presses Universitaires de Bruxelles, Brussels
2. Gandhi SK (1977) *Semiconductor Power Devices.* Wiley, New York
3. Leturcq Ph (1978) *Physique des Composants Actifs à Semiconducteurs.* Dunod, Paris
4. Blicher A (1981) *Field-Effect and Bipolar Transistor Physics.* Academic Press, New York
5. Sze SM (1981) *Physics of Semiconductor Devices,* (2nd ed.). Wiley, New York
6. Oxner ES (1982) *Power FET's and Their Applications.* Prentice-Hall, Englewood Clifs, N.J
7. Kubat M (1984) *Power Semiconductors.* Springer, Berlin
8. Baliga BJ (1987) *Modern Power Devices.* Wiley, New York
9. Taylor PD (1987) *Thyristor Design and Realization.* Wiley, New York
10. Williams BW (1987) *Power Electronics. Devices, Drivers and Applications* Wiley, New York

Junctions and diodes

11. Choo SC (1970) Effect of carrier lifetime on the forward characteristics of high-power devices. *IEEE Trans. Electron Devices,* **17**(9): 647–652
12. Kao YC and Davis JR (1970) Correlations between reverse recovery time and lifetime of p–n junction driven by a current ramp. *IEEE Trans. Electron Devices,* **17**(9): 652–656
13. Kao YC (1970) The design of high-voltage high-power silicon junction rectifiers. *IEEE Trans. Electron Devices,* **17**(9): 657–660
14. Schroen WH (1970) Characteristics of a high-current high-voltage Shockley diode. *IEEE Trans. Electron Devices,* **17**(9): 694–705 (42 references)
15. Polgar P, Mouyard A, Shiner B (1970) A high-current metal–semiconductor rectifier. *IEEE Trans. Electron Devices,* **17**(9): 725–731
16. Miller SE, Reynolds JE, Washburn JR (1971) Design of high-power Zener diode and its energy absorption capability. *IEEE Trans. Gen. Ind. Appl.,* **7**(2): 208–211

17. Bulucea CD, Prisecaru DC (1973) The calculation of the avalanche multiplication factor in silicon P–N junction taking into account the carrier generation (thermal or optical) in the space-charge region. *IEEE Trans. Electron Devices*, **20**(8): 692–701
18. Smith WB, Pontius DH, Budenstein PP (1973) Second breakdown and damage in junction devices. *IEEE Trans. Electron Devices*, **20**(8): 731–744
19. Chowdhuri P (1973) Transient-voltage characteristics of silicon power rectifiers. *IEEE Trans. Ind. Appl.*, **9**(5): 582–592
20. Kannam PJ (1976) Design concepts of high energy punchthrough structures. *IEEE Trans. Electron Devices*, **23**(8): 879–882
21. Naito M, Matsuzaki H, Ogawa T (1976) High current characteristics of asymmetrical p–i–n diodes having low forward voltage drops. *IEEE Trans. Electron Devices*, **23**(8): 945–949
22. Bhattacharyya AB, Kumar R (1976) Avalanche breakdown characteristics of punchthrough diodes. *IEEE Trans. Electron Devices*, **23**(9): 1016–1023
23. Miller MD (1976) Differences between Platinum- and Gold-doped silicon power devices. *IEEE Trans. Electron Devices*, **23**(12): 1279–1283
24. Munoz-Yague A, Leturcq Ph (1978) High-level behaviour of power rectifiers: a quantitative analysis of the forward voltage drop. *IEEE Trans. Electron Devices*, **25**(1): 42–49
25. Nakagawa A, Kurata M (1981) Computer-aided design consideration on low-loss p–i–n diodes. *IEEE Trans. Electron Devices*, **28**(3): 231–237
26. Maertens RP, Nijs JF, Van Overstraeten RJ, Jain SC (1982) The DC current–voltage characteristics of diodes under high-injection conditions. *IEEE Trans. Electron Devices*, **29**(5): 922–928
27. Chuang CT (1983) On the current–voltage characteristics of n^+–p–p^+ diodes. *IEEE Trans. Electron Devices*, **30**(12): 1709–1716
28. Amemiya Y, Mizushim Y (1984) Bipolar-mode Schottky contact and applications to high-speed diodes. *IEEE Trans. Electron Devices*, **31**(1): 35–42
29. Shimuzu Y, Naito M, Murakami S, Teresawa Y (1984) High-speed low-loss p–n diode having a channel structure. *IEEE Trans. Electron Devices* **31**(9): 1314–1319

Bipolar Transistors

30. Bailey RL (1970) Large-signal nonlinear analysis of a high-power high-frequency junction transistor. *IEEE Trans. Electron Devices*, **17**(2): 108–119
31. Hower PL, Reddi VGK (1970) Avalanche injection and second breakdown in transistors. *IEEE Trans. Electron Devices*, **17**(4): 320–335
32. Denning R, Moe DA (1970) Epitaxial π–ν p–n–p high-voltage power transistors. *IEEE Trans. Electron Devices*, **17**(9): 711–716
33. Wang PP (1971) Thermal instability and second breakdown in power transistors. *IEEE Trans. Aerosp. Electron. Syst.*, **7**(6): 1195–1200
34. Wang PP (1971) Temperature analysis of current crowded effect in power transistors. *IEEE Trans. Aerosp. Electron. Syst.*, **7**(6): 1200–1204
35. Hower PL (1973) Optimum design of power transistor switches. *IEEE Trans. Electron Devices*, **20**(4): 426–435
36. Popescu C (1974) The second breakdown in reverse biased transistor as an electrothermal switching. *IEEE Trans. Electron Devices*, **21**(7): 428–436
37. Balthasar PP (1976) The integrated power switch. *IEEE Trans. Ind. Appl.*, **12**(2): 179–191
38. Oettinger FF, Blackburn DL, Rubin S (1976) Thermal characterization of power transistors. *IEEE Trans. Electron Devices*, **23**(8): 831–838
39. Beatty BA, Krishna S, Adler MS (1976) Second breakdown in power transistors due to avalanche injection. *IEEE Trans. Electron Devices*, **23**(8): 851–857
40. Hower PL (1976) Application of a charge-control model to high-voltage power transistors. *IEEE Trans. Electron Devices*, **23**(8): 863–870
41. Wheatley CF, Einthoven WG (1976) On the proportioning of chip area for multistage Darlington power transistors. *IEEE Trans. Electron Devices*, **23**(8): 870–878
42. Koyanagi K, Hane K, Suzuki T (1977) Boundary conditions between current mode and thermal mode second breakdown in epitaxial planar transistors. *IEEE Trans. Electron Devices*, **24**(6): 672–678
43. Beatty BA, Krishna S, Adler MS (1977) Second breakdown in power transistors due to avalanche injection. *IEEE Trans. Ind. Electron. Control Instrum.*, **24**(4): 306–312

44. Hower PL, Einthoven WG (1978) Emitter current crowding in high-voltage transistors. *IEEE Trans. Electron Devices*, **25**(4): 465–471
45. Kondo HL, Yukimoto Y (1980) A new bipolar transistor – GAT. *IEEE Trans. Electron Devices*, **27**(2): 373–379
46. Sasayama T (1980) A method to evaluate the second breakdown resistance of power transistors. *IEEE Trans. Instrum. Meas.*, **29**(1): 48–51
47. Chen DY, Jackson B (1981) Turn-off characteristics of power transistors using emitter-open turn-off. *IEEE Trans. Aerosp. Electron. Syst.*, **17**(3): 386–391
48. Bennett WP, Kumbatovic RA (1981) Power and energy limitations of bipolar transistors imposed by thermal-mode and current-mode second-breakdown mechanisms. *IEEE Trans. Electrons Devices*, **28**(10): 1154–1162
49. Evans PD, Saied BM (1981) Fault-current control in power-conditioning units using power transistors. *Proc. Inst. Electr. Eng., Part B*, **128**(6): 335–337
50. Chen DY (1982) Application of transistor emitter-open turn-off scheme to high voltage power inverters. *IEEE Trans. Ind. Appl.*, **18**(4): 411–415
51. Auckland DW, Foo CF, Shuttleworth R (1982) Control method for compensation of switching delays in transistors. *Proc. Inst. Electr. Eng., Part B*, **129**(4): 199–204
52. Evans PD, Saied BM (1982) Protection methods for power-transistors circuits. *Proc. Inst. Electr. Eng., Part B*, **129**(6): 359–363
53. Chen DY, Lee FC, Blackburn DL, Berning DW (1983) Reverse-bias second breakdown of high power Darlington transistors. *IEEE Trans. Aerosp. Electron. Syst.*, **19**(6): 840–846
54. Van der Broeck HW, VanWyck JD, Schoeman JJ (1985) On the steady-state and dynamic characteristics of bipolar transistor power switches in low-loss technology. *Proc. Inst. Electr. Eng., Part B*, **132**(5): 251–259
55. Jovanic MM, Lee FC, Chen DY (1986) Nondestructive characterization of RBSOA of high-power bipolar transistors. *IEEE Trans. Aerosp. Electron. Syst.*, **22**(2): 138–145
56. Bonkowski RB (1986) A technique for increasing power transistor switching frequency. *IEEE Trans. Ind. Appl.*, **22**(2): 240–243
57. Hwang K, Navon DH, Tang TW, Hower PL (1986) Second breakdown prediction by two-dimensional analysis of BJT turn-off. *IEEE Trans. Electron Devices*, **33**(7): 1067–1071
58. Harada K, Sakamoto H, Shoyama M (1986) On the high-speed turn-off of a power transistor by a small saturable core. *IEEE Trans. Power Electron.*, **1**(3): 175–180
59. Villa FF (1986) Improved second breakdown in integrated bipolar power transistors. *IEEE Trans. Electrons Devices*, **33**(12): 1971–1976
60. Rockot JH (1987) Losses in high-power bipolar transistor. *IEEE Trans. Power Electron.*, **2**(1): 72–80
61. Bhat BN, Kumar MJ, Ramasubramanian V, George P (1987). The effects of collector lifetime on the characteristics of high-voltage power transistors operating in the quasi-saturation region. *IEEE Trans. Electron Devices*, **34**(5): 1163–1169
62. Curran P, Sang S (1987) Nonplanar multiple-epitaxy bipolar power integrated-circuit process. *IEEE Trans. Electron Devices*. **34**(8): 1823–1830
63. Widlar RJ (1987) Turn-off processes in high-voltage n–p–ν–n switches. *IEEE Trans. Electron Devices*, **34**(9): 2013–2022

Power MOSFETs

64. Devine ML (1970) AC performance of FET analogue switch. *Proc. Inst. Electr. Eng.*, **117**(7): 1205–1210
65. Reiser M (1973) A two-dimensional numerical FET model for DC, AC and large-signal analysis. *IEEE Trans. Electron Devices*, **20**(1): 35–45
66. Salama CAT, Oakes JG (1978) Nonplanar power Field-Effect Transistors. *IEEE Trans. Electron Devices*, **25**(10): 1222–1228
67. Lisiak KP, Berger J (1978) Optimization of nonplanar power MOS Transistors. *IEEE Trans. Electron Devices*, **25**(10): 1229–1234
68. Fong E, Pitzer DC, Zeman RJ (1980) Power DMOS for high-frequency and switching applications. *IEEE Trans. Electron Devices*, **27**(2): 322–330
69. Coen RW, Tsang DW, Lisiak KP (1980) A high-performance planar power MOSFET. *IEEE Trans. Electron Devices*, **27**(2): 340–343

70. Temple VAK, Love RP, Gray PV (1980) A 600-Volt MOSFET designed for low on-resistance. *IEEE Trans. Electron Devices*, **27**(2): 343–349
71. Lane WA, Salama CAT (1980) Epitaxial power transistors. *IEEE Trans. Electron Devices*, **27**(2): 349–355
72. Sun SC, Plummer JD (1980) Modeling of the on-resistance of LMOS, VDMOS and VMOS power transistors. *IEEE Trans. Electron Devices*, **27**(2): 356–367
73. Yoshida I, Okabe T, Katsueda M, Ochi S, Nagata M (1980) Thermal stability and secondary breakdown in planar power MOSFET's. *IEEE Trans. Electron Devices*, **27**(2): 395–398
74. Adler MS, Westbrook SR (1982) Power semiconductor switching devices – A comparison based on inductive load. *IEEE Trans. Electron Devices*, **29**(6): 947–952
75. Hu C, Chi MH (1982) Second breakdown of vertical power MOSFET's. *IEEE Trans. Electron Devices*, **29**(8): 1287–1293
76. Tamer AA, Rauch K, Moll JL (1983) Numerical comparison of DMOS, VMOS and UMOS power transistors. *IEEE Trans. Electron Devices*, **30**(1): 73–76
77. Chi MH, Hu C (1983) The operation of power MOSFET in reverse mode. *IEEE Trans. Electron Devices*, **30**(12): 1825–1828
78. Board K, Byrne DJ, Stowers M (1984) The optimization of the on-resistance in vertical DMOS power devices with linear and hexagonal surface geometries. *IEEE Trans. Electron Devices*, **31**(1): 75–80
79. Love RP, Gray PV Adler MS (1984) A large-area power MOSFET designed for low conduction losses. *IEEE Trans. Electron Devices*, **31(6)**: 817–820
80. Hu C, Chi MHA, Patel VM (1984) Optimum design of power MOSFET's. *IEEE Trans. Electron Devices*, **31**(12): 1693–1700
81. Darwish MN, Board K (1984) Optimization of breakdown voltage and on-resistance of VDMOS transistors *IEEE Trans. Electron Devices*, **31**(12): 1769–1773
82. Wheatley CF, Ronan HR (1987) Switching waveforms of the L^2 FET: a 5-V gate-drive power MOSFET. *IEEE Trans. Electron Devices*, **2**(2): 81–89
83. Blackburn DL (1987) Turn-off failure of power MOSFET. *IEEE Trans. Power Electr.*, **2**(2): 136–143
84. Ueda D, Takagi H, Kano G (1987) An ultra-low on-resistance power MOSFET fabricated by using a fully self-aligned process. *IEEE Trans. Electron Devices*, **34**(4): 926–930

FET–Bipolar Combinations

85. Chen DY, Chandrasekaran S, Chin SA (1984) A new FET–bipolar combinational power semiconductor switch. *IEEE Trans. Aerosp. Electron. Syst.*, **20**(2): 104–110
86. Baliga BJ, Adler MS, Love RP, Gray PV, Zommer ND (1984) The Insulated Gate Transistor: a new three-terminal MOS-controlled bipolar power device. *IEEE Trans. Electron Devices*, **31**(6): 821–828
87. Chen DY, Chin SA (1984) Bipolar-FET combinational power transistors for power conversion applications. *IEEE Trans. Aerosp. Electron. Syst.*, **20**(5): 659–664
88. Baliga BJ (1984) Switching speed enhancement in Insulated Gate Transistors by electron irradiation. *IEEE Trans. Electron Devices*, **31**(12): 1790–1795
89. Chen DY, Chin SA (1984) Design considerations for FET-gated power-transistors. *IEEE Trans. Electron Devices*, **31**(12): 1834–1837
90. Adler MS (1986) A comparison between BIMOS device types. *IEEE Trans. Electron Devices*, **33**(2): 286–293
91. Yilmaz H, Benjamin JL, Dyer RF, Chen LS, Ron van Dell W, Pifer GC (1986) Comparison of punch-through and non-punch-through IGT structures. *IEEE Trans. Ind. Appl.*, **22**(3): 466–470
92. Sukumar V, Chen DY (1986) IGT/COMFET latching characteristics and application to brushless DC motor drive. *IEEE Trans. Aerosp. Electron. Syst.*, **22**(5): 540–544
93. Fossum JG, Mc Donald RJ (1986) Charge-control analysis of the COMFET turn-off transient. *IEEE Trans. Electron Devices*, **33**(9): 1377–1382
94. Baliga BJ (1986) Power integrated circuits. A brief overview. *IEEE Trans. Electron Devices*, **33**(12): 1936–1938
95. Sin JKO, Salama CAT, Hou LZ (1986) The SINFET – A Schottky injection MOS-gated power transistor. *IEEE Trans. Electron Devices*, **33**(12): 1940–1947

96. Tanaka T, Yasuda Y, Ohayashi M (1986) A new MOS-gated bipolar transistor for power switches. *IEEE Trans. Electron Devices*, **33**(12): 2041–2045
97. Lutteke G, Raets HC (1987) 220-V mains 500-kHz class-E converter using a BIMOS. *IEEE Trans. Power Electron.*, **2**(3): 186–193
98. Hefner AR, Blackburn DL (1987) A performance trade-off of the Insulated Gate bipolar Transistor: Buffer layer versus lifetime reduction. *IEEE Trans. Power Electron.*, **2**(3): 194–207

Field-Controlled Thyristors

99. Nishizawa JI, Terasaki T, Shibata J (1975) Field-Effect Transistor versus analog transistor (Static Induction Transistor). *IEEE Trans. Electron Devices*, **22**(4): 185–197
100. Houston DE, Krishna S, Piccone DA, Finke RJ, Sun YS (1976) A field terminated diode. *IEEE Trans. Electron Devices*, **23**(8): 905–911
101. Wessels BW, Baliga BJ (1978) Vertical channel field-controlled thyristors with high gain and fast switching speed. *IEEE Trans. Electron Devices*, **25**(10): 1261–1265
102. Plummer JD, Scharf BW (1980) Insulated-Gate planar thyristors. *IEEE Trans. Electron Devices*, **27**(2): 380–387
103. Baliga BJ (1981) Temperature dependence of Field-Controlled Thyristor characteristics. *IEEE Trans. Electron Devices*, **28**(31): 257–264
104. Baliga BJ (1982) Breakover phenomena in Field-Controlled Thyristors. *IEEE Trans. Electron Devices*, **29**(10): 1579–1587
105. Baliga BJ (1983) The dv/dt capability of Field-Controlled Thyristors. *IEEE Electron Devices*, **30**(6): 612–616
106. Fisher CA, Paxman DH, Slatter JAG (1984) Design and performance of a new Static Induction Thyristor – The gate V-groove p–i–n diode. *IEEE Trans. Electron Devices*, **31**(9): 1299–1308
107. Nishizawa, JI, Muraoka K, Kawamura Y, Tamamushi (1986) A low-loss high-speed switching device: the 2500-V 300-A Static Induction Thyristor. *IEEE Trans. Electron Devices*, **33**(4): 507–514
108. Nakamura Y, Tadano H, Takigawa M, Igarashi I, Nishizawa JI (1986) Very high speed static induction thyristor. *IEEE Trans. Ind. Appl.*, **22**(6): 1000–1006
109. Temple VAK (1986) MOS-controlled thyristors – A new class of power devices. *IEEE Trans. Electron Devices*, **33**(10): 1609–1618
110. Nishizawa JI, Yukimoto Y, Kondou H, Harada M, Pan H (1987) A double-gate type Static Induction Thyristor. *IEEE Trans. Electron Devices*, **34**(6): 1396–1406

Thyristors

111. Chauprade R (1970) Evolution des circuits de commande des convertisseurs de puissance à thyristors. *Rev. Gén. Electr.*, **79**(7): 577–589
112. Ruhl HJ (1970) Spreading velocity of the active area boundary in a thyristor. *IEEE Trans. Electron Devices*, **17**(9): 672–680
113. Ikeda S, Tsuda S, Waki Y (1970) The current pulse ratings of thyristors. *IEEE Trans. Electrons Devices*, **17**(9): 690–693
114. Somos I, Piccone DE (1970) Plasma spread in high-power thyristors under dynamic and static conditions. *IEEE Trans. Electron Devices*, **17**(9): 680–687
115. Gaudry M (1970) Les thyristors. *Rev. Gén. Electr.*, **79**(11): 897–901
116. Dubois R (1970) Les thyristors: organes de commutation. Commutation à la fermeture. *Rev. Gén. Electr.*, **79**(11): 902–906
117. Berlioux R (1970) Les thyristors: organes de commutation en courant continu. *Rev. Gén. Electr.*, **79**(11): 907–913
118. Peter JM (1971) Les limites en di/dt des thyristors. *Rev. Gén. Electr.*, **80**(6): 503–508
119. Cornu J, Lietz M (1972): Numerical investigation of the thyristor forward characteritics. *IEEE Trans. Electron Devices*, **19**(8): 975–981
120. Balenovich JD, Gillot DM, Motto JW (1973) Thyristor high-frequency ratings by current testing and computer simulation. *IEEE Trans. Ind. Appl.*, **9**(2): 227–235

121. Terasawa Y (1973) Observation of turn-on action in a gate-triggered thyristor using a new microwave technique. *IEEE Trans. Electron Devices*, **20**(8): 714–721
122. Assalit HB, Studtmann GH (1974) Description of a technique for the reduction of thyristor turn-off time. *IEEE Trans. Electron Devices*, **21**(7): 416–420
123. Ruhl HJ, Shafer PO (1975) A new rating system for high-current high-frequency thyristors. *IEEE Trans. Ind. Appl.*, **11**(5): 540–545
124. Revankar GN, Srivastava PK (1975) Turn-off model of an SCR. *IEEE Trans. Ind. Electron. Control Instrum.*, **22**(4): 507–510
125. Matteson FM, Ruhl HJ, Shafer PO, Wolley ED (1976) The recovered charge characteristics of high power thyristors. *IEEE Trans. Ind. Appl.*, **12**(3): 305–311
126. Newell WE (1976) A design tradeoff relationship between thyristor ratings. *IEEE Trans. Ind. Appl.*, **12**(4): 397–405
127. Hartmann K (1976) Improvement of thyristor turn-on by calculation of the gate–cathode characteristics before injection. *IEEE Trans. Electron Devices*, **23**(8): 912–916
128. Munoz-Yague, Leturcq Ph (1976) Optimum design of thyristor gate–emitter geometry. *IEEE Trans. Electron Devices*, **23**(8): 917–924
129. Schlegel ES (1976) A technique for optimizing the design of power semiconductor devices. *IEEE Trans. Electron Devices*, **23**(8): 924–927
130. Dannhäuser F, Voss P (1976) A quasi-stationary treatment of the turn-on delay of one-dimensional thyristors – Part 1: Theory. *IEEE Trans. Electron Devices*, **23**(8): 928–936. – Part 2: Experiments, *ibid.*, 936–939
131. Jaecklin AA (1976) The first dynamic phase at turn-on of a thyristor. *IEEE Trans. Electron Devices*, **23**(8): 940–944
132. Kado-Sysoev AF, Reshetin VP, Tchashnikov IG, Schuman VB (1976) Avalanche injection in high-speed thyristors. *IEEE Trans. Electron Devices*, **23**(11): 1203–1211
133. Cornick JAF, Ramsbottom MJ (1976) Behaviour of thyristors when turned on by gate current *Proc. Inst. Electr. Eng.*, **123**(12): 1365–1367
134. Adler MS (1978) Accurate calculations of the forward drop and power dissipation in thyristors. *IEEE Trans. Electron Devices*, **25**(1): 16–22
135. Roulston DJ, Nakla MR (1979) Efficient modeling of thyristor static characteristics from device fabrication data. *IEEE Trans. Electron Devices*, **26**(2): 143–147
136. Burkes TR, Craig JP, Hagler MO, Kristiansen M, Portnoy WM (1979) A review of high-power switch technology. *IEEE Trans. Electron Devices*, **26**(10) 1401–1411
137. Adler MS, Temple VAK (1980) The dynamics of the thyristor turn-on process. *IEEE Trans. Electron Devices*, **27**(2): 483–494
138. Adler MS (1980) Details of the plasma-spreading process in thyristors. *IEEE Trans. Electron devices*, **27**(2): 495–502
139. Fukui H, Naito M, Terasawa Y (1980) One-dimensional analysis of reverse recovery and dv/dt triggering characteristics for a thyristor. *IEEE Trans. Electron Devices*, **27**(3): 596–602
140. Jaecklin A (1982) Two-dimensional model of a thyristor turn-on channel. *IEEE Trans. Electron Devices*, **29**(10): 1529–1535
141. Temple VAK (1983) Optimizing carrier lifetime profile for improved trade-off between turn-off time and forward drop. *IEEE Trans. Electron Devices*, **30**(7): 782–790
142. Woodhouse ML (1987) Voltage and current stresses on HVDC valves. *IEEE Trans. Power Delivery*, **2**(1): 199–206
143. Hudgins JL, Portnoy WM (1987) High di/dt pulse switching of thyristors. *IEEE Trans. Power Electron.*, **2**(2): 143–148
144. Hudgins JL, Portnoy WM (1987) Gating effects on thyristor anode current di/dt. *IEEE Trans Power Electron.*, **2**(2): 149–153

Special thyristors

145. Kokosa RA, Tuft BR (1970) A high-voltage high-temperature reverse conducting thyristor. *IEEE Trans. Electron. Devices*, **17**(9): 667–672
146. Oka H, Gamo H (1973) Electrical characteristics of high-voltage high-power fast-switching reverse-conducting thyristor and its applications for chopper use. *IEEE Trans. Ind. Appl.*, **9**(2): 236–247

147. Ogawa T, Kamel T, Morita K (1974) Electrical characteristics of ultrahigh-voltage thyristors and related problems. *IEEE Trans. Ind. Appl.*, **10**(1): 112–115
148. Shimizu J, Oka H, Funakawa S, Gamo H, Iida T, Kawakami A (1976) High-voltage high-power gate-assisted turn-off thyristor for high-frequency use. *IEEE Trans. Electron Devices*, **23**(8): 883–887
149. Schlegel ES (1976) Gate-Assisted Turnoff Thyristors. *IEEE Trans. Electron Devices*, **23**(8): 888–892
150. Temple VAK, Ferro AP (1976) High-power dual amplifying gate light triggered thyristors. *IEEE Trans. Electron Devices*, **23**(8): 893–898
151. Silber D, Winter W, Fullmann M (1976) Progress in light activated power thyristors. *IEEE Trans. Electron Devices*, **23**(8): 899–904
152. De Bruyne P, Van Iseghem PM, Vlasak T, Frech R, Skrabo B (1978) Les thyristors de puissance à conduction inverse ou l'intégration judicieuse d'un diode. *Rev. Gén. Electr.*, **87**(5): 359–364
153. Williams BW (1978) State-space computer triac model. *Proc. Inst. Electr. Eng.* **125**(5): 413–415
154. Iida T, Iwamoto H, Oka H, Funakawa S (1980) New DC chopper circuits using fast-switching reverse-conducting thyristors for low-voltage DC motor control. *IEEE Trans. Ind. Appl.*, **16**(1): 111–118
155. Temple VAK (1980) Development of a 2.6-KV light-triggered thyristor for electric power systems. *IEEE Trans. Electron Devices*, **27**(3): 583–591
156. Kapoor AK, Henderson HT (1980) A new planar injection-gated bulk switching device based upon deep impurity trapping. *IEEE Trans. Electron Devices*, **27**(7): 1268–1274
157. Hayashi H, Mamine T, Matsuhita T (1981) A high-power gate-controlled switch (GCS) using new lifetime control method. *IEEE Trans. Electron Devices*, **28**(3): 245–251
158. Kapoor AK, Henderson HT (1981) Injection-gated DI diode with gate-controlled holding voltage. *IEEE Trans. Electron Devices*, **28**(5): 557–560
159. Temple VAK (1981) Comparison of light triggered and electrically triggered thyristor turn-on. *IEEE Trans. Electron Devices*, **28**(7): 860–865
160. Temple VAK (1982) Thyristor devices for electric power systems. *IEEE Trans. Power Appar. Syst.*, **101**(7): 2286–2291
161. Jaecklin AA (1982) Turn-on phenomena in optically and electrically fired thyristors. *IEEE Trans. Electron Devices*, **29**(10): 1552–1560
162. Kobayashi S, Takahashi T, Tanabe S, Yoshino T, Horiuchi T, Senda T (1983) Performance of high-voltage light-triggered thyristor valve. *IEEE Trans. Power Appar. Syst.*, **102**(8): 2784–2791
163. Chu CK, Spisak PB, Walczak DA (1987) High-power asymmetrical thyristors. *IEEE Trans. Power Electron.*, **2**(2): 98–100
164. Ekström A, Eklund L (1987) HVDC thyristor valve development. *IEEE Trans. Power Electron.* **2**(3): 177–185
165. Przybysz JX, Miller DL, Lesli SG, Kao YC (1987) High d*i*/d*t* light-triggered thyristors. *IEEE Trans. Electron Devices*, **34**(10): 2192–2199

Gate Turn-Off thyristors

166. Naito M, Nagano T, Fukui H, Terasawa Y (1979) One-dimensional analysis of turn-off phenomena for a Gate Turn-Off thyristor. *IEEE Trans. Electron Devices*, **26**(3): 226–231
167. Azuma M, Kurata M, Takigami K (1981) 2500-V 600-A Gate Turn-Off thyristor (GTO). *IEEE Trans. Electron Devices*, **28**(3): 270–274
168. Shimizu Y, Naito M, Odamura M, Terasawa Y (1981) Numerical analysis of turn-off characteristics for a Gate Turn-Off thyristor with a shorted anode emitter. *IEEE Trans. Electron Devices*, **28**(9): 1043–1047
169. Nakagawa A, Ohashi H (1984) A study on GTO turn-off failure mechanism – A time- and temperature-dependant 1-D model analysis. *IEEE Trans. Electron Devices*, **31**(3): 273–279
170. Silard AP (1984) Switching characteristics of 45-A double-interdigitated GTO thyristors. *IEEE Trans. Electron Devices*, **31**(9): 1230–1237
171. Adler MS, Owyang KW, Baliga BJ, Kokosa RA (1984). The evolution of power device technology. *IEEE Trans. Electron Devices*, **31**(11): 1570–1591 (83 references)
172. Yatsuo T, Nagano T, Fukui H, Okamura M, Sakurada S (1984) Ultrahigh-voltage high-current Gate Turn-Off thyristors. *IEEE Trans. Electron Devices*, **31**(12): 1681–1686

173. Gough PA, Slatter JA (1984) A model for the GTO thyristor during switching-off. *IEEE Trans. Electron Devices*, **31**(12): 1796–1803
174. Fukui H, Yaginuma T (1985) Two-dimensional numerical analysis of turn-off process in a GTO under inductive load. *IEEE Trans. Electron Devices*, **32**(9): 1830–1834
175. Hashimoto O, Kirihata H, Watanabe M, Nishiura A, Tagami S (1986) Turn-on and turn-off characteristics of a 4.5-kV 3000-A Gate Turn-Off thyristor. *IEEE Trans. Ind. Appl.*, **22**(3): 478–482
176. Ho EYY, Sen PC (1986) Effect of gate-drive circuits on GTO thyristor characteristics. *IEEE Trans. Ind. Electron.*, **33**(3): 325–331
177. Harada K, Sakamoto H, Shoyama M (1987) On the effective turn-off of GTO by a small saturable core. *IEEE Trans. Power Electron.*, **2**(1): 20–27
178. Hayashi Y, Suzuki T, Ishibashi S, Sueoka T (1987) A consideration on turn-off failure of GTO with amplifying gate. *IEEE Trans. Power Electron.*, **2**(2): 90–97
179. Silard AP, Turtudau FI, Margarit MN, Kosa BB (1987) High-power double interdigitated (TIL) GTO/GTA thyristors. *IEEE Trans. Electron devices*, **34**(8): 1807–1814

Thermal analysis, Cooling

180. Tserng HQ, Plumbee HR (1970) The forward voltage technique measurement to measure junction temperature operating triacs. *IEEE Trans. Electron Devices*, **17**(9): 755–761
181. Tserng HQ, Plumbee HR (1970) Temperature measurement of AC operating triacs using a gate trigger technique. *IEEE Trans. Electron Devices*, **17**(9): 761–765
182. Wenthen FT (1970) Computer-aided thermal analysis of power semiconductor devices. *IEEE Trans. Electron Devices*, **17**(9): 765–770
183. Linsted RD, Surty RJ (1972) Steady-state junction temperatures of semiconductor chips. *IEEE Trans. Electron Devices*, **19**(1): 41–44
184. Cornick JAF, Ramsbotton MJ (1972) Instantaneous temperature rise in thyristors under invertor and chopper operating conditions. *Proc. Inst. Electr. Eng.*, **119**(8): 1141–1148
185. Golden FB (1972) Liquid cooling of power transistors. *IEEE Trans. Ind. Appl.*, **8**(5): 601–606
186. Le Ponner J, Peter JM (1972) Résistance et impédance thermique des triacs. *Rev. Gén. Electr.*, **81**(11): 711–719
187. Mc Laughlin MH, Vonzastrow EE (1975) Power semiconductor equipment cooling methods and application criteria. *IEEE Trans. Ind. Appl.*, **11**(5): 546–555
188. Newell WE (1976) Dissipation in solid-state device – The magic of I^{1+N}. *IEEE Trans. Ind. Appl.*, **12**(4): 386–396
189. Newell WE (1976) Transient thermal analysis of solid-state power devices – Making a dreaded process easy. *IEEE Trans. Ind. Appl.*, **12**(4): 405–420
190. Matsumara S, Harumoto Y, Kawagoe E, Tominaga S, Osawa Y, Yamamoto Y, Marutani T (1983) Development of HVDC thyristor valve insulated and cooled by compressed gas. *IEEE Trans. Power Appar. Syst.*, **102**(9): 3243–3253
191. Leturcq Ph, Dorkel JM, Napieralski A, Lachiver E (1987) A new approach to thermal analysis of power devices. *IEEE Trans. Electron. Devices*, **34**(5): 1147–1156

Chapter 3

DC–DC Converter Circuits: An Overview

The previous chapters reviewed commutations, supplies and loads, as well as power semiconductor switches. The remarks concerning these are relevant to all static converters. This chapter is devoted to the various types of DC–DC converters. In the following chapters, there will be a further and more detailed analysis of the most commonly used circuits.

The structure of a DC–DC converter (or chopper) can be adequately defined by

- the nature of the "sources" between which the converter controls the power transfer;
- the presence or absence of an internal energy-storage element and, if present, an indication of its nature;

which leads to the choice of

- the configuration of its elements,
- the types of semiconductor device used,
- the control law of the semiconductor devices used.

In this overview of DC–DC converters it is assumed that

- the "sources" are perfect (see Chap. 1)
- the "switches" are perfect,
- the commutations are instantaneous.

Starting from a circuit diagram showing mechanical switches, we will, firstly, show how each switch is supposed to behave and, secondly, deduce the type of semiconductor device which could suitably replace them.

Before describing the main types of DC–DC converter with direct energy transfer and those including a storage element, we must show how "switches" can be classified.

3.1 "Switches" Classification

3.1.1 Unidirectional "Switches"

A semiconductor device alone is always unidirectional in current. Its choice depends on what it is required to do, i.e. mainly on the type of commutation it must carry out.

3.1.1.1 Naturally Commutated Converters

Only diodes or thyristors may be used in converters where *all* the commutations are natural.

- *Diode* D is an *uncontrolled* "switch"; its operational mode depends totally on the circuit into which it is embedded. It can allow a forward current to flow or block a reverse voltage. Figure 3.1a provides a schematic diagram of its characteristic:

  branch OB: ON state,

  branch OC: reverse blocking state.

- *Thyristor* T is a "switch" *controlled at turn-on.* In schematic diagram form (Fig. 3.1b) its characteristic is made up of three branches:

  OA: forward blocking state,

  OB: ON state,

  OC: reverse blocking state.

  The thyristor can be turned on by the gate, which allows the change from OA to OB. Firing is shown by a single-lined arrow.

  The conventional thyristor cannot be turned off by the gate. Only when the current becomes zero does the thyristor turn off, provided that the operating point remains for a sufficient period on branch OC.

3.1.1.2 Force-Commutated Converters

Controlled turn-on and turn-off "switches" are needed in force-commutated converters, i.e. where *some* commutations are forced.

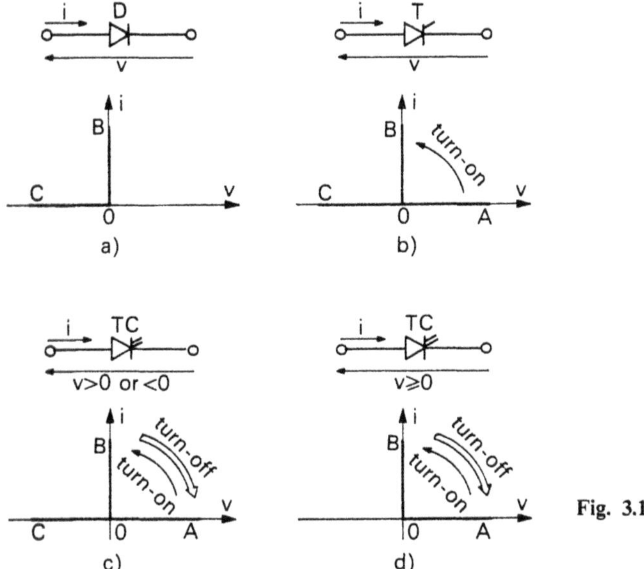

Fig. 3.1

Figure 3.1c represents such a switch by means of a "thyristor" to which a second gate has been added. TC denotes this switch in the circuit diagrams. It is called a controlled turn-on/turn-off switch. Its characteristic has three branches:

OA: forward blocking state,

OB: ON state,

OC: reverse blocking state.

It allows the passage from OA to OB *and* from OB to OA. Turn-off is shown by a double-lined arrow.

Such a characteristic is obtained by use of a thyristor provided with an auxiliary commutation circuit, or by a symmetrical GTO thyristor, i.e. capable of blocking the same voltage in reverse or forward bias.

Asymmetrical GTO thyristors can only block a reduced reverse voltage. Transistors can only withstand a very low reverse voltage; a diode must be connected in series with them in order to obtain the characteristic of Fig. 3.1c.

However, as will be seen, it is often the case that controlled turn-on/turn-off switches do not have to block negative voltages. TC will once more denote those switches whose characteristic shows only two branches (Fig. 3.1d), if they allow for the change from OA to OB and vice versa.

3.1.2 Bidirectional "Switches"

- In many DC–DC and DC–AC converters, a diode D must be connected in antiparallel with the controlled turn-on/turn-off device (Fig. 3.2a).

  If TC and D are both used during the same operational mode, the function of TC is to turn on or off the positive current *i* at suitable moments. Diode D must allow current *i* to flow when that current reverses its polarity, during the time interval when this "switch" must be on.

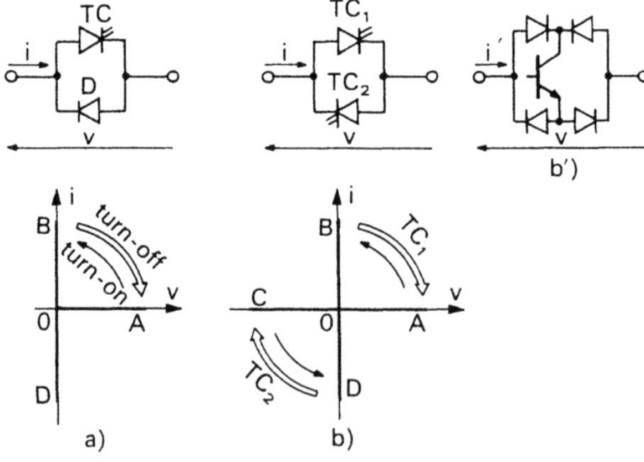

Fig. 3.2

In other cases, the diode current settles or disappears suddenly as the result of the turn-off/turn-on of a different controlled "switch" from the one across which it is connected. In such cases, the diode ensures that the current continues to flow. Finally, in some converters, TC can be used for one operating mode and D for another.

The schematic characteristic in the diagram has a negative-current branch OD, but negative-voltage branch OC has disappeared since it is incompatible with the presence of a diode. If using an asymmetrical GTO thyristor or a transistor, there is no need to add a diode in series with TC to produce this component.

- Some DC–DC converters (or frequency changers, see Vol. 2, Appendix) require bidirectional "switches" capable of controlling the turn-on and the turn-off of the current in both directions.

  Such "switches" can be composed of two controlled turn-on/turn-off devices TC_1 and TC_2 connected in antiparallel (Fig. 3.2b). Each of these must be able to block a positive or negative voltage across it.

If two transistors are used, a diode must be connected in series with each of them. Therefore, the use of a single transistor in a four-diode bridge is generally preferred (Fig. 3.2b′)

3.1.3 Remarks

a) When indicating the function to be carried out by a "switch", it is not enough to show (in the voltage–current axes system) the branches of characteristics on which it will have to be able to operate. The switchings from ON to OFF or vice versa have to be indicated.

   If this is not indicated, no difference should appear between a conventional thyristor (Fig. 3.1b) and a thyristor with forced commutation circuit (Fig. 3.1c). The bidirectional controlled turn-on/turn-off switch (Fig. 3.2b) would be mistaken for the two thyristors connected in antiparallel as used in AC regulators (Fig. 3.3).

b) In presenting the DC–DC converter structures, the commutations are assumed to be instantaneous; this, however, is not the case. To each controlled

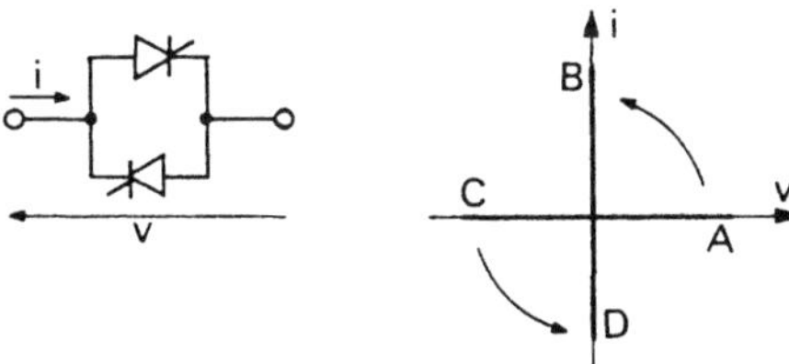

Fig. 3.3

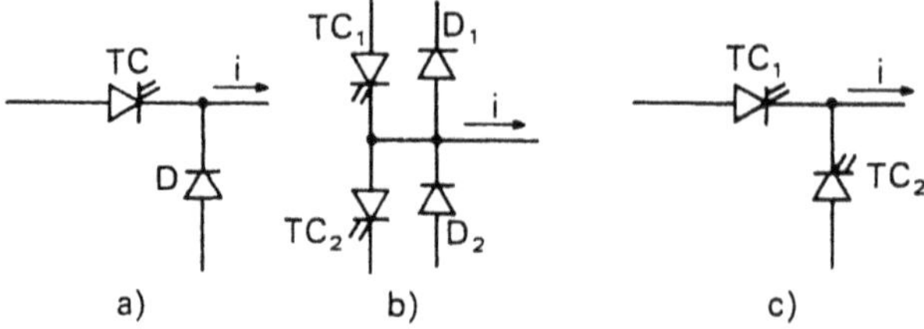

Fig. 3.4

turn-off semiconductor switch the diagram must connect a device capable of ensuring the continuous flow of current i.

Such a device is frequently a diode (Fig. 3.4a); current i switching off in TC is linked to the current rise in D and conversely.

In many reversible DC–DC and DC–AC converters, there are two switches made up of a controlled semiconductor device and a diode in antiparallel (Fig. 3.4b). Current transfer takes place between TC_1 and D_2 or vice versa, and between TC_2 and D_1 or vice versa.

In some configurations, the commutation must take place between two controlled semiconductor switches (Fig. 3.4c). In such cases, the controls must overlap. TC_1 can only be turned off when TC_2 is turned on and therefore acts in this case as a diode (and vice versa).

3.2 Directly Linked DC–DC Converters with Two "Switches"

3.2.1 Common Properties

3.2.1.1 Directly Linked DC–DC Converters

- A DC–DC converter is said to be directly linked if, between its input and its output, there is no energy storage element (inductor or capacitor). It comprises switches only.

  As the losses in the latter are not taken into account, the instantaneous powers at input, ui, and output, $u'i'$, are equal. The same obviously applies to the average values:

$$ui = u'i'$$

$$(ui)_{\text{mean}} = (u'i')_{\text{mean}} \,. \tag{3.1}$$

- The "switches" connect or separate input and output sources; they chop the input waveforms. If the converter is supplied from a current source, the current will be chopped at output. If the input voltage is constant, the output voltage is chopped. As the generator and the load are assumed to be perfect, *a directly linked DC–DC converter can only connect two sources of a different type*, a voltage source and a current source.

- For a given number of switches, the converter structure is imposed by the reversibility of the sources.

  We will first consider DC–DC converters with two "switches", before going on to discuss the bridge configuration with four "switches". We will only consider the most common types of circuit.

3.2.1.2 Configuration and Transformation Ratio of Converters with Two Switches

- All double-"switch" directly linked DC–DC converters have the same operational configuration, whether they connect a DC voltage generator to a DC current load or a DC current generator to a DC voltage load (Fig. 3.5):
  - the first "switch" K_1 enables the input and the output to be connected;
  - the second, K_2, short-circuits the current source (or generator or load) when K_1 is open.

The two switches must be in complementary states, so that the current source is never in open circuit nor the voltage source short-circuited (at least between commutations; see Remark 3.1.3b).

- U is used to denote the voltage across the voltage source (generator or load). This voltage remains constant during the operating cycle T of the converter. It can be positive or negative if a voltage-reversible source is used.
  - I' is used to denote the current flowing through the current source (generator or load). This current remains constant during the operating cycle T, but can be positive or negative if a current-reversible source is used.
  - i is the chopped current relative to voltage U.
  - u' is the chopped voltage relative to current I'.
  - α_1 denotes the duty ratio of K_1. This switch is turned on for $\alpha_1 T$ during each cycle T;
  - α_2 is the duty ratio of K_2. As both switches are complementary,

$$\alpha_1 + \alpha_2 = 1 \tag{3.2}$$

If, for $0 < t < \alpha_1 T$, switch K_1 is turned on:

$$u' = U; \quad i = I'.$$

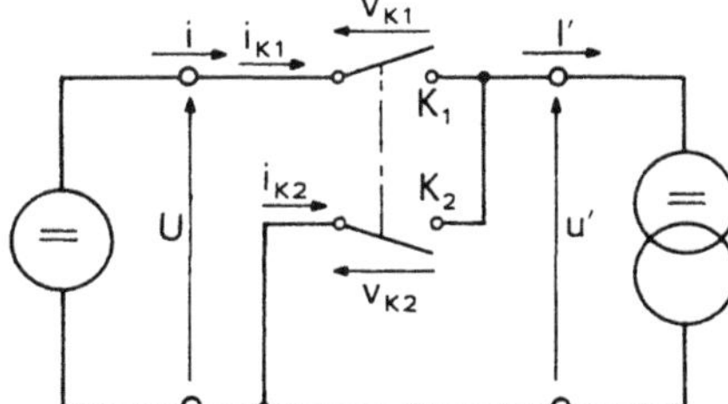

Fig. 3.5

During $\alpha_1 T < t < T$, K_2 is turned on:

$$u' = 0; \quad i = 0.$$

Thus the average values U' and I of the chopped voltage and current are

$$U' = \alpha_1 U; \quad I = \alpha_1 I'$$

giving

$$\frac{U'}{U} = \frac{I}{I'} = \alpha_1. \tag{3.3}$$

α_1 acts as the transformation ratio with regard to the voltages and currents, as does the ratio of the number of turns in a transformer.

3.2.2 Non–Reversible DC–DC Converters

3.2.2.1 Buck Converter (Fig. 3.6)

The buck converter controls the power flow between
a *voltage generator U*
and a *current load I'*.

- The turn-on of K_1 (state 1) makes i_{K_1} equal to I',
v_{K_2} equal to $-U$.
- The turn-on of K_2 (state 2) makes v_{K_1} equal to $+U$,
i_{K_2} equal to I'.

The branches of the (v–i) characteristics used by both switches are given in Fig. 3.6, below the operational diagram. It can be seen that K_1 must be controlled at turn-off and turn-on (but must not block a reverse voltage). K_2 can be merely a diode.

- We can thus derive the basic configuration of the buck converter shown on the figure. This converter is made up of a controlled turn-on/off semiconductor switch TC and a diode D.
  Underneath have been plotted the waveforms of
  output voltage u',
  input current i, equal to current i_T in TC,
  voltage v_T across TC,
  current i_D through diode D,
  voltage v_D across the latter,
  as well as the conduction diagram of both semiconductor devices.

- We check that commutation D–TC is natural (provided by the voltage source U): when TC is turned on, $-U$ is applied across the diode. However, commutation TC–D is forced: switching off TC applies $+U$ across it. In Fig. 3.6, the two types of commutation are shown by N and F.

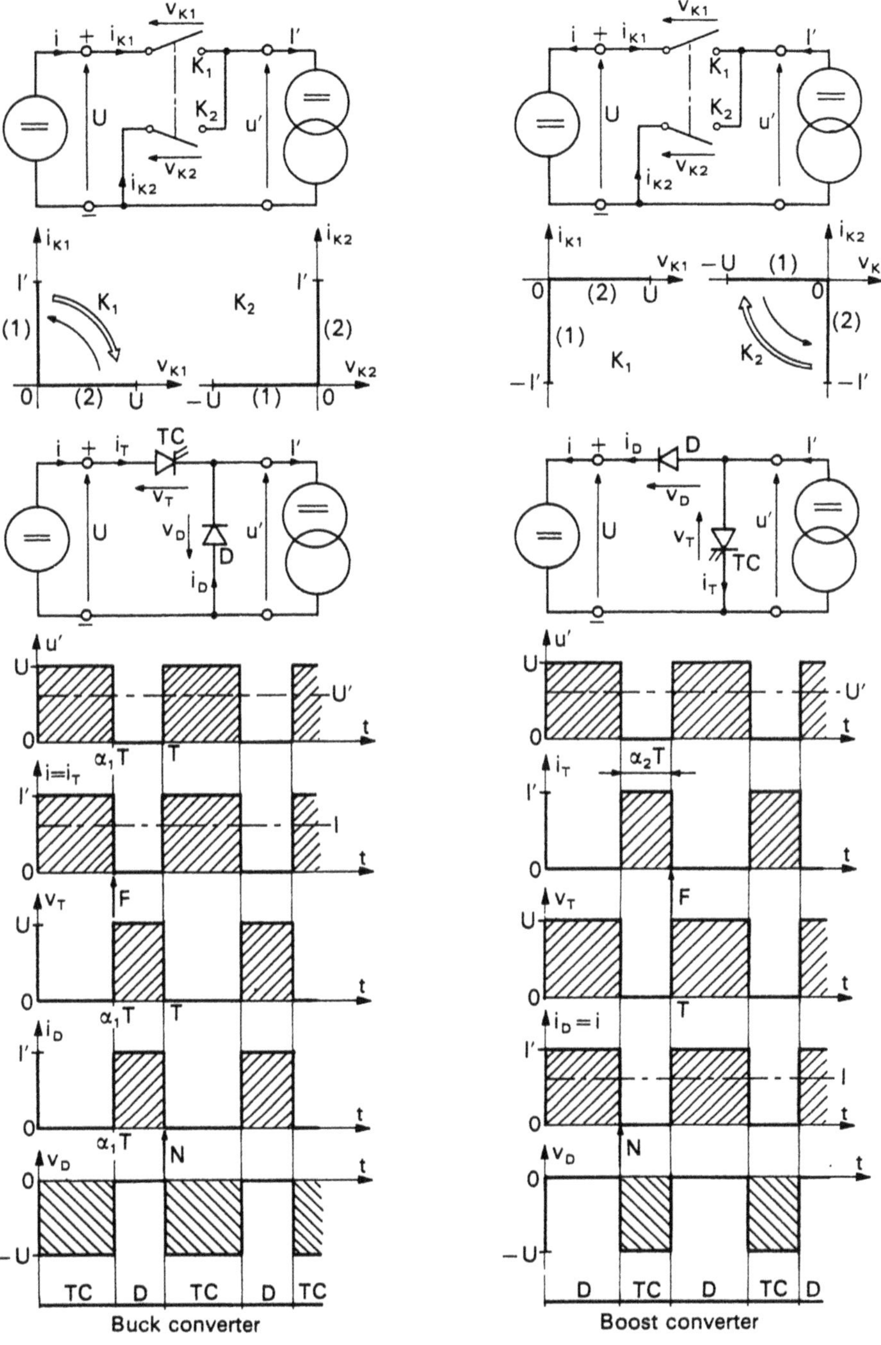

Fig. 3.6

Fig. 3.7

- Since U' equals $\alpha_1 U$, the average value of output voltage U' can be varied from U to zero, by varying the duty ratio α_1 from 1 to zero.
  This converter is sometimes called a series *chopper*, because the controlled semiconductor switch is serially connected with the generator. It is also known as a *step-down chopper* because it reduces – to a variable extent – the output voltage compared to the input voltage.

3.2.2.2 Boost Converter (Fig. 3.7)

The boost converter enables the current supplied
by a *current generator*
to a *voltage load*
to be varied.

In order to keep the same circuit topology for all double-switch directly linked choppers, the generator has been placed on the right side and the load on the left side, at the top of Fig. 3.7. The directions of currents I' and i have been reversed; the first is positive, the second is positive or zero.

- Turning on K_1 (state 1: $u' = U$; $i = I'$) gives

$$i_{K_1} = -I'; \quad v_{K_2} = -U\,.$$

- Turning on K_2 (stage 2: $u' = 0$; $i = 0$) gives

$$i_{K_1} = -I'; \quad v_{K_1} = +U\,.$$

The voltage–current characteristics of the two switches show that K_2 must be controlled at turn-on/turn-off, that K_1 can be merely a diode and that the direction of both positive currents in the semiconductor switches is the reverse of that for i_{K_1} and i_{K_2}.

- We can then consider the basic configuration of the converter with switch TC connected across the current generator and diode D connected between the input and the output.
  Underneath the diagram are shown the waveforms of
  output voltage u'
  current i_T through TC,
  voltage v_T, equal to u', across TC,
  current i_D, equal to i', through diode D,
  voltage v_D across D.

The conduction diagram of the two semiconductor switches has also been drawn. The type of commutation – forced (F) or natural (N) – is shown.

- Control is carried out by acting on the duty ratio α_2 of TC conduction intervals. The average values, U' of voltage u' and I of current i, are still given by

$$U'/U = I/I' = \alpha_1 = 1 - \alpha_2$$

If α_2 is made to vary from 1 to zero, U' is made to vary from zero to U and I from zero to I'.

This converter is sometimes called a *parallel chopper* because the controlled turn-on/turn-off switch is connected across the input. It is also called *step-up chopper*, because output voltage U is higher than the average input voltage U'.

3.2.2.3 Remark

There is no basic difference between the buck and the boost converters.

If, in the diagram (Fig. 3.6) representing a buck converter, the current source is moved and connected,

- no longer between terminal A and the negative terminal of the voltage source (in broken lines in Fig. 3.8a),
- but between terminal A and the positive terminal of the voltage source (in mixed lines),

the diagram of a boost converter can be obtained (Fig. 3.8b).

It is clear that the position of the diode on the outward or inward wire of the current i has no effect on the operation.

This remark on the possibility of changing from one structure to another will be used in the analysis of commutation circuits (Sect. 5.6.1).

3.2.3 Reversible DC–DC Converters

3.2.3.1 General Remarks

The structures with two "switches" only allow energy exchange to be reversed between *two sources which have the same type(s) of reversibility*.

Indeed the basic configuration (Fig. 3.5) shows that

- current i of the voltage source is sometimes equal to current I' of the current source, and sometimes zero:

  reversal of i means that I' is also reversed and vice versa;
- chopped voltage u' is sometimes equal to voltage U of the voltage source, and some times zero:

  reversal of u' means that U is also reversed and vice versa.

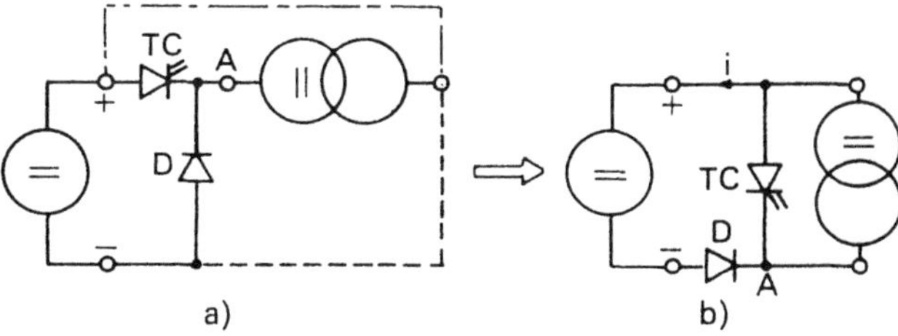

Fig. 3.8

The DC–DC converter can be characterized by indicating the reversibilities common to the voltage and current sources:

- the voltage-reversible converter connects two voltage-reversible sources;
- the current-reversible converter connects two current-reversible sources;
- the current- and voltage-reversible converter connects two sources both of which are current- and voltage-reversible.

3.2.3.2 Presentation of the Three Structures

Switches K_1 and K_2 are shown in the general diagram at the top of Fig. 3.9. Indicating the type(s) of reversibility to be carried out enables the branches of the current–voltage characteristics on which these switches must operate to be shown. The types of semiconductor devices to be used can be deduced from this.

- *Voltage-reversible chopper* (Fig. 3.9a)

Current I' is always positive.
When voltage U is positive,

$$K_1 \text{ closed, } K_2 \text{ open (state 1): } i_{K_1} = I' > 0; \quad v_{K_2} = -U < 0,$$

$$K_1 \text{ open, } K_2 \text{ closed (state 2): } v_{K_1} = U > 0; \quad i_{K_2} = I' > 0.$$

When voltage U is negative,

$$K_1 \text{ closed, } K_2 \text{ open (state 3): } i_{K_1} = I' > 0; \quad v_{K_2} = -U > 0,$$

$$K_1 \text{ open } K_2 \text{ closed (state 4): } v_{K_1} = U < 0; \; i_{K_2} = I' > 0.$$

Switches K_1 and K_2 must be made of two controlled turn-on/turn-off semiconductor devices capable of blocking reverse voltages, TC_1 and TC_2, as shown in the basic diagram.

- *Current-reversible chopper* (Fig. 3.9b)

Voltage U is always positive.

When I' is positive, the switches must ensure state 1 and 2 (see above) and the change in one or other direction between these states.

When I' is negative,

$$K_1 \text{ closed, } K_2 \text{ open (state 5): } i_{K_1} = I' < 0; \quad v_{K_2} = -U < 0,$$

$$K_1 \text{ open, } K_2 \text{ closed (state 6): } v_{K_1} = U > 0; \quad i_{K_2} = I' < 0.$$

Switches K_1 and K_2 must be made up of controlled turn-on/turn-off semiconductor devices, TC_1 and TC_2, each of them having a diode connected in antiparallel across it, as shown in the basic diagram. Note that TC_1 and TC_2 do not have to block a reverse voltage.

It is clear that this chopper is a combination of a buck and a boost converter.

- *Current- and voltage-reversible chopper* (Fig. 3.9c)

Voltage U and current I' can change polarity separately or simultaneously.

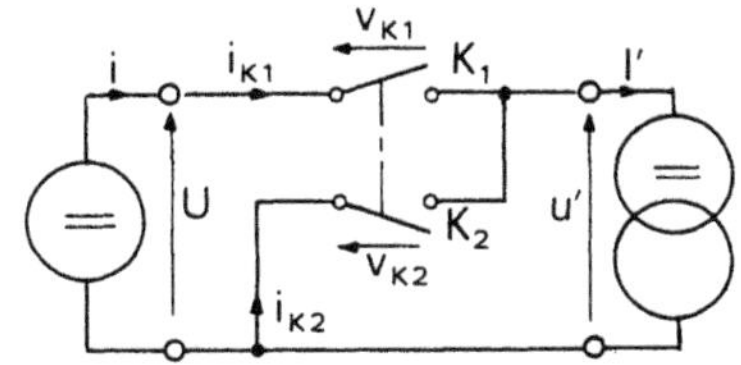

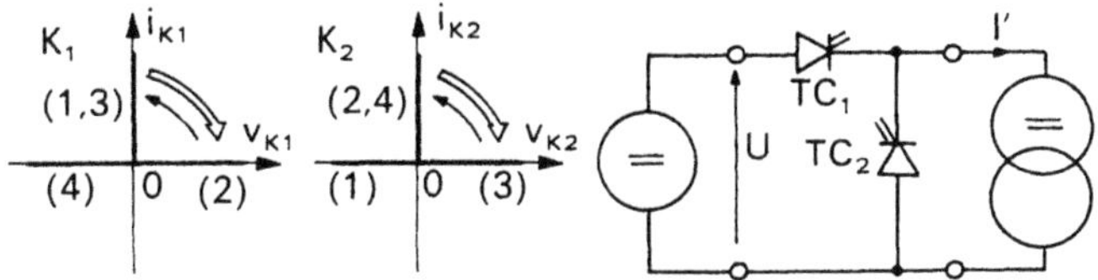

a) Voltage-reversible chopper

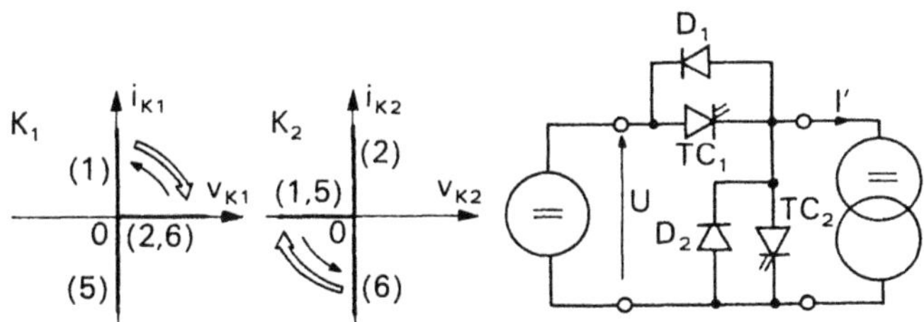

b) Current-reversible chopper

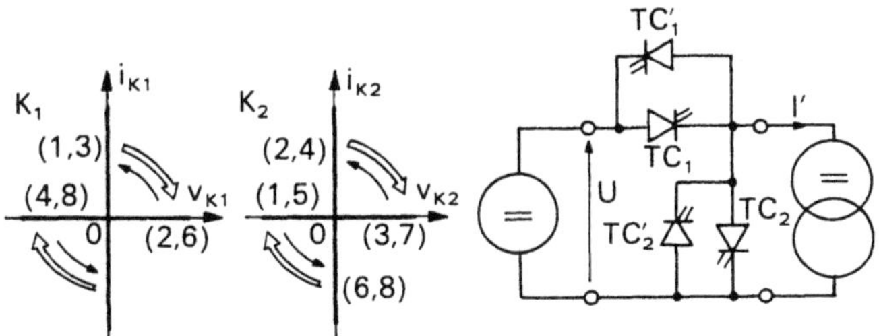

c) Current- and voltage-reversible chopper

Fig. 3.9

When U and I' are positive,
states 1 and 2 are to be found.
When U is negative and I' positive,
states 3 and 4 are to be found.
When U is positive and I' negative,
states 5 and 6 are to be found.

When U and I' are negative,

K_1 closed, K_2 open (state 7): $i_{K_1} = I' < 0$; $v_{K_2} = -U > 0$.

K_1 open, K_2 closed (state 8): $v_{K_1} = U < 0$; $i_{K_2} = I' < 0$.

Each switch K_1 and K_2 must be made up of two controlled turn-on/turn-off semiconductor switches, TC_1 and TC'_1, TC_2 and TC'_2, connected in anti-parallel, as shown in the diagram. The four semiconductor devices must be able to block a reverse voltage.

• *Remark*

In the current-reversible chopper (Fig. 3.9b), the controlled semiconductor device TC_1 and diode D_2 (like TC and D in the buck converter) have their cathodes connected together and to the same terminal of current source I'. There is no difficulty in transferring current I' from one device to another, during the commutation following the turn-on/turn-off command of TC_1. Moreover, the device which begins to conduct allows the reverse current peak to flow from the one which is being turned off.

The same case applies to the controlled device TC_2 and diode D_1. Their anodes are connected together (as with TC and D of the boost converter).

However, there are only controlled semiconductor devices in voltage-reversible (Fig. 3.9a) or voltage- and current-reversible (Fig. 3.9c) choppers. There must be an overlap interval between the turn-off command of the device to be turned off and the turn-on command of the device to which the current which flowed through the previous device must be transferred.

3.2.3.3 Analysis of the Current-Reversible Chopper

Among the three reversible DC–DC converters shown in Fig. 3.9, the current-reversible converter is the most widely used. For example, using a storage battery supply, it enables a DC motor to be supplied under variable voltage, when I' is positive (motoring operation in quadrant 1 of the torque-speed plane) and when I' is negative (recuperative braking operation in quadrant 2).

• *Theoretical waveforms*

The basic diagram of this chopper is represented again at the top of Fig. 3.10. Below are shown, for I' positive and then I' negative, the waveforms of

the output voltage u',
the input current i,
the current i_{T1} through TC_1,
the voltage v_{T1} across TC_1, equal to $-v_{D1}$,
the current i_{D1} through diode D_1,
the current i_{T2} through TC_2,
the voltage v_{T2} across TC_2, equal to $-v_{D2}$,
the current i_{D2} through diode D_2.

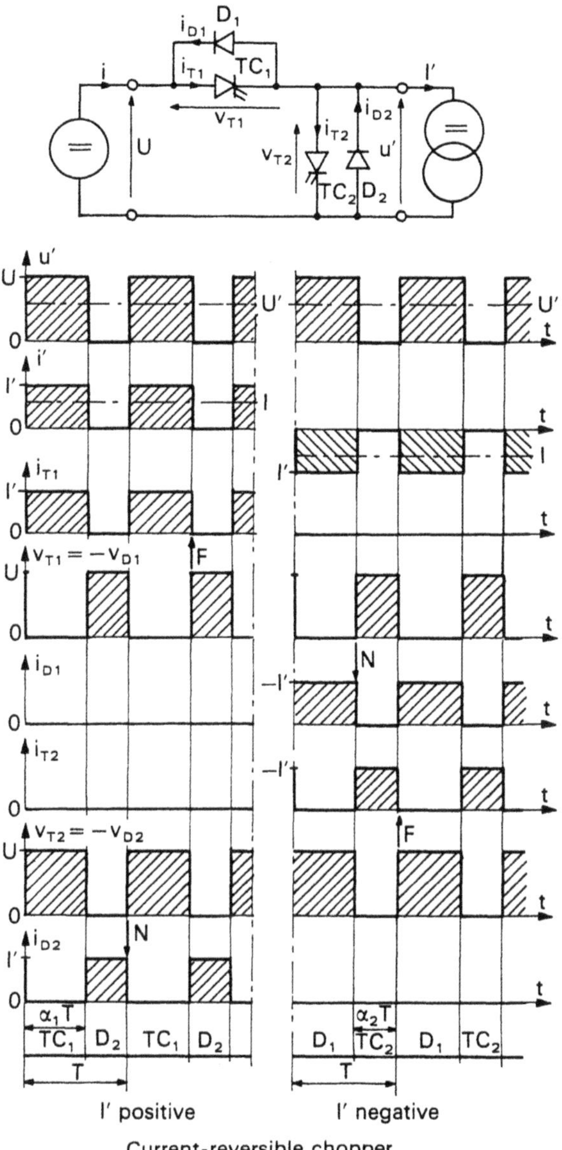

Fig. 3.10

The type of commutation and the conduction diagram of the four devices are shown.

- *When current I' is positive*, the voltage source operates as a generator, the current source as a load. The operation is regulated by the control of TC_1.

When TC_1 is turned on, it lets I' flow; when TC_1 is turned off, I' goes via D_2.

TC_1 turned on: $u' = U$; $i = I'$;

$i_{T_1} = I'$; $i_{D_1} = 0$; $v_{T_1} = -v_{D_1} = 0$;

$i_{T_2} = i_{D_2} = 0$; $v_{T_2} = +U$; $v_{D_2} = -U$.

TC_1 turned off: $u' = 0$; $i = 0$

$i_{T_1} = i_{D_1} = 0$; $v_{T_1} = +U$; $v_{D_1} = -U$;

$i_{D_2} = I'$; $i_{T_2} = 0$; $v_{T_2} = -v_{D_2} = 0$.

TC_1 and D_2 conduct alternately, as if they were operating as a buck converter.

- *When current I' is negative*, energy is transferred from the current source to the voltage source. It is controlled by TC_2 which lets through $-I'$ during $(1-\alpha_1)T$ for each cycle and transfers it to D_1 for the remaining time.

TC_1 turned on: $u' = 0$; $i = 0$;

$i_{T_1} = i_{D_1} = 0$; $v_{T_1} = +U$; $v_{D_1} = -U$;

$i_{T_2} = -I'$; $i_{D_2} = 0$; $v_{T_2} = -v_{D_2} = 0$.

TC_1 turned off: $u' = U$; $i = I'$;

$i_{D_1} = -I'$; $i_{T_1} = 0$; $v_{T_1} = -v_{D_1} = 0$;

$i_{T_2} = i_{D_2} = 0$; $v_{T_2} = +U$; $v_{D_2} = -U$.

Owing to the alternating conduction of TC_2 and D_1, the circuit operates as a boost converter.

- *Remark on the complementary control*

Using a constant voltage source U, this chopper allows a variable mean voltage U' to be delivered to a current source. The flow of the current is controlled by acting upon α_1 when current I' is positive, upon α_2 when I' is negative.

$$U' = \alpha_1 U \quad \text{or} \quad (1-\alpha_2)U.$$

- Because of the finite value of the current source inductance, current i' in the latter actually varies above and below its mean value I'. This has no effect on the relation between U'/U and α_1 or α_2, except in the following case:
  - current source i' has an EMF – e.g. armature circuit of a DC motor,
  - and the mean value I' of the current is low. Current i' then falls periodically to zero. For a given α_1 or α_2, the ratio U'/U shows a considerable variation, depending on I'.

This phenomenon – caused by *discontinuous conduction* – is to be found in rectifiers. (Vol. 1, Chap. 6, § 2 and Chap. 7 § 6). In the case of choppers, it will be examined in more detail in the next chapter.

- For the moment, it can be noted that these rapid variations of U', when I'

moves towards zero, disappear in the case of current reversible choppers, if complementary control is chosen:

Instead of turning on TC_1 during $\alpha_1 T$ or TC_2 during $(1 - \alpha_1)T$, a complementary control strategy is adopted for both switches.

TC_1 is closed during $\alpha_1 T$, TC_2 is closed during the remainder $(1 - \alpha_1)T$ of the duty cycle.

I' may therefore move from a positive value to a negative one or vice versa, without voltage U' showing any dicontinuity.

3.3 Full-Bridge DC–DC Converters

In the different types of chopper (or single-phase inverter), the four-switch or full-bridge type offers the greatest number of possibilities. It enables each of the two output terminals to be connected to each of the two input terminals or to separate them.

Since it is a directly linked chopper, either the input or output source should be a voltage one and the other should be a current one . But the energy transfer between *two different reversibility sources* can be controlled.

Figure 3.11 provides the operational diagram of a four-switch DC–DC converter. K_1 and K'_1 on the one hand, K_2 and K'_2 on the other hand, must be complementary-controlled to avoid the voltage source short-circuiting and to ensure the current source is never in open circuit.

The type of switches used depends on the nature and the reversibility of the sources placed at the input and ouput. The full-bridge chopper is naturally reserved for the "conversions" which cannot be achieved using the choppers with two switches.

3.3.1 Most Usual Configuration

The most frequent use of the full-bridge chopper occurs in cases where there is a link between

a *current-reversible voltage source*

$$(U > 0; \quad i > 0 \quad \text{or} \quad < 0)$$

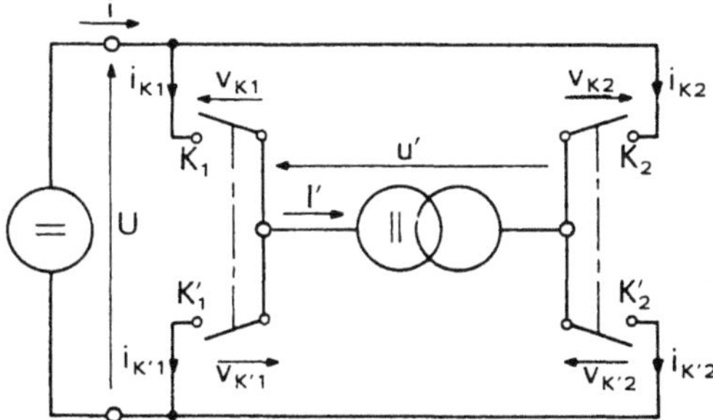

Fig. 3.11

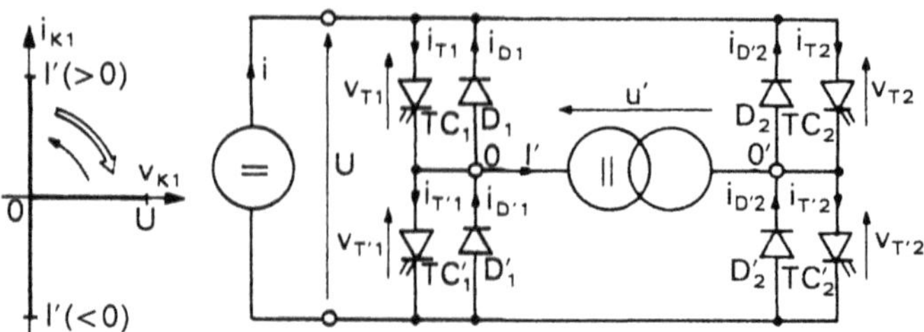

Fig. 3.12

and a *current- and voltage-reversible current course*

$$(I' > 0 \quad \text{or} \quad < 0; \quad u' > 0 \quad \text{or} \quad < 0).$$

This occurs, for example, in cases where a storage battery (voltage source, current-reversible but not voltage-reversible) has to be connected to the armature of a DC motor (current source, current- and voltage-reversible), thus enabling the machine to be used to the maximum of its possibilities.

In the operational diagram (Fig. 3.11), the various states imposed on switch K_1 are shown:

When K_1 is closed, $i_{K_1} = I'$, positive if I' is positive,
negative if I' is negative.

When K_1 is open, $v_{K_1} = U$, positive.

This accounts for the three branches (Fig. 3.12) of the v–i characteristics on which K_1 has to operate. For the other three switches, the same applies. Each switch must therefore be made up of a controlled turn-on/turn-off semiconductor device and a diode in antiparallel. This gives the basic diagram shown in Fig. 3.12.

The "switches" can be controlled in two ways:

- by use of two different laws, according to whether the mean value U' of the output voltage is positive or negative;
- by use of a single control law.

3.3.1.1 Sequential Control (Fig. 3.13)

During an operational cycle T of the converter, the turn-on/turn-off of only one semiconductor device is acted on. The waveforms are the same as those found in Fig. 3.10.

- *To obtain a positive output voltage*, the ON state of TC'_2 is constantly controlled. If current I' is positive, it leaves terminal O′ of the current source via TC_2; otherwise, it leaves it via D'_2.
  - *When I' is positive*, TC_1 is the switching semiconductor device:

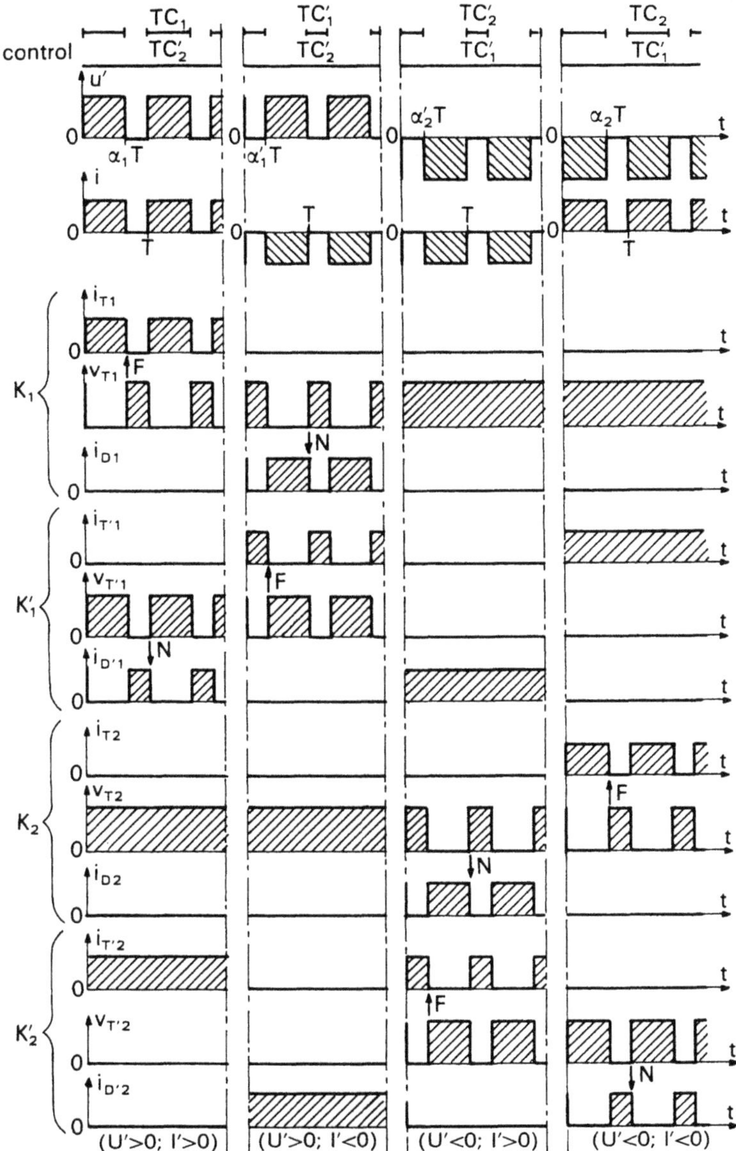

Fig. 3.13

- when TC_1 is on, it lets I' through:

$$i = i_{T_1} = I' = i_{T_2'}; \quad u' = U;$$

- when TC_1 is off, I' goes via D_1';

$$i = 0; \quad I' = i_{T_2'} = i_{D_1'}; \quad u' = 0.$$

If $\alpha_1 T$ denotes the time duration of TC_1 on interval during each cycle:

$$\frac{U'}{U} = \frac{I}{I'} = \alpha_1 . \tag{3.4}$$

- *When I' is negative*, switching is carried out by TC'_1:
  - when TC'_1 is on, $-I'$ goes through it:

    $$i = 0; \quad i_{D'_2} = -I' = i_{T'_1}; \quad u' = 0;$$

  - when TC'_1 is off, $-I'$ goes via D_1:

    $$i_{D'_2} = -I' = i_{D_1} = -i; \quad u' = U.$$

  If α'_1 is the relative duration of the on period of TC'_1:

  $$\frac{U'}{U} = \frac{I}{I'} = 1 - \alpha'_1 . \tag{3.4'}$$

- *To obtain a negative output voltage*, the ON state of TC'_1 is constantly controlled. If current I' is negative, it goes via TC'_1; if it is positive, it goes via D'_1.
  - *When I' is positive*, TC'_2 is on for $\alpha'_2 T$ during each duty cycle T:
    - when TC'_2 is on, it lets I' flow:

      $$i = 0; \quad i_{D'_1} = I' = i_{T'_2}; \quad u' = 0;$$

    - when TC'_2 is off, I' goes via D_2:

      $$i_{D'_1} = I' = i_{D_2} = -i; \quad u' = -U.$$

    Thus

    $$-\frac{U'}{U} = -\frac{I}{I'} = 1 - \alpha'_2 . \tag{3.4''}$$

  - *When I' is negative*, TC_2 is on for $\alpha_2 T$ during each cycle T.
    - when TC_2 is on, it lets $-I'$ flow:

      $$i = i_{T_2} = -I' = i_{T'_1}; \quad u' = -U;$$

    - when TC_2 is off, $-I'$ goes via D'_2:

      $$i = 0; \quad -I = i_{T'_1} = i_{D'_2}; \quad u' = 0.$$

    Thus,

    $$-\frac{U'}{U} = -\frac{I}{I'} = \alpha_2 . \tag{3.4'''}$$

In Fig. 3.13, for each operating mode

$$(U' > 0;\ I' > 0),\quad (U' > 0;\ I' < 0),\quad (U' < 0;\ I' > 0)$$
$$\text{and } (U' < 0;\ I' < 0),$$

we have shown the periods during which the controlled semiconductor devices receive the turn-on command.

We have then plotted the waveforms of
the output voltage u',
the input current i,
the voltages v_{T_1}, $v_{T'_1}$, v_{T_2}, $v_{T'_2}$ across the "switches",
the currents i_{T_1} and i_{D_1}, i_{T_1}, and $i_{D'_1}$, i_{T_2} and i_{D_2}, $i_{T'_2}$ and $i_{D'_2}$ through the eight semiconductor devices.

The peak value of the voltage pulses is equal to U, that of the current pulses to $|I'|$.

- *Remarks*

– To make control easier at low values of $|I'|$, especially when this current falls to zero, a complementary control of the switching semiconductor devices can be used, i.e.
  - to make $\alpha_1 + \alpha'_1 = 1$, when U' is positive;
  - to make $\alpha_2 + \alpha'_2 = 1$, when U' is negative.

– The roles devoted to the controlled semiconductor devices can be changed by taking
  - when U' is positive,
    TC_1 as permanently turned-on device,
    TC_2 and TC'_2 as switching devices;
  - when U' is negative,
    TC_2 as permanently turned-on device,
    TC_1 and TC'_1 as switching devices.

3.3.1.2 Continuous Control (Fig. 3.14)

The following procedure avoids any discontinuity in the control:
- for each cycle T, the turn-on of TC_1 and TC'_2 is controlled for a period equal to $\alpha_1 T$;
- and the turn-on of TC_2 and TC'_1 is controlled for the remainder of the cycle.

When applied simultaneously to both half-bridges, this complementary control enables voltage u' to be fixed independently of the polarity of current I'.

For $0 < t < \alpha_1 T$, the turn-on of TC_1 and TC'_2 is controlled:

$$u' = U$$

$$\text{if } I' > 0,\quad i = i_{T_1} = I' = i_{T'_2};$$

$$\text{if } I' < 0,\quad i_{D'_2} = -I' = i_{D_1} = -i.$$

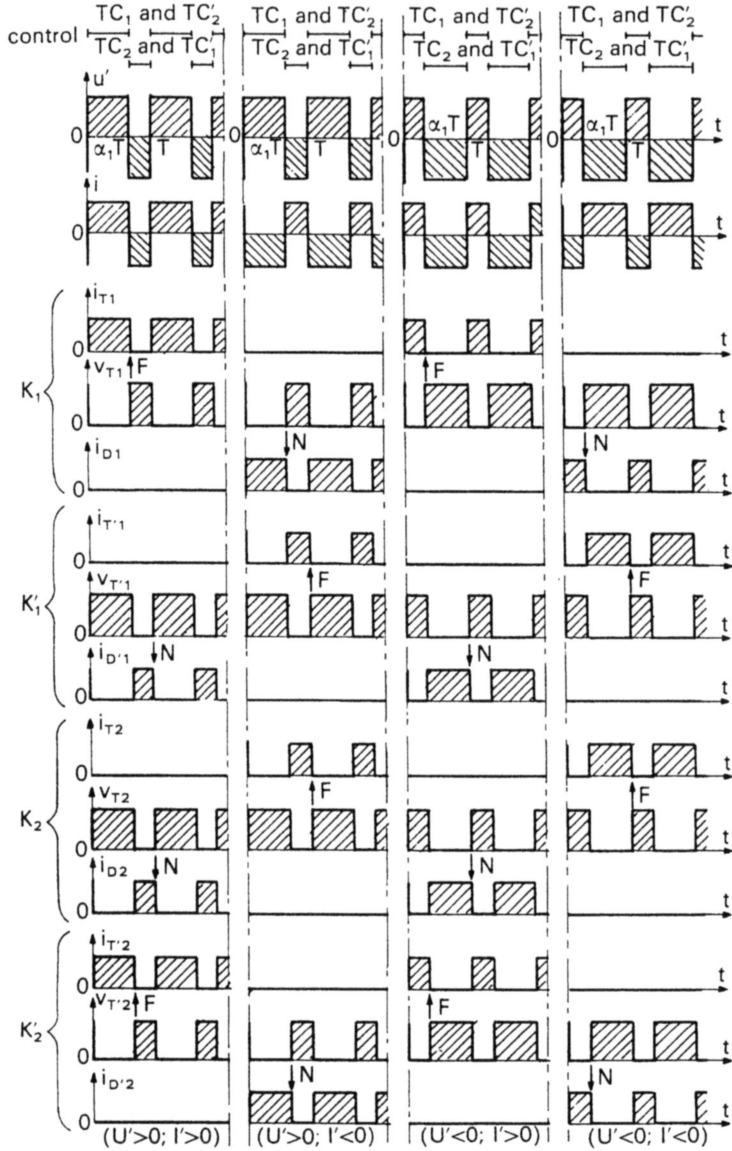

Fig. 3.14

For $\alpha_1 < t < T$, the turn-on of TC_2 and TC'_1 is controlled:

$$u' = -U$$

$$\text{if } I' > 0, \quad i_{D'_1} = I' = i_{D_2} = -i;$$

$$\text{if } I' < 0, \quad i = i_{T_2} = -I' = i_{T'_1}.$$

The relationship between the mean values U' and I and the constant values U and I' is the same for all operation modes:

$$\frac{U'}{U} = \frac{I}{I'} = \frac{\alpha_1 T - (1 - \alpha_1)T}{T}$$

$$\frac{U'}{U} = \frac{I}{I'} = 2\alpha_1 - 1 . \tag{3.5}$$

If α_1 is varied from 1 to zero, U' is varied from $+U$ to $-U$.

Figure 3.14 gives, for the four operational modes, the waveforms of the output voltage u', the input current, the currents through the semiconductor devices and the voltages across them. The diagrammatic representation of the control signals, at the top of the figure, shows that the control sequence always remains the same. Only the duration of the signals varies.

3.3.1.3 Comparison of the Two Control Methods

The change from sequential control to continuous control has two drawbacks and two advantages.

- The first drawback is that the *harmonics content* of voltage u' and current i is higher. Instead of being made up of unidirectional pulses (u' equal to $+U$ or zero, if U' is positive; u' equal to $-U$ or zero, if U' is negative), the output voltage u' is made up of pulses of alternating polarities (u' equals $+U$ or $-U$). The same applies to the current i. The voltage source and the current source are disturbed to a greater extent.
- The second drawback arises from the *number of commutations*. These are shown by small arrows in Figs. 3.13 and 3.14. It is indicated whether they are natural (N) or forced (F). It can be seen that the number of commutations per cycle is doubled as well as the corresponding losses.
- The first advantage arises from the fact that all the "switches" turn off and turn on once per cycle: the waveform of the voltage across them is, significantly, the same at all times and for all "switches". This is of great help in *checking on* the operation and enables action to be taken more rapidly, if any problems arise.
- The second advantage comes from the *single control law*. When the current of the current source is reversed or when the voltage across it has to be reversed, the control cycle of the controlled switches need not be changed. Only the width of the control signals need be varied. This avoids any dead time when the operational mode is being changed and improves the response speed.

Although the steady-state performances are less satisfactory, the second type of control is chosen when very fast reversals are needed.

3.3.2 Remarks on Choppers with Two Controlled Switches Series-Connected Across the DC Voltage Source

The chopper with two current-reversible "switches" in Fig. 3.10 comprises two controlled turn-off/turn-on semiconductor devices, series-connected across voltage U, each being shunted by an antiparallel diode. The full-bridge chopper (Fig. 3.12) consists of two identical assemblies, parallel-connected under voltage U.

- Both of these structures can operate as *inverters.*

- In the case of the first one, a DC voltage source U with a mid-point O′ is needed (Fig. 3.15a) and the complementary control of the two "switches" with a duty ratio α_1 equal to 1/2 must be used. The voltage between points O and N is then equal to $+U$ for a half-cycle and to zero during the other half. The voltage between point O and the mid-point O′ is equal to $U/2$ during one half-cycle, and to $-U/2$ during the other. It can be applied to a voltage-reversible AC current source.
- In the case of the full-bridge topology, with α_1 equal to 1/2 and the control shown in Fig. 3.14, voltage u' is equal to $+U$ during one half-cycle and to $-U$ during the other. A voltage-reversible AC current source can be connected between points O and O′ (Fig. 3.15b).

The two inverter diagrams in Fig. 3.15 are those most frequently used when designing voltage-source inverters.

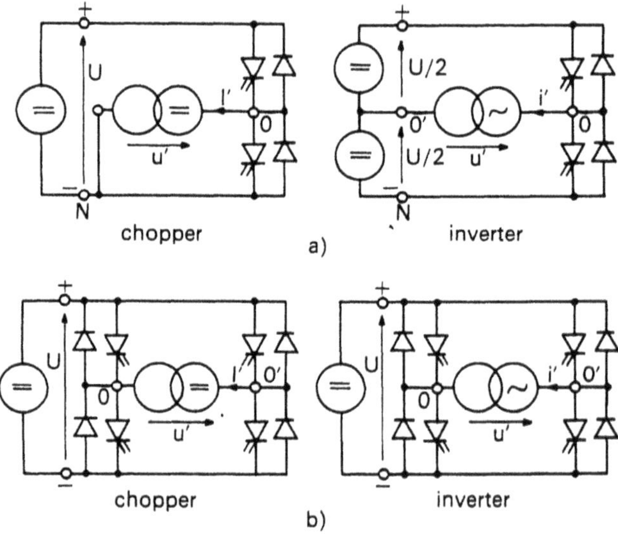

Fig. 3.15

- The same *commutations* have to be carried out in the chopper and inverter operating modes, since I' and i' can be either positive or negative.

  We will study the current transfers between these series-connected switches under a DC voltage in the next volume, which deals with inverters.

  It is, however, worth noting that series-connection of the controlled semiconductor devices may make it easier to block them. This applies especially when thyristors are concerned: turning on one controls the turn-off of the other and vice versa.

- In general, *complementary control* is chosen to constantly settle voltage u' across the current source by using voltage U of the voltage source. This obviously requires a certain number of precautions to avoid the short-circuiting of voltage source U by the simultaneous conduction of the two series-connected switches.

3.3.3 Other Types of Full–Bridge DC–DC Converters

By taking into account the five possible types of DC current type generator, load or source:

generator,
load,
current-reversible source,
voltage-reversible source,
current- and voltage-reversible source,

and the five similar types of DC voltage type generator, load or source, 25 different combinations are possible. We could therefore expect to find 25 types of chopper.

In fact, a generator can only be connected to a load and vice versa. A source can only use its reversibility (-ies) if it is connected via the chopper to another source which is itself reversible. Table 3.1 shows, horizontally, the current type sources and, vertically, the voltage type sources. Only 11 boxes correspond to possible converters and the others are crossed.

The buck converter corresponds to 2.1, and the boost converter to 1.2. Using two switches, it is possible to fill the boxes on the main diagonal, where the chopper must connect two sources of different type but with the same reversibility (-ies): boxes 3.3, 4.4 and 5.5 correspond to the structures in Figs. 3.9b, 3.9a and 3.9c. These five converter types can obviously be built using full-bridge structures, but this brings unnecessary complications.

Full-bridge structures must be used to fill the other boxes outside the main diagonal since, in such cases, two sources with different reversibilities must be linked. The DC–DC converter studied previously corresponds to box 5.3. We shall now give a rapid description of the structures corresponding to boxes 4.3 and 5.4.

Table 3.1

Voltage type	Current type				
	Generator	Load	Current-reversible source	Voltage-reversible source	Current- and volt.-revers. source
Generator	1.1	2.1 chopper Fig. 3.6	3.1	4.1	5.1
Load	1.2 chopper Fig. 3.7	2.2	3.2	4.2	5.2
Current-reversible source	1.3	2.3	3.3 chopper Fig. 3.9b	4.3 chopper Fig. 3.16	5.3 chopper Fig. 3.12
Voltage-reversible source	1.4	2.4	3.4	4.4 chopper Fig. 3.9a	5.4 chopper Fig. 3.17
Current- and voltage-reversible source	1.5	2.5	3.5	4.5	5.5 chopper Fig. 3.9c

Note that not all the boxes can be filled, if the aim is to achieve a rational use of semiconductor devices.

3.3.3.1 Chopper Connecting a Current-Reversible Voltage Source to a Voltage-Reversible Current Source (Fig. 3.16)

Since Ui equals $u'I'$, the input current and the output voltage must change their polarity simultaneously.

- To obtain u' and i alternately positive and zero,
  - first K_1 and K_2' are closed (state 1):

$$u' = U; \quad i = I'$$

$$i_{K_1} = I'; \quad v_{K_1'} = U; \quad v_{K_2} = U; \quad i_{K_2'} = I'$$

  - then K_1 and K_2 are closed (state 2):

$$u' = 0; \quad i = 0$$

$$i_{K_1} = I'; \quad v_{K_1'} = U; \quad i_{K_2} = -I'; \quad v_{K_2'} = U$$

- To obtain u' and i alternately negative and zero,

• first K'_1 and K_2 are closed (state 3):

$$u' = -U; \quad i = -I'$$

$$v_{K_1} = U; \quad i_{K'_1} = -I'; \quad i_{K_2} = -I'; \quad v_{K'_2} = U$$

• then K'_1 and K'_2 are closed (state 4):

$$u' = 0; \quad i = 0$$

$$v_{K_1} = U; \quad i_{K'_1} = -I'; \quad v_{K_2} = U; \quad i_{K'_2} = I'.$$

On the operational diagram common to all full-bridge choppers, the polarities which can be attributed to u' and i are shown. The branches of the v–i characteristics on which the four switches must operate are then indicated. It can be seen that

K_1 and K'_2 must be controlled turn-on/turn-off devices,

K_2 and K'_1 can be simple antiparallel-connected diodes.

$$i_D = -i_K; \quad v_D = -v_K$$

This gives the circuit configuration.

The controlled semiconductor device TC_1 makes the reversal possible (u' and i change polarity during the periods when they are not zero, i.e. the mean values, U' and I, change polarity). The controlled device TC'_2 takes charge of the variations in the ratios $|U'/U|$ or $|I/I'|$.

If TC'_2 is conducting for αT during each cyle T,

$$U'/U = I/I' = \alpha, \qquad \text{if } U' \text{ and } I \text{ are positive,}$$

$$-U'/U = -I/I' = 1 - \alpha, \qquad \text{if } U' \text{ and } I \text{ are negative.}$$

It can be seen that the commutations TC_1–D'_1 and $TC'_2 - D_2$ must be forced and commutations D'_1–TC_1 and D_2–TC'_2 are natural.

3.3.3.2 Chopper Connecting a Voltage-Reversible Voltage source to a Current- and Voltage-Reversible Current Source (Fig. 3.17)

Since the voltage source is not current-reversible, its current i is positive when the converter connects the two sources and zero when it disconnects them. Equating input and output instantaneous powers gives:

$$u' = \frac{i}{I'} U.$$

If I' is positive, u' equals $+U$ or zero. If I' is negative, u' equals $-U$ or zero.

– *For a positive* I', K_1 and, alternately, K_2 and K'_2 are closed:

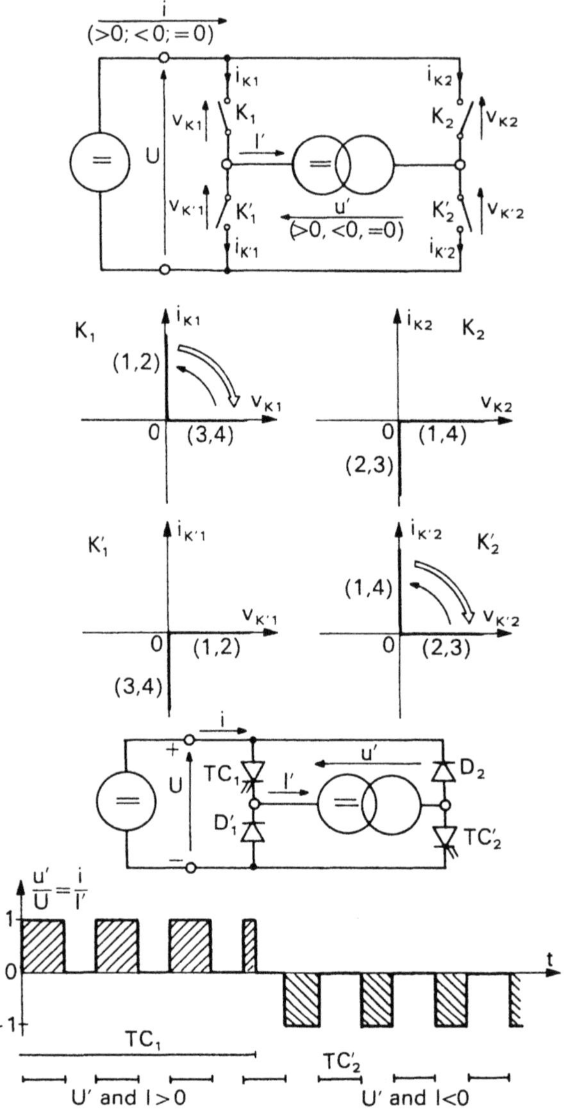

Fig. 3.16

- When K_1 and K'_2 are closed,

$$i = i_{K_1} = I' = i_{K'_2} > 0; \quad u' = U \qquad \text{(state 1 or 1')}$$

$$\text{if } U \text{ is positive, } v_{K'_1} = U > 0; \quad v_{K_2} = U > 0 \qquad \text{(state 1)}$$

$$\text{if } U \text{ is negative, } v_{K'_1} = U < 0; \quad v_{K_2} = U < 0. \qquad \text{(state 1')}$$

i
(>0; =0)
i_{K1}
i_{K2}
v_{K1}
K_1
K_2
v_{K2}
U
(>0; <0)
I′
(>0; <0)
u′
(>0; <0; =0)
$v_{K'1}$
K'_1
K'_2
$v_{K'2}$
$i_{K'1}$
$i_{K'2}$
(1,1′,2,2′)
(3,3′,4,4′)
(3′)
(3)
(4,4′)
(1′)
(1)
(2,2′)
(3,3′)
(1,1′)
(1′,2′,4′)
(1,2,4)
(2′,3′,4′)
(2,3,4)
T_1
TC_1
T_2
TC_2
TC'_1
TC'_2
u′
i
t
U>0
U<0
I′ positive
I′ negative

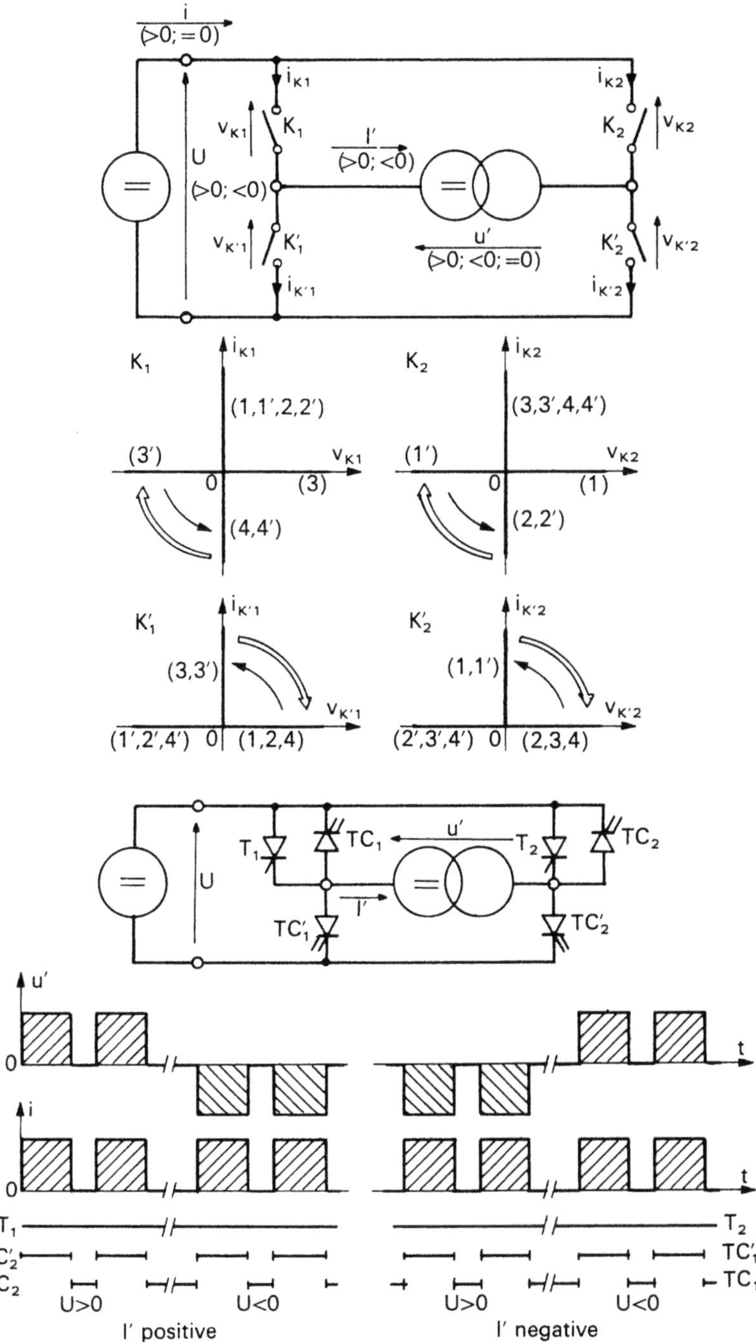

Fig. 3.17

• When K_1 and K_2 are closed,

$$i = 0; \quad i_{K_1} = I' = -i_{K_2} > 0; \quad u' = 0 \qquad \text{(state 2 or 2')}$$

$$\text{if } U \text{ is positive, } v_{K_1'} = U > 0; \quad v_{K_2'} = U > 0 \qquad \text{(state 2)}$$

$$\text{if } U \text{ is negative, } v_{K_1'} = U < 0; \quad v_{K_2'} = U < 0. \qquad \text{(state 2')}$$

– *For a negative* I', K_2 and, alternately, K_1 and K'_1 are closed:

• When K_2 and K'_1 are closed:

$$i = i_{K_2} = -I' = i_{K_1'} > 0; \quad u' = U \qquad \text{(state 3 or 3')}$$

$$\text{if } U \text{ is positive, } v_{K_2'} = U > 0; \quad v_{K_1} = U > 0 \qquad \text{(state 3)}$$

$$\text{if } U \text{ negative, } v_{K_2'} = U < 0; \quad v_{K_1} = U < 0. \qquad \text{(state 3')}$$

• When K_2 and K_1 are closed,

$$i = 0; \quad i_{K_2} = -I' = -i_{K_1'} > 0; \quad u' = 0 \qquad \text{(state 4 or 4')}$$

$$\text{if } U \text{ is positive, } v_{K_2'} = U > 0; \quad v_{K_1'} = U > 0 \qquad \text{(state 4)}$$

$$\text{if } U \text{ is negative, } v_{K_2'} = U < 0; \quad v_{K_1'} = U < 0. \qquad \text{(state 4')}$$

The representation of these various states on the current–voltage characteristics of the four switches shows that

- switch K_1 must be capable of turning on and off under negative voltage (going from state 3′ to states 4 or 4′, and vice versa); it must allow a positive current (states 1 or 1′, 2 or 2′) to flow; finally it must be capable of blocking a positive voltage (state 3). This switch is built using a conventional thyristor T_1 and a controlled turn-on/turn-off device TC_1, connected in antiparallel;
- switch K_2 plays a similar role and is built using a thyristor T_2 and a controlled turn-on/turn-off semiconductor device TC_2 connected in antiparallel;
- switch K'_1 must be capable of switching on or off a positive current under a positive voltage and of blocking a negative voltage. It is thus a controlled turn-on/turn-off semiconductor device TC'_1;
- switch K'_2, which plays a similar role to K'_1, is also a controlled turn-on/turn-off device TC'_2;
- all the controlled semiconductor devices, TC_1, TC'_1 TC_2, TC'_2, must be capable of blocking a reverse voltage.

In Fig. 3.17, below the operational diagram of the full-bridge choppers, there is shown, successively, the switching function to be carried out by the switches, the waveforms of voltage u' and current i and the control periods of the various semiconductor devices.

For a positive I' (T_1 always on and chopping by TC'_2 and TC_2), if α is the relative ON-time duration of TC'_2,

$$\frac{U'}{U} = \frac{I}{I'} = \alpha .$$

For a negative I' (T_2 always on and chopping by TC'_1 and TC_1), if α is the relative ON-time duration of TC'_1,

$$-\frac{U'}{U} = -\frac{I}{I'} = \alpha .$$

3.4 Indirectly Linked DC–DC Converters

Directly linked DC–DC converters only allow energy transfer control between sources of different natures – one of which is a current source and the other a voltage source. In order to be able to link two sources of the same nature, the converter must contain an energy storage element:

- This element is charged during a part of the cycle by connecting it to the input.
- During the other part, it is discharged by connecting it to the output.

This enables energy to be transferred from input to output without connecting these directly.

- When both sources are *voltage sources*, the storage element is an *inductor*. Sources which oppose the voltage discontinuities can withstand the current discontinuities, corresponding to their connection or disconnection with this inductor. The latter acts as an intermediate current source.

  When the converter is connected between two *current sources*, the storage element to which these sources are alternately connected must be a *capacitor*. The latter acts as an intermediate voltage source.

- In presenting the structures of indirectly linked DC–DC converters, we will also assume that

  the sources connected to the input and output are perfect;
  the "switches" are perfect;
  the commutations are instantaneous.

  Losses in storage element will not be taken into account, but *the finite value of the inductor or the capacitor* will.[1]

This has no effect on the general relations and does not make them more complicated to establish. It enables the operation to be presented more clearly by showing the charge and the discharge of the storage element.

It will simply be assumed that conduction is continuous, i.e. current in a "switch" is not interrupted during its normal conducting period.

- The instantaneous input power, ui, is no longer equal to the instantaneous output power, $u'i'$, since the converter contains an energy storage element.

[1] If the intermediate sources are assumed to be perfect, (infinite inductor or capacitor) the ripple in the current through the inductor or in the voltage across the capacitor disappear. In Figs. 3.18–3.25, the pulses are of constant amplitude.

However, as losses are not taken into account, the average power is the same at input and output.

$$UI = U'I'. \tag{3.6}$$

We will once again base our analysis on the operational diagrams, in order to determine the nature of the switches to be used. Our analysis concerns only diagrams with two switches.

3.4.1 Non-Reversible Chopper with Inductive Energy Storage

The non-reversible chopper (Fig. 3.18) only allows energy to be transferred from the voltage generator U to the voltage load U'.

- The operational diagram at the top of Fig. 3.18 illustrates the principle of the converter with inductive energy storage:
  - When K_1 is closed, the voltage generator U increases the current i_L in the inductor L which stores energy.
  - When K_2 is closed, inductor L discharges into the voltage load U', thus supplying it with current.

Both switches must be complementarily controlled to avoid the two voltage sources short-circuiting, or leaving the inductor (current source) in open circuit.

- The branches of the v–i characteristics used, together with the commutations to be carried out, show that K_1 must be replaced by a controlled turn-on/turn-off semiconductor device and that K_2 can be replaced by a simple diode. This gives the basic diagram with TC_1 and D_2.
- If, during each cycle T, the controlled device TC_1 is conducting for αT:
  - when TC_1 is conducting,

$$i = i_{T_1} = i_L; \quad i' = 0; \quad v_{D_2} = -(U + U')$$

$$U = L\frac{di_L}{dt} \text{ shows that } i_L \text{ increases linearly;}$$

  - when D_2 is conducting,

$$i_L = i_{D_2} = i'; \quad i = 0; \quad v_{T_1} = +(U + U')$$

$$U' = -L\frac{di_L}{dt} \text{ shows that } i_L \text{ decreases linearly.}$$

In steady-state operation, current i_L has the same average value I_L during its increase as during its decrease. This gives the average value I of the input current i and that I' of the output current i':

$$I = \alpha I_L; \quad I' = (1 - \alpha)I_L.$$

Since UI equals $U'I'$, the equation linking the ouput and input values is

$$\frac{U'}{U} = \frac{I}{I'} = \frac{\alpha}{1-\alpha}. \tag{3.7}$$

By varying α from 1 to zero, it is theoretically possible to vary these ratios from infinity to zero. Therefore this chopper is frequently referred as the buck–boost converter.

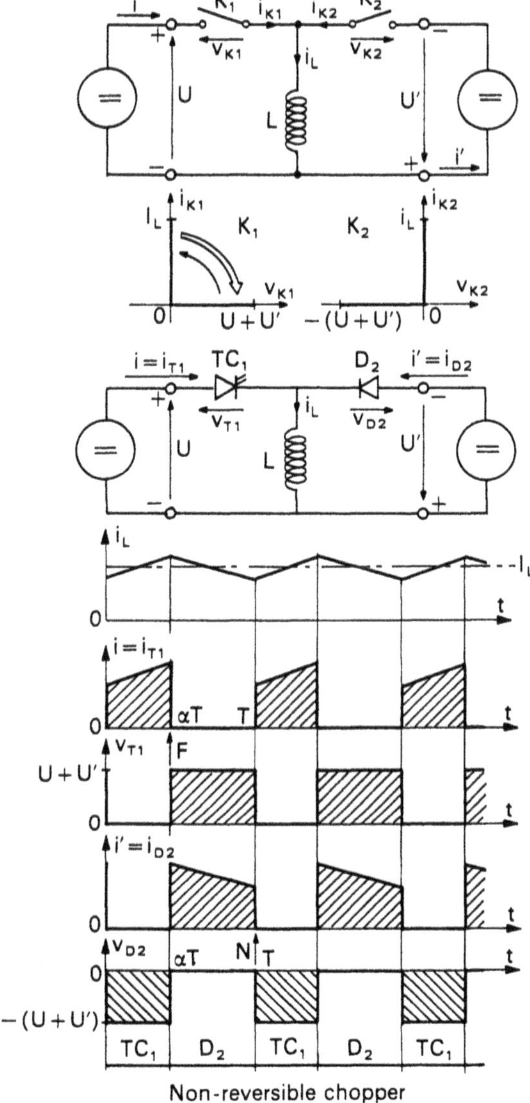

Non-reversible chopper

Fig. 3.18

– At the bottom of Fig. 3.18 can be found the waveforms of
current i_L in the inductor,
input current i, equal to current i_{T_1} though TC_1,
voltage v_{T_1} across TC_1,
output current i' equal to current i_{D_2} through D_2,
voltage v_{D_2} across D_2,
as well as the conduction diagram of both switches. It can be clearly seen that commutation TC_1–D_2 is forced and commutation D_2–TC_1 is natural.

3.4.2 Reversible Choppers with Inductive Energy Storage

Two-"switch" choppers with inductive energy storage use the same diagram as shown at the top of Fig. 3.18. The only change is the semiconductor devices used.

The choppers with two switches only allow sources which have the *same type (s) of reversibility to be connected.*[1] Thus, current i in the inductor flows in the same direction throughout the operational cycle. This current is equal to the input current i when K_1 is turned on and the output current i' when K_2 is turned on. The average values I and I' of the input and output currents necessarily have the same polarity; they can only be reversed together. Since UI equals $U'I'$, voltages U and U' can also only be reversed together.

As in the case of double-"switch", directly-linked choppers, these converters can be characterized by the type(s) of reversibility of the sources which they connect.

Whatever the reversibilities, if αT is the on-time duration of K_1 during each cycle, in all cases:

$$\frac{U'}{U} = \frac{I}{I'} = \frac{\alpha}{1-\alpha}.$$

3.4.2.1 Voltage-Reversible Chopper (Fig. 3.19)

As in the case of the non-reversible chopper, K_1 ensures the closing and the opening of the current i_L path when voltages U and U' are positive. K_2 must only allow i_L to continue to flow when K_1 is open.

When U and U' are negative, the right-hand source acts as a generator and the left-hand one as a load: K_2 must act as did K_1 previously, and vice versa.

The two switches must be replaced by two controlled turn-on/turn-off devices, TC_1 and TC_2 able of blocking reverse voltages. They must have complementary controls.

– For positive U and U',

[1] A four-switch or full-bridge structure would be required to connect two sources of different reversibilities.

- when TC_1 is on:

$$i = i_{T_1} = i_L; \quad i' = 0; \quad v_{T_2} = -(U + U') < 0$$

$$i_L, \text{ given by } \quad L\frac{di_L}{dt} = U, \text{ increases;}$$

- when TC_1 is off:

$$i = 0; \quad i_L = i' = i_{T_2}; \quad v_{T_1} = +(U + U') > 0$$

$$i_L, \text{ given by } \quad L\frac{di_L}{dt} = -U', \text{ decreases.}$$

– For negative U and U',

- when TC_2 is on:

$$i' = i_{T_2} = i_L; \quad i = 0; \quad v_{T_1} = +(U + U') < 0$$

$$i_L, \text{ given by } \quad L\frac{di_L}{dt} = -U', \text{ increases;}$$

- when TC_2 is off:

$$i = 0; \quad i_L = i = i_{T_1}; \quad v_{T_2} = -(U + U') > 0$$

$$i_L, \text{ given by } \quad L\frac{di_L}{dt} = U, \text{ decreases.}$$

3.4.2.2 Current-Reversible Chopper (Fig. 3.20)

The operational diagram remains that which appears at the top of the previous figure.

When i_L is positive, i and i' are alternately positive and zero:

K_1 must be able to switch on or off a positive current under a positive voltage,

K_2 must be able to let a positive current flow or block a reverse voltage.

When i_L is negative, i and i' are negative or zero in turn:

K_1 must be able to let a negative current flow or block a positive voltage;

K_2 must be able to switch on or off a negative current under a negative voltage.

Switch K_1 must thus be replaced by a controlled turn-on/turn-off semiconductor device TC_1, with a diode D_1 connected in antiparallel. Similarly switch K_2 must be made up by TC_2 in antiparallel with D_2; however, since all the current and voltage stresses are reversed, these two semiconductor devices must be connected in the direction shown in the basic diagram.

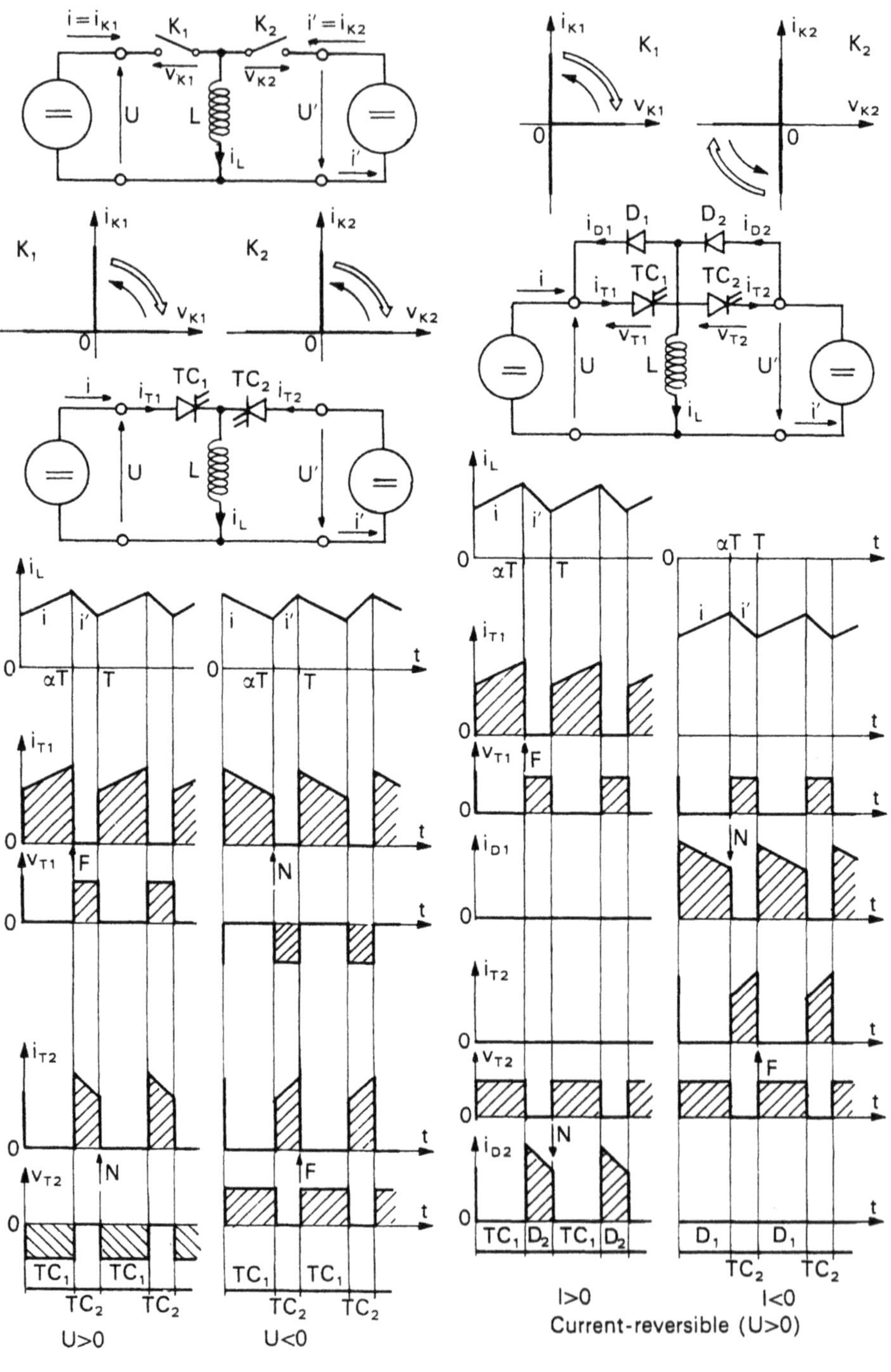

Fig. 3.19

Fig. 3.20

K1
K2
TC1
TC'1
TC2
TC'2
TC1
TC'1
TC2
TC'2
U
U'
L
i
i'
iL
(I > 0; U > 0)
(I>0; U<0)
(I<0; U>0)
(I<0; U<0)
Current- and voltage-reversible

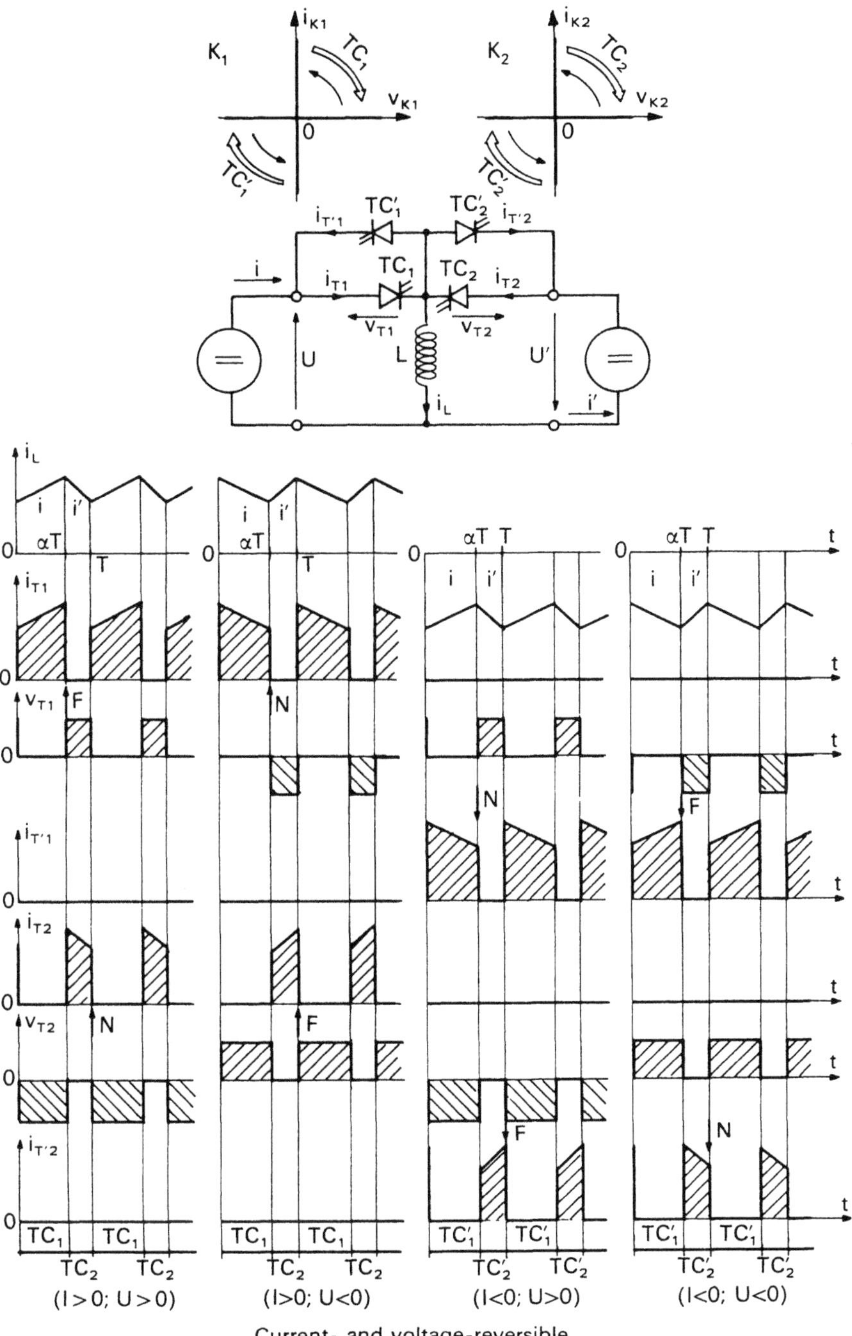

Fig. 3.21

3.4.2.3 Current- and Voltage-Reversible Chopper (Fig. 3.21)

The stresses applied on the two switches are the sum of those which correspond to the voltage reversibility ($I > 0$; $U > 0$ or < 0) and the current reversibility ($U > 0$; $I > 0$ or < 0). To those should be added the stresses arising from operating with U and I simultaneously negative.

Switch K_1 must be able to switch on or off a positive current under a positive voltage and switch on or off a negative current under a negative voltage. The same applies to K_2. This gives the basic diagram in Fig. 3.21. Each switch comprises two semiconductor devices connected in antiparallel TC_1 and TC'_1, TC_2 and TC'_2. Each device is turn-on/turn-off controlled and capable of blocking reverse voltages.

For each of the four operational modes, the device which uses its ability to turn-on and turn-off is located on the generator side. The other acts as a free-wheeling diode, enabling i_L to flow when the first is turned off.

The semiconductor device which has to control commutations is

$$TC_1, \quad \text{for} \quad U > 0 \quad \text{and} \quad I > 0,$$

$$TC'_2, \quad \text{for} \quad U > 0 \quad \text{and} \quad I < 0,$$

$$TC'_1, \quad \text{for} \quad U < 0 \quad \text{and} \quad I < 0,$$

$$TC_2, \quad \text{for} \quad U < 0 \quad \text{and} \quad I > 0.$$

- *Remark*:

In the non-reversible chopper (Fig. 3.18) and the current-reversible chopper (Fig. 3.20), the commutations appear between a controlled device and a diode. These semiconductor devices have either their cathodes or their anodes connected together. The current i_L transfer and the flow of the reverse recovery current of the device turned off, pose no problems.

In the other two choppers (Figs. 3.19 and 3.21) the commutations appear between two controlled semiconductor devices. A suitable overlapping of their controls is necessary in order to enable the transfer of i_L and the reverse recovery current flow of the device which has to be blocked.

3.4.3 Non-Reversible Chopper with Capacitive Energy Storage

- Often named after their inventor – the *Cúk converters* – the choppers with capacitive energy storage enable energy to be transferred between two *current sources*. This dual aspect makes them similar to choppers with inductive energy storage connected between two voltage sources.

The operational diagram at the top of Fig. 3.22 is common to all choppers with two "switches" and capacitive energy storage. Capacitor C is shown series-connected between the two current sources. The two "switches" K_1 and

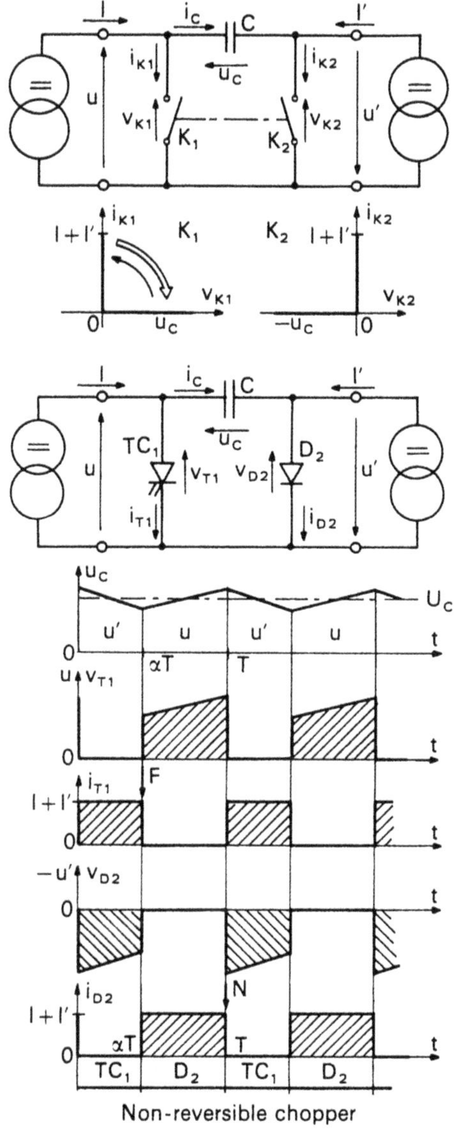

Fig. 3.22

K_2 must be complementarily controlled,

- so that a current source is not placed in open circuit;
- so that the capacitor, which acts as an intermediate voltage source, is not short-circuited.

- The remainder of Fig. 3.22 concerns the non-reversible chopper, connected between current generator I and current load I'.

– When K_2 is closed and K_1 open, the capacitor is charged by the generator at constant current I. Voltage u_C across it increases linearly:

$$u = u_C; \quad u' = 0$$

$$i_C = I; \quad \frac{\mathrm{d}u_C}{\mathrm{d}t} = \frac{I}{C}$$

$$v_{K_1} = u_C > 0; \quad i_{K_2} = I + I' > 0.$$

– When K_2 is open and K_1 closed, the capacitor supplies the constant current I' load. Voltage u_C decreases linearly:

$$u = 0; \quad u' = u_C$$

$$i_C = -I'; \quad \frac{\mathrm{d}u_C}{\mathrm{d}t} = -\frac{I'}{C}$$

$$i_{K_1} = I + I' > 0; \quad v_{K_2} = -u' < 0.$$

The branches used on the v–i characteristics show that K_2 ($v_{K_2} < 0$ or $i_{K_2} > 0$) may consist of a diode and that K_1 ($v_{K_1} > 0$ or $i_{K_1} > 0$; switching from one branch to the other in one direction and then in the other) has to be made with a controlled turn-on/turn-off semiconductor device TC_1. The latter does not have to block any reverse voltage.

Below the basic diagram are shown the waveforms of

voltage u_C across the capacitor,
input voltage u, equal to voltage v_{T_1} across TC_1,
current i_{T_1} through TC_1,
voltage v_{D_2} across D_2, equal to output voltage u' with opposite sign,
current i_{D_2} through diode D_2.

- Whatever the nature of the "switches", in steady-state operation, the mean value U_C of the voltage across the capacitor is the same during charging as during discharging.

  If αT is the on-time duration of K_1 during each cycle T, the mean values U and U' of the input u and output u' voltages are given by

$$U' = U_C \frac{\alpha T}{T}; \quad U = U_C \frac{(1-\alpha)T}{T}.$$

  Since UI equals $U'I'$, between the input and output values, this gives the equation

$$\frac{U'}{U} = \frac{I}{I'} = \frac{\alpha}{1-\alpha}. \tag{3.8}$$

3.4.4 Reversible Choppers with Capacitive Energy storage

The values of voltage u and u' are, in turn, u_C and zero; their mean values U and U' must have the same polarity, since inverting one of them assumes the other will also be inverted. With UI equal to $U'I'$, the inversion of I assumes that of I' also.

Like all choppers with two "switches", the capacitive storage chopper only allows for reversible energy exchanges between two sources which have the same type(s) of reversibility.

3.4.4.1 Voltage-Reversible Chopper (Fig. 3.23)

In the case of positive U, U' and U_C, the switches act as shown for the non-reversible chopper.

In the case of negative U, U' and U_C, current source I' acts as a generator, and current source I as a load. The roles of the switches are also reversed. K_2 has to switch on and off current $I + I'$ under a positive voltage; K_1 has to allow a positive current to flow when K_2 is open, and to block a negative voltage when K_2 is closed.

If the stresses related to the two operating modes are added, it can be seen that K_1 and K_2 must both be replaced by controlled turn-on/turn-off semiconductor devices, TC_1 and TC_2, both of which are capable of blocking reverse voltages.

3.4.4.2 Current-Reversible Chopper (Fig. 3.24)

When currents I and I' are positive, the operating mode is that of the non-reversible chopper: a diode D_2 and a controlled semiconductor device TC_1 are required.

When currents I and I' are negative, the right-hand source operates as a generator and the left-hand one as a load. Since the roles of the two sources and the current directions are reversed, a diode D_1 must be connected in antiparallel across TC_1, while a controlled turn-on/turn-off semiconductor device TC_2 must be connected in antiparallel across D_2.

3.4.4.3 Current- and Voltage-Reversible Chopper (Fig. 3.25)

Each of the switches K_1 and K_2 must be replaced by two controlled turn-on/turn-off devices, TC_1 and TC'_1, TC_2 and TC'_2, capable of blocking a reverse voltage and connected in antiparallel.

For each of the four operating modes, the turn-on and turn-off ability of only one semiconductor device is used; one of the three remaining ones must be adequately controlled in order to enable it to act as a diode.

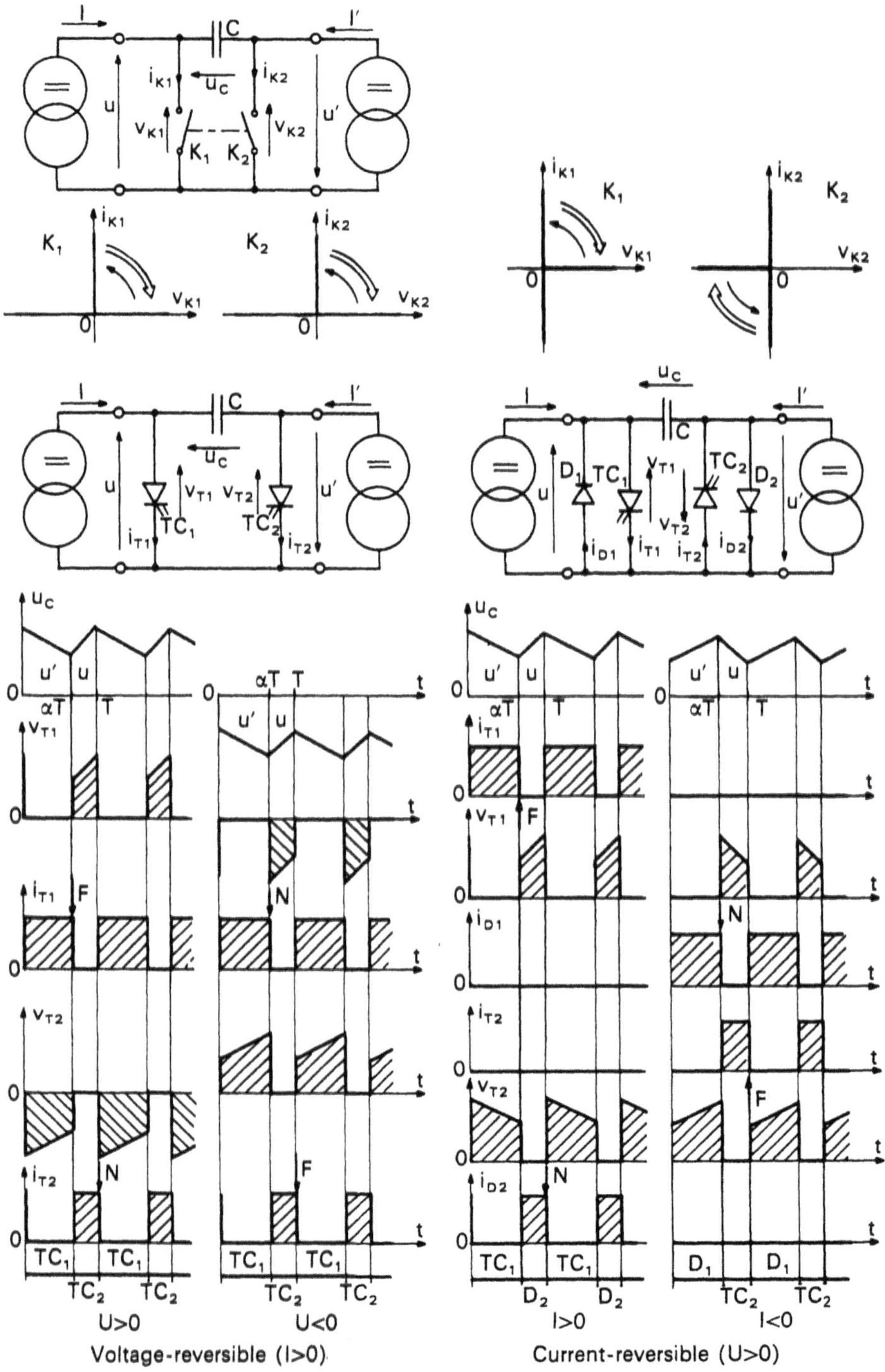

Fig. 3.23

Fig. 3.24

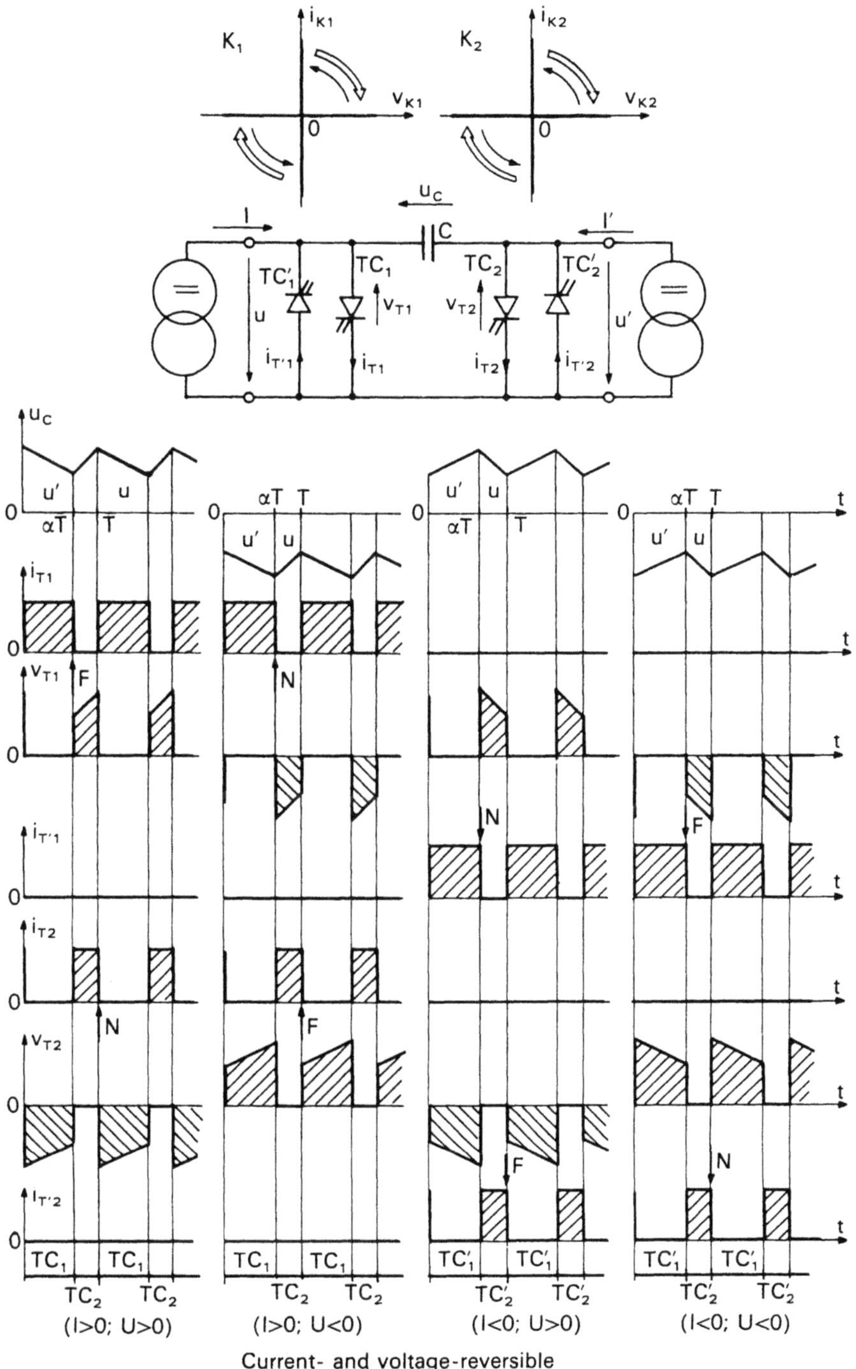

Current- and voltage-reversible

Fig. 3.25

The turning on and turning off ability of the following device is used:

$$\begin{aligned} &TC_1, \quad \text{for} \quad U > 0 \quad \text{and} \quad I > 0,\\ &TC'_2, \quad \text{for} \quad U > 0 \quad \text{and} \quad I < 0,\\ &TC_2, \quad \text{for} \quad U < 0 \quad \text{and} \quad I > 0,\\ &TC'_1, \quad \text{for} \quad U < 0 \quad \text{and} \quad I < 0. \end{aligned}$$

The remarks on the commutations, made in the section devoted to choppers with inductive energy storage, applies to choppers with capacitive storage which have the same reversibilities.

3.4.5 Remarks on Choppers with Intermediate Energy Storage Stage

- In a chopper with capacitive energy storage, the current in the conducting switch is the sum of the generator and load currents. If the current supplied by the former to the latter has to vary from zero to I, the current rating of the switches must be taken as equal to $2I$.

  In the case of the choppers with inductive energy storage, the voltages of the two sources are added across the blocked switch. The voltage ratings of the switches must be doubled.

- For all choppers with intermediate energy storage and using only two switches, the point which is common to both sources corresponds to the + terminal of one and the − terminal of the other. This can be a problem in certain applications where the two − terminals must be grounded.

*

* *

Choppers with inductive or capacitive energy storage are rarely used at medium or high power ratings. They are, however, often used in switch-mode power supplies which deal with lower power ratings.

Indeed the higher the operating frequency and the smaller the power ratings, the smaller may be the inductor (or the capacitor) without its role as an intermediate current (or voltage) source being greatly affected. At high power ratings, the limitations of the semiconductor devices used leads to the frequency being reduced. Because of the power increase and frequency decrease, the inductor (or capacitor) would have to be given a value leading to prohibitive cost and excessive size.

It should be noted that, in the switch-mode power supplies, coupled coils are often used. These act as both inductor and transformer. The operating modes of the converter can, in such cases, show considerable differences from those described in this chapter.

This explains the plan for the following chapters.

- For high- and medium-power applications, we will develop the analysis of *directly linked choppers* (Chap. 4) emphasizing the most common uses: buck converter, boost converter, current-reversible chopper, full-bridge chopper with current- and voltage-reversible output. The load considered is the active RLE type.
- Such choppers often make use of the thyristor. Chapter 5 is thus devoted to the *forced commutation of thyristors* in the four directly linked choppers mentioned above.
- *Switch-mode power supplies* (Chap. 6) normally concern another field of application. The load is likened to a pure voltage-fed resistance. This explains the new quantitative analysis of the buck and boost converters, and that of converters with intermediate energy storage. This is followed by an analysis of the operational mode and the characteristics of circuits using transformers.

These three chapters in no way lay claim to solving all the problems raised by DC–DC converters. They confine themselves to those which can arise in the most frequent types of application. Moreover the methods used can be transferred to other such structures.

Chapter 4

Operation and Characteristics of Directly Linked Choppers

In the previous chapter, we presented the principle of directly linked DC–DC converters either with two switches or in full-bridge topology. They enable the energy transfer to be controlled between a DC voltage generator and a DC current receptor, or between a DC current generator and a DC voltage receptor. To make matters simpler, we have used the word "source" to cover the functions of both generator and receptor.

For this simplified presentation, the current and voltage sources are assumed to be perfect. The main intention of this chapter is to analyze the *consequences of source imperfection* on the operation and characteristics of directly linked choppers

In the first part, we examine the *effects of current source imperfection*, i.e. the finite value of its inductance.

The second part concerns the *correction of the voltage source*, i.e. the filter placed between the converter and the voltage source in order to correct the effects of the non-zero value of this source's inductance.

In both cases, our analysis is limited to steady-state operation.

Two series of notes have been added to these first two parts:
- on the *multiphase* choppers,
- on using a chopper in *traction drives*.

4.1 Effects of Current Source Imperfection

The analysis of the influence of current "source" nature is similar in many respects to that which was put forward in the case of rectifiers (Vol. 1, Chap. 6, § 5 and 6). It makes use of the same phenomena, the same calculation procedure and the same parameters.

We will therefore attempt to shorten this study by
- immediately examining the RLE circuit instead of RL and then RLE;
- making it applicable to the most usual types of directly linked choppers, instead of considering each type individually.

Figure 4.1 shows the diagram used in the major part of this analysis.

- *The current "source"* consists of a resistance R and an inductance L series-connected with an EMF E. Apart from the source inductance itself, induc-

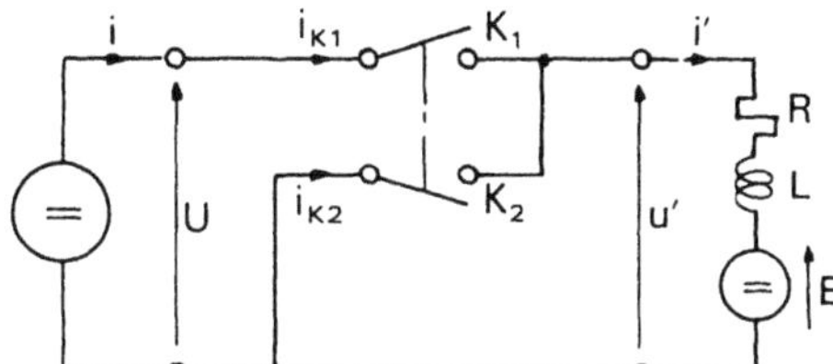

Fig. 4.1

tance L usually comprises a smoothing inductance intended to keep the ripple of current i' within certain limits.

Two parameters are required to characterize the current source. We will use:

$$Q = \frac{L\omega}{R}; \quad m = \frac{E}{U}. \tag{4.1}$$

The first parameter shows the importance of inductance L; it is the ratio of the impedance presented by the inductance at switching angular frequency ω to resistance R.

The second parameter shows the value of the EMF E, by relating it to the value U of the voltage source. The average value U' of voltage u' varies between zero (K_1 always open) and U (K_1 always closed); as there is little difference between E and U' if the losses in R are low, we shall limit our analysis to cases where

$$0 \leqslant m \leqslant 1.$$

- For the chopper, we use the configuration with *two switches*.

If these switches – K_1 and K_2 – are current-reversible, a current i' flows permanently through the RLE circuit: the chopper conduction is said to be continuous.

If the switches K_1 and K_2 are unidirectional in current (as is the case of the buck or boost converter), the flow of current i' can be either permanent or intermittent, depending on the values of Q, of m, and of duty ratio α. In cases where it is intermittent, the conduction will be said to be discontinuous.

Firstly, we will analyze the continuous conduction, with the specific aim of finding the limits of this operating mode when the switches are unidirectional. We will then analyze discontinuous conduction. Finally, we will show the effect on results obtained when a full-bridge structure is considered instead of the structure with two switches.

4.1.1 Continuous Conduction: Straightforward Calculation

When switch K_1 is closed and switch K_2 open, voltage u' is equal to voltage U of the voltage source. When switch K_2 is closed and switch K_1 open, voltage u' is zero.

If the relation between the duration of the K_1 on-time period and the operating period is taken to be duty ratio α, the average value U' of voltage u' is

$$U' = \alpha U \tag{4.2}$$

whatever the energy transfer direction.

4.1.1.1 Current Expressions

The instant of K_1 turn-on is taken as zero time.

- For $0 < \omega t < 2\pi\alpha$, K_1 is closed:

$$i' = i_{K_1} = i\,; \quad i_{K_2} = 0$$

from $L\mathrm{d}i'/\mathrm{d}t + Ri' + E = U$, it can be deduced that

$$i' = (1 - m)\frac{U}{R}\left[1 - \exp\left(-\frac{\omega t}{Q}\right)\right] + i'_0 \exp\left(-\frac{\omega t}{Q}\right) \tag{4.3}$$

by denoting the value of i' for $\omega t = 0$ by i'_0.

- For $2\pi\alpha < \omega t < 2\pi$, K_2 is closed:

$$i' = i_{K_2}\,; \quad i = i_{K_1} = 0$$

from $L\mathrm{d}i'/\mathrm{d}t + Ri' + E = 0$, it can be deduced that

$$i' = -m\frac{U}{R}\left[1 - \exp\left(-\frac{\omega t - 2\pi\alpha}{Q}\right)\right] + i'_1 \exp\left(-\frac{\omega t - 2\pi\alpha}{Q}\right) \tag{4.4}$$

by denoting the value of i' for $\omega t = 2\pi\alpha$ by i'_1.

As current i' cannot suffer any discontinuity, the value of i'_1 is obtained by substituting $2\pi\alpha$ for ωt in Eq. (4.3):

$$i'_1 = (1 - m)\frac{U}{R}\left[1 - \exp\left(-\frac{2\pi\alpha}{Q}\right)\right] + i'_0 \exp\left(-\frac{2\pi\alpha}{Q}\right). \tag{4.5}$$

The value of i'_0 can easily be found by using the periodicity of current i' in steady state. Noting that, for $\omega t = 2\pi$, current i' recovers its value i'_0, this gives

$$i'_0 = \frac{U}{R}\left[\frac{\exp(2\pi\alpha/Q) - 1}{\exp(2\pi/Q) - 1} - m\right]. \tag{4.6}$$

By substituting this value of i'_0 into Eqs. (4.3) and (4.5) and then into Eq. (4.4), the two expressions of current i' during a cycle can be obtained:

- for $0 < \omega t < 2\pi\alpha$,

$$i' = \frac{U}{R}\left[1 - m - \frac{\exp(2\pi/Q) - \exp(2\pi\alpha/Q)}{\exp(2\pi/Q) - 1}\exp\left(-\frac{\omega t}{Q}\right)\right]; \tag{4.7a}$$

- for $2\pi\alpha < \omega t < 2\pi$,

$$i' = \frac{U}{R}\left[-m + \frac{\exp(2\pi/Q)\left[\exp(2\pi\alpha/Q) - 1\right]}{\exp(2\pi/Q) - 1}\exp(-\omega t/Q)\right]. \tag{4.7b}$$

Figure 4.2 shows the three types of possible waveforms for current i':

a) current i' is always positive: the DC current source is a receptor;
b) current i' is alternately positive and negative: the DC current source acts alternately as a receptor and a generator;
c) current i' is always negative: the DC current source is a generator.

4.1.1.2 Characteristics

- *Average value of current i'*

– The average value I' of current i' can easily be determined by denoting that average voltage value across inductance L is zero. In these conditions, the

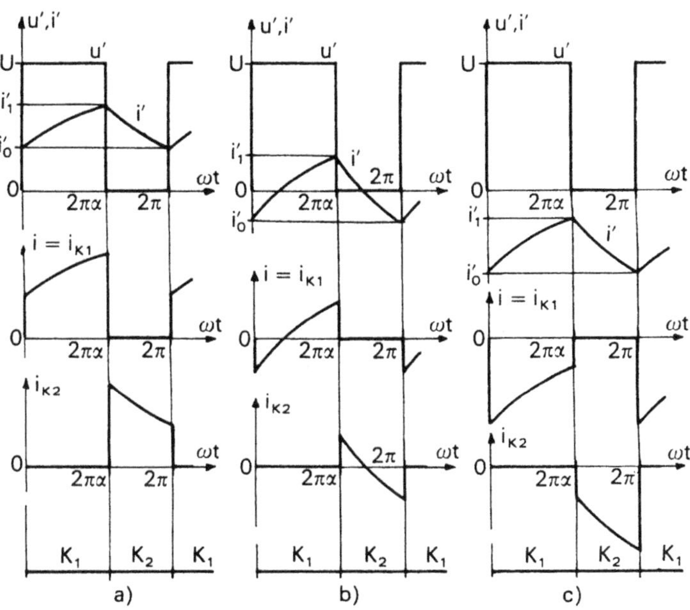

Fig. 4.2

equation

$$u' = L\frac{di'}{dt} + Ri' + E$$

gives the following relation between the average values:

$$U' = RI' + E. \tag{4.8}$$

On the other hand, since the impedance of the voltage source U and the voltage drop in the conducting switches are not taken into account, the average value of the output voltage of the chopper, in continuous conduction, is

$$U' = \alpha U.$$

By transferring this to Eq. (4.8), the result is

$$I' = \frac{\alpha U - E}{R} = \frac{U}{R}(\alpha - m). \tag{4.9}$$

- Figure 4.3 shows that I' can be determined by the intersection between the chopper output characteristic $U'(I')$ and the characteristic $U'(I')$ of the current "source" connected across its terminals. The former is an horizontal straight line of equation $U' = \alpha U$, and the latter a straight line of equation $U' = E + RI'$.

  The figure shows the intersections of the horizontal lines corresponding to $\alpha_1, \alpha_2, \alpha_3$ respectively with line corresponding to value E_1 of the EMF E. It can be seen how I' can be varied by acting on α, for a given E.

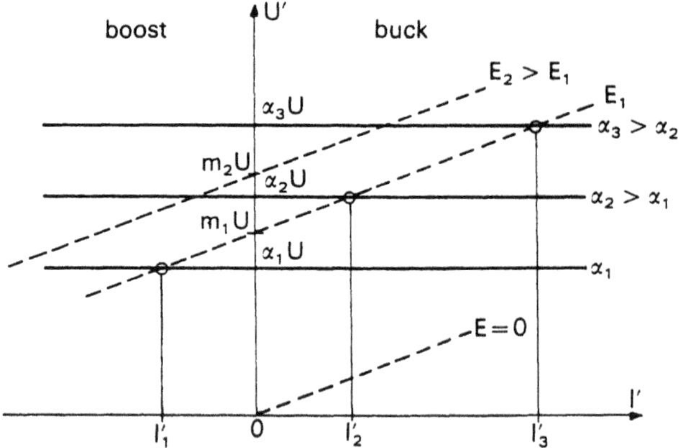

Fig. 4.3

When E increases, I' decreases. If I' has to be kept constant, α must be increased.

- If α is higher than m, current I' is positive: the energy is transferred from the voltage source to the current source, and the chopper operates as a buck converter. If α is less than m, I' is negative and the operation is that of a boost converter.

  If E is zero, current I' cannot be reversed and the only possible operation is as a buck converter.

- *Current i' ripple*

Equations (4.7a) and (4.7b) show that current i' increases when ωt goes from 0 to $2\pi\alpha$ and decreases when ωt goes from $2\pi\alpha$ to 2π. The minimum thus corresponds to ωt equal to zero and the maximum to ωt equal to $2\pi\alpha$:

$$i'_{min} = i'_0 = \frac{U}{R}\left[\frac{\exp(2\pi\alpha/Q) - 1}{\exp(2\pi/Q) - 1} - m\right] \tag{4.10a}$$

$$i'_{max} = i'_1 = \frac{U}{R}\left[\frac{\exp(2\pi/Q) - \exp((2\pi - 2\pi\alpha)/Q)}{\exp(2\pi/Q) - 1} - m\right] \tag{4.10b}$$

The peak-to-peak ripple $\Delta i'$, equal to $i_{max} - i_{min}$, is given by

$$\Delta i' = \frac{U}{R}\frac{[\exp(2\pi\alpha/Q) - 1][\exp((2\pi - 2\pi\alpha)/Q) - 1]}{\exp(2\pi/Q) - 1}. \tag{4.11}$$

Current i' ripple is independent of E and m; Fig. 4.4 shows its variations as a function of α for different values of Q.

For a given value of Q, $\Delta i'$ is maximum for α equal to 0.5. The value of this maximum is thus

$$\Delta i'_{max} = \frac{U}{R}\frac{\exp(\pi/Q) - 1}{\exp(\pi/Q) + 1} = \frac{U}{R}\tanh\left(\frac{\pi}{2Q}\right) \tag{4.12}$$

For high values of Q (in fact for $Q > 4$), the hyperbolic tangent may be replaced by its argument, thus giving

$$\Delta i'_{max} \simeq \frac{U}{R}\frac{\pi}{2Q} = \frac{\pi}{2}\frac{U}{L\omega}. \tag{4.12'}$$

This equation allows for the determination of the minimal value which must be given to inductance L for the currrent ripple to be less than a maximum imposed:

$$L_{min} = \frac{\pi}{2\omega}\frac{U}{\Delta i'_{max}}.$$

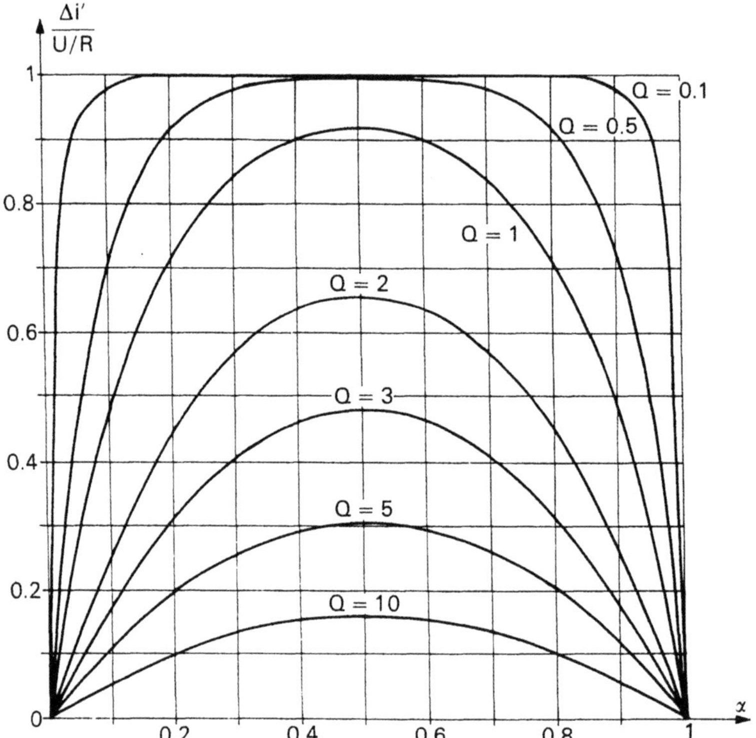

Fig. 4.4

High values of Q are needed to obtain small or medium current ripple. If, for example, for the rated I' current, RI' equals $5U/100$ and if $\Delta i'_{max}$ is limited to 20% of rated I',

$$\Delta i'_{max} = 0.2 \times \frac{5}{100}\frac{U}{R} = \frac{1}{100}\frac{U}{R}$$

which requires, according to Eq. (4.12′),

$$Q = \frac{\pi}{2}\frac{U/R}{\Delta i'_{max}} = 157.$$

- *Average values of current i and i_{K_1}*

– We denote by I the average value of current i, which is equal to current i_{K_1} flowing through switch K_1 and delivered by the constant voltage source U. The power provided by this source is given by the average value of Ui and equals UI.

If current i' were perfectly smoothed, the power received by the current source, i.e. the average value of $u'i'$, would be equal to $U'I'$.

Power conservation would give

$$I = \frac{U'}{U} I' = \alpha I' = \frac{U}{R}(\alpha^2 - \alpha m).$$

– In order to determine more precisely the value of I, it should be denoted that current i is equal to i' during interval $(0, 2\pi\alpha)$ and is zero for the remainder of the cycle:

$$I = \frac{1}{2\pi} \int_0^{2\pi\alpha} i' \, \mathrm{d}\omega t$$

or, taking the expression (4.7a) of current i' into account,

$$I = \frac{U}{R} \left\{ (1 - m)\alpha - \frac{Q}{2\pi} \frac{\exp((2\pi - 2\pi\alpha)/Q) - 1}{\exp(2\pi/Q) - 1} [\exp(2\pi\alpha/Q) - 1] \right\}. \tag{4.13}$$

Figure 4.5 gives the variations of the difference between I and $\alpha I'$ as a function of α, for different values of Q; this difference is referred to U/R.

This difference depends only on Q and on α; it is independent of I'. Its relative value can be ignored, except for the very low values of Q and I (this can be seen, for example, in the characteristics of Fig. 4.17).

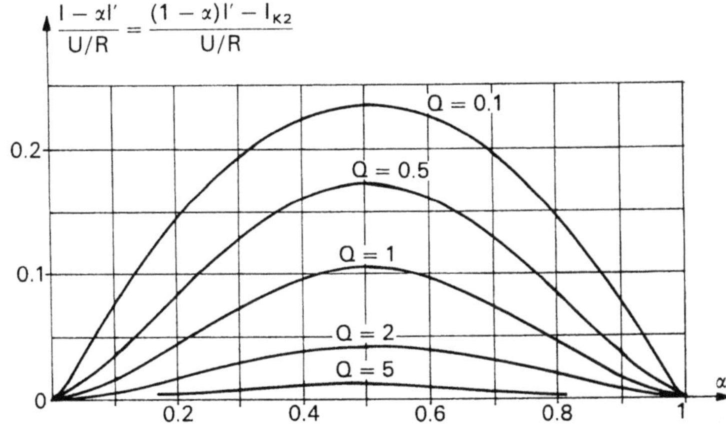

Fig. 4.5

For normal values of Q and medium or high current, this difference can be ignored. This gives

$$I \simeq \frac{U'}{U} I' = \frac{U}{R}(\alpha^2 - \alpha m). \tag{4.14}$$

• *Average value of current* i_{K_2}

If current i' were perfectly smoothed, the average value I_{K_2} of the current flowing through switch K_2 would be

$$I_{K_2} = (1 - \alpha)I' = \frac{U}{R}(\alpha - m)(1 - \alpha).$$

More precisely, since current i_{K2} is zero for ωt between zero and $2\pi\alpha$, equal to i' for ωt between $2\pi\alpha$ and 2π, we have

$$I_{K_2} = \int_{2\pi\alpha}^{2\pi} i' \mathrm{d}\omega t$$

Taking Eq. (4.7b) into account, this becomes

$$I_{K_2} = \frac{U}{R}\left\{-m(1 - \alpha) + \frac{Q}{2\pi}\frac{\exp((2\pi - 2\pi\alpha)/Q) - 1}{\exp(2\pi/Q) - 1}[\exp(2\pi\alpha/Q) - 1]\right\}.$$

A comparison with Eq. (4.13) indicates that the difference between I_{K_2} and $(1 - \alpha)I'$ is equal, except for the sign, to that between I and $\alpha I'$. This result is predictable since, necessarily,

$$I_{K_2} + I = I'.$$

As a function of α and for different values of Q' the curves in Fig. 4.5 give

$$\frac{U}{R}(I - \alpha I') \text{ as well as } \frac{U}{R}[(1 - \alpha)I' - I_{K_2}].$$

• *Remarks concerning switched currents*

The commutation of current i' from one switch to the other takes place for $\omega t = 0$ and for $\omega t = 2\pi\alpha$. The minimum and maximum values of currents i_{K_1} and i_{K_2} are thus equal to the minimum and maximum values of current i'.

When i' is positive, the commutation from K_1 to K_2, for $\omega t = 2\pi\alpha$, is a forced commutation and the commutation from K_2 to K_1, for $\omega t = 0$, is natural. When current i' is negative, the reverse is true: commutation K_1–K_2 is natural and commutation K_2–K_1 must be forced.

In both cases, forced commutation occurs at the maximum value of current i' taken at its absolute value. The switching ability of the controlled turn-on/turn-off switch must therefore correspond to the peak value of $|i'|$.

4.1.1.3 Continuous Conduction Limit

When the switches are unidirectional in current, the previous analysis can only be applied when the polarities of the currents agree with the switch non-reversibility.

• *Buck converter*

With the buck converter (Figure 4.6), current i' cannot be negative. Thus, the conduction is continuous only if the minimum value i'_0 of this current is positive or zero (Fig. 4.2a). Eq. (4.10a) shows that, for this to happen, the following is necessary:

$$\frac{\exp(2\pi\alpha/Q)-1}{\exp(2\pi/Q)-1} \geqslant m. \tag{4.15}$$

The difference which is required between α and m for conduction to be continuous depends on α and Q; condition (4.15) can be written as

$$\alpha \geqslant \frac{Q}{2\pi} \ln\left(m\left[\exp\left(\frac{2\pi}{Q}\right)-1\right]+1\right). \tag{4.15'}$$

To show that continuous conduction requires a minimum value of I' and to draw boundary curves in the I', U' plane, we can replace

$$\alpha - m \text{ by } \frac{I'}{U/R}, \quad \alpha \text{ by } \frac{U'}{U} \quad \text{and thus} \quad m \text{ by } \frac{U'}{U} - \frac{I'}{U/R};$$

in which case, Eq. (4.15) becomes:

$$\frac{I'}{U/R} \geqslant \frac{U'}{U} - \frac{\exp((2\pi/Q)U'/U)-1}{\exp(2\pi/Q)-1}. \tag{4.15''}$$

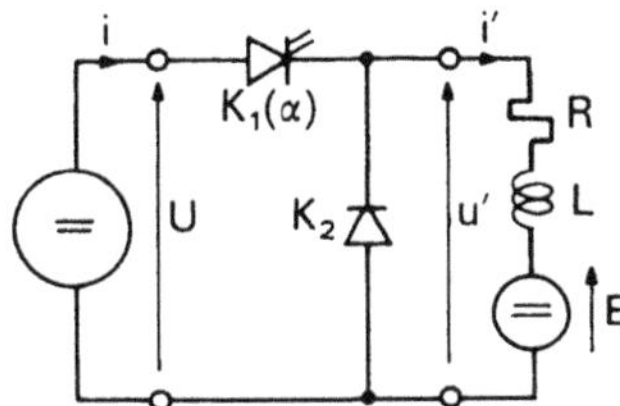

Fig. 4.6

• *Boost converter*

By keeping the same polarity conventions for the currents and voltages (Fig. 4.7), current i' cannot become positive with the boost converter. Conduc-

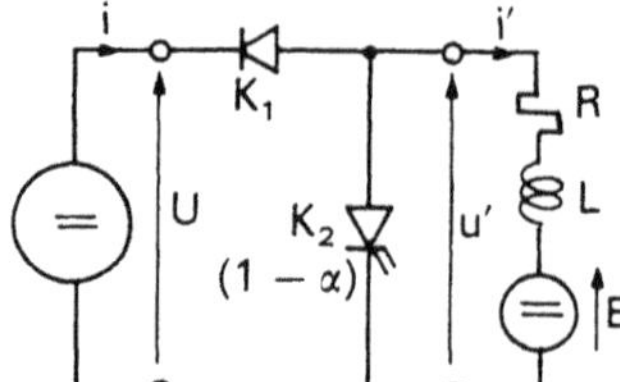

Fig. 4.7

tion is continuous if the maximum i'_1 of current i' is negative or zero (Fig. 4.2c). Following Eq. (4.10b), this assumes that

$$\frac{\exp(2\pi/Q) - \exp((2\pi - 2\pi\alpha)/Q)}{\exp(2\pi/Q) - 1} \leqslant m. \tag{4.16}$$

To calculate the limit below which, for a given m and Q, the value of α must be kept in order to obtain continuous conduction, condition (4.16) can be written as

$$\alpha \leqslant 1 - \frac{Q}{2\pi} \ln\left\{\exp\left(\frac{2\pi}{Q}\right) - m\left[\exp\left(\frac{2\pi}{Q}\right) - 1\right]\right\}. \tag{4.16'}$$

To draw the boundary curves in the I', U' plane, we can also write:

$$\frac{I'}{U/R} \leqslant \frac{U'}{U} - \frac{\exp(2\pi/Q) - \exp(2\pi/Q - (2\pi/Q)U'/U)}{\exp(2\pi/Q) - 1}. \tag{4.16''}$$

- *Boundary curves*

- Using Eqs. (4.15″) and (4.16″), we have drawn in Fig. 4.8, for different values of Q', the curves which separate continuous conduction from discontinuous conduction in the current I'–voltage U' plane. Average current I' is referred to U/R and average voltage U' to U.

The right-hand part of the figure corresponds to the buck converter ($I' > 0$). Continuous conduction occurs when the operating point is to the right side of the boundary curve.

The left-hand part corresponds to the boost converter ($I' < 0$). Continuous conduction occurs when the operating point is to the left side of the boundary curve.

- Using Eqs. (4.15′) and (4.16′), we have also drawn, in Fig. 4.9, for various values of Q, the boundary between continuous and discontinuous conduction in the m, α axis system.

The part which corresponds to the buck converter ($\alpha > m$) is located above the straight line segment given by equation $\alpha = m$. For a given value of Q, discontinuous conduction occurs when the point of coordinates m, α lies between the curve drawn for this value of Q and the segment of equation $\alpha = m$.

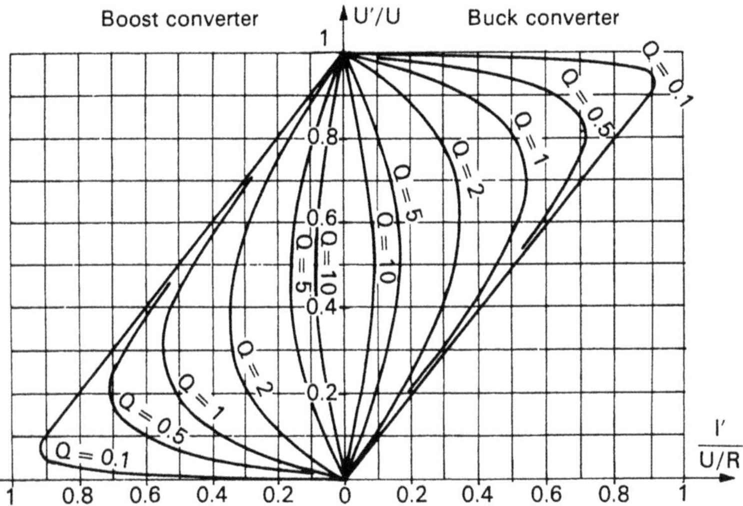

Fig. 4.8

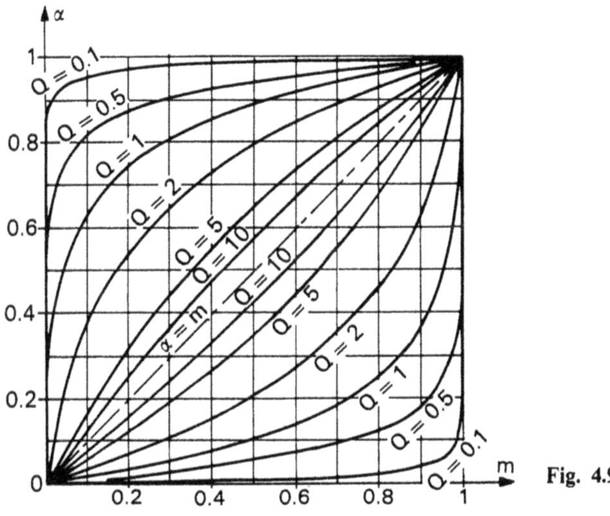

Fig. 4.9

In the case of the boost converter, the region which has to be used corresponds to α less than m. For a given Q, discontinuous conduction occurs when the point of coordinates m, α lies above the boundary curve.

- *Remarks on the current-reversible chopper*

In the case of the current-reversible chopper (Fig. 4.10), the continuous conduc-

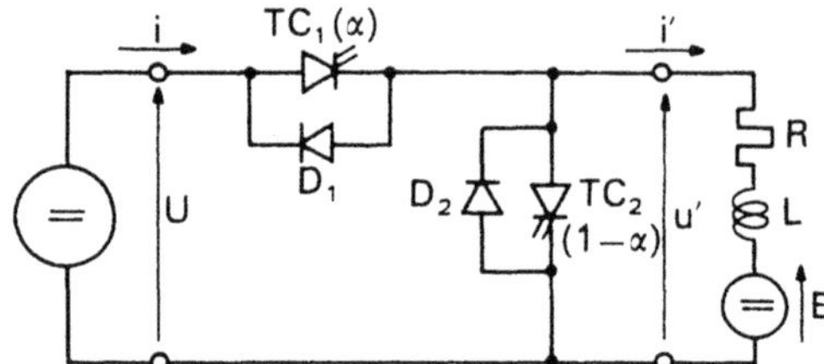

Fig. 4.10

tion areas would still be delimited as in Figs. 4.8 and 4.9, if only one switch was being controlled for each operating mode:

TC_1 for step-down chopper operation;
TC_2 for step-up chopper operation.

- For small values of $|I'|$ located between the continuous conduction zones of the step-down and of the step-up choppers, continuous conduction only occurs if TC_1 and TC_2 are *complementarily* controlled (see Sect. 3.2.3.3):
  TC_1 turn-on is controlled for $0 < \omega t < 2\pi\alpha$,
  and TC_2 turn-on is controlled for $2\pi\alpha < \omega t < 2\pi$.

Figure 4.11 shows the turn-on control time intervals, the waveforms of voltage u' and current i' and the semiconductor device conduction periods for
a) i' always positive,
b) i' either positive or negative,
c) i' always negative.

It is in the second case that complementary control is required to keep the conduction continuous. A similar problem arises in the case of rectifiers (see Vol. 1, Chap. 9, § 3 and 4), during the crossing of the low-current zone, when the rectified current has to be rapidly reversed. In the latter case it was shown that the continuity of characteristics could also be ensured by use of a complementary control and by accepting a circulating current strong enough to avoid discontinuous conduction.

- The complementary control of two switches, which are directly series-connected in the same direction under voltage U, could raise the problem of this source being short-circuited by the simultaneous conduction of the two switches.

  Figure 4.11 shows that, in all cases, current i' commutations occur between a controlled semiconductor device and a diode, and never between the two controlled devices.

  A sufficient time interval must thus be left between the TC_1 (or TC_2) turn-off signal command and the TC_2 (or TC_1) turn-on signal command, so as to be sure that the former is turned off before the latter is turned on.

- *Remarks on the other reversible choppers*

- The *voltage-reversible* chopper with two switches (Fig. 4.12) behaves in the same way as the step-down chopper. TC_1 ensures the intermittent flow of the

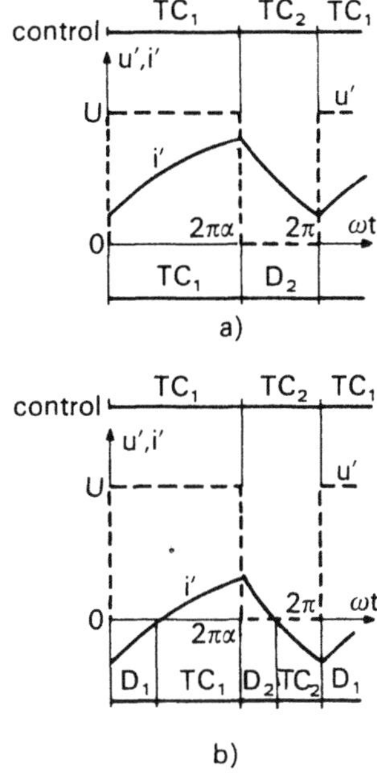

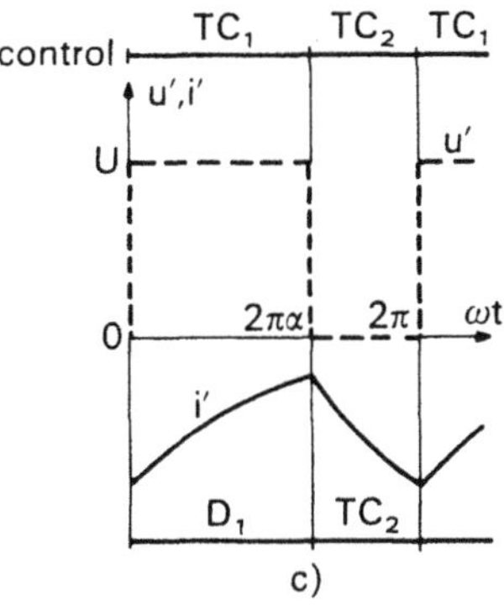

Fig. 4.11

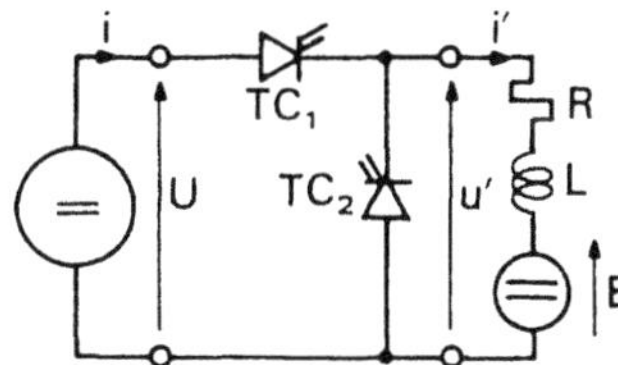

Fig. 4.12

current delivered by the voltage source. TC_2 short-circuits the current source when TC_1 is turned off.

The two conduction modes may be separated by use of the same curves as for the step-down chopper.

In order to ensure current i' commutations, both semiconductor devices must be controlled so that their conduction intervals overlap, at least during the commutations. This is in no way dangerous since one device is connected in the forward direction and the other in the reverse direction with respect to voltage U.

- The *current- and voltage-reversible* chopper with two switches (Fig. 4.13) can use the complementary control.

  To avoid the short-circuiting of the voltage source by the simultaneous conduction of TC_1 and TC_2 (when U is positive) or TC_2' and TC_1' (when U is

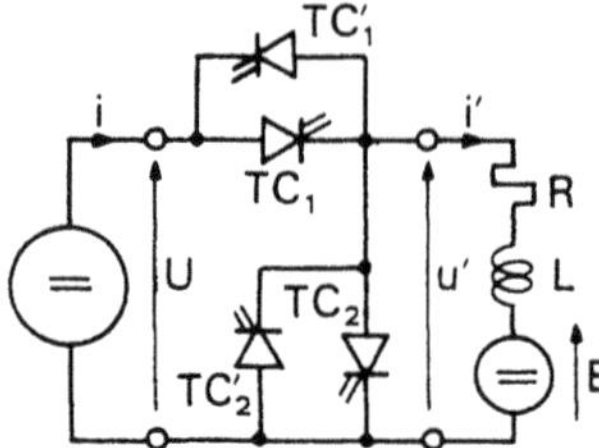

Fig. 4.13

negative), the same precautions must be taken as for the current-reversible chopper. However, there must be some overlapping of the control signals of TC_1 and TC_2', of TC_1' and TC_2 to allow for current i' commutations.

4.1.2 Continuous Conduction: "First-Harmonic" Method

As in the case of rectifiers (see Vol. 1, Chap. 6, § 2.2. and 2.4), it is useful to show the interest of the first-harmonic method as a means of evaluating approximately but quickly the effects of the current source imperfection.

In continuous conduction, voltage u' is equal to U for $0 < \omega t < 2\pi\alpha$, and to zero for $2\pi\alpha < \omega t < 2\pi$. In addition to its average value U', equal to αU, its harmonic spectrum contains harmonics at angular frequency $\omega, 2\omega, \ldots, p\omega, \ldots$

The harmonic with angular frequency $p\omega$ has an amplitude equal to

$$\frac{2}{p\pi} U \sin p\pi\alpha.$$

Current i' is linked to voltage u' by the equation

$$Ri' + L\frac{\mathrm{d}i'}{\mathrm{d}t} + E = u'.$$

In making a rapid evaluation of the current i' ripple and its consequences, only the average value of this current and its first harmonic may be taken into account.

The average value is

$$I' = \frac{U' - E}{R} = \frac{U}{R}(\alpha - m).$$

The amplitude of the term with pulsation ω is

$$I_1'\sqrt{2} = \frac{1}{\sqrt{R^2 + L^2\omega^2}}\frac{2}{\pi} U \sin \pi\alpha = \frac{2}{\pi}\frac{U}{R}\frac{1}{\sqrt{1 + Q^2}} \sin \pi\alpha.$$

- *The current i′ ripple* is equal to twice the amplitude of the component with angular frequency ω:

$$\Delta i' = \frac{4}{\pi}\frac{U}{R}\frac{1}{\sqrt{1+Q^2}}\sin\pi\alpha .$$

For a given value of Q, when α varies, $\Delta i'$ is maximum for α equal to 1/2 and is then given by

$$\Delta i'_{max} = \frac{4}{\pi}\frac{U}{R}\frac{1}{\sqrt{1+Q^2}};$$

If Q is high enough for $R\sqrt{1+Q^2}$ to be approximated by $L\omega$, the maximum ripple is given by

$$\Delta i'_{max} = \frac{4}{\pi}\frac{U}{L\omega} = 1.27\frac{U}{L\omega};$$

this value can be compared to $(\pi/2)U/L\omega = 1.57\,U/L\omega$ given by Eq. (4.12′).

- *The limit of continuous conduction* is reached when the amplitude of term with angular frequency ω becomes equal to the average value.
  - For the step-down chopper, I' is positive. Conduction is continuous if

$$\frac{2}{\pi}\frac{U}{R}\frac{1}{\sqrt{1+Q^2}}\sin\pi\alpha \leqslant \frac{U}{R}(\alpha-m)$$

or

$$\alpha - m \geqslant \frac{2}{\pi}\frac{\sin\pi\alpha}{\sqrt{1+Q^2}}.$$

  - For the step-up chopper, I' is negative. Its absolute value is equal to $(m-\alpha)(U/R)$. Conduction is continuous if

$$\frac{2}{\pi}\frac{U}{R}\frac{1}{\sqrt{1+Q^2}}\sin\pi\alpha \leqslant \frac{U}{R}(m-\alpha)$$

or

$$\alpha - m \leqslant -\frac{2}{\pi}\frac{\sin\pi\alpha}{\sqrt{1+Q^2}}.$$

For Q equal to 10 and $m = 0.5$, for example, straightforward calculation gives the following boundaries:

step-down chopper, $\alpha \geqslant 0.5 + 0.077$
step-up chopper, $\alpha \leqslant 0.5 - 0.077$.

The "first-harmonic" method gives:

step-down chopper, $\alpha \geqslant 0.5 + 0.062$
step-up chopper, $\alpha \leqslant 0.5 - 0.062$.

This example and all analogous calculations show that the "first-harmonic" method gives a clear idea of the size of the current i ripple and its effects, but that it considerably underestimates them.

4.1.3 Discontinuous Conduction: Straightforward Calculation

4.1.3.1 Current Analysis

When the circulation of current i' in RLE circuit becomes discontinuous, the case of the step-down chopper must be distinguished from that of the step-up chopper, since the current i' does not fall to zero during the same time interval.

- *Step-down chopper* (Fig. 4.14)

For given values of Q and α, if the value of m is higher than the value corresponding to the boundary between continuous and discontinuous conduction, current i' falls to zero during the on-period of switch K_2. As the latter has only unidirectional conduction capability, i' remains zero until K_1 is turned on again.

During the period when i' is zero, the load is in open circuit: voltage u' across it is equal to its EMF E;

– For $0 < \omega t < 2\pi\alpha$, switch K_1 is closed and we have:

$$u' = U = Ri + L\frac{di}{dt} + E.$$

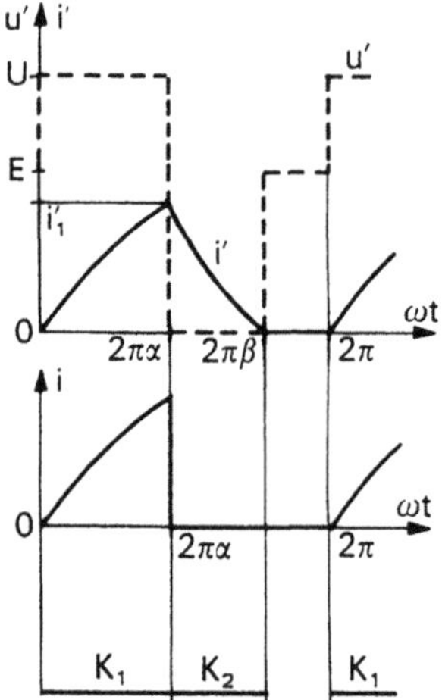

Fig. 4.14

Since current i starts from zero, it is expressed in the following way during this period:

$$i' = \frac{U}{R}(1-m)\left[1-\exp\left(-\frac{\omega t}{Q}\right)\right]. \tag{4.17}$$

At the end of this period, the value of i' has reached i'_1:

$$i'_1 = \frac{U}{R}(1-m)\left[1-\exp\left(-\frac{2\pi\alpha}{Q}\right)\right]. \tag{4.18}$$

This value corresponds to the maximum value of i'. Since the minimum is zero, i'_1 gives the ripple $\Delta i'$ if current i'.

– From $\omega t = 2\pi\alpha$, voltage u' is zero. Current i' is expressed as:

$$i' = \frac{U}{R}\left\{\left[\exp\left(\frac{2\pi\alpha}{Q}\right)-1+m\right]\exp\left(-\frac{\omega t}{Q}\right)-m\right\}. \tag{4.19}$$

It falls to zero, for ωt equal to $2\pi\beta$, β being given by the equation

$$\beta = \frac{Q}{2\pi}\ln\left[\frac{\exp(2\pi\alpha/Q)-1+m}{m}\right]. \tag{4.20}$$

– For $2\pi\beta < \omega t < 2\pi$, i' is zero and u' is equal to E.

• *Step-up chopper* (Fig. 4.15)

For given values of Q and α, when the value of m falls below the value

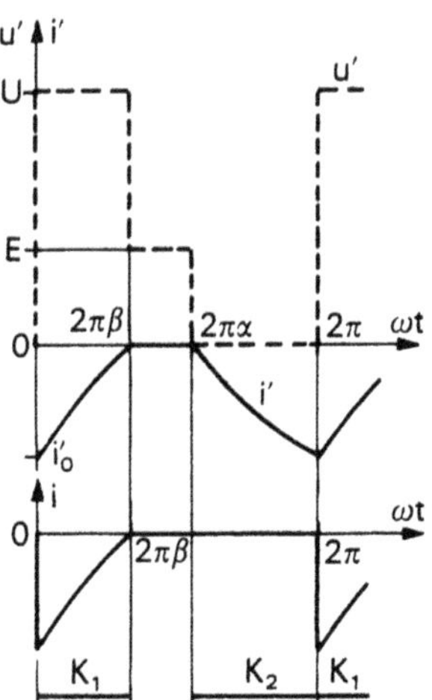

Fig. 4.15

corresponding to boundary between continuous and discontinuous conduction, current i' becomes zero during the on-period of K_1; it remains at zero until K_2 is turned on.

- For $2\pi\alpha < \omega t < 2\pi$, voltage u' is zero. Current i', equal to zero at the beginning of this period, is expressed by

$$i' = -m\frac{U}{R}\left[1 - \exp\left(\frac{-\omega t + 2\pi\alpha}{Q}\right)\right]. \tag{4.21}$$

Current i' reaches its minimum for $\omega t = 2\pi$ and is then equal to

$$i'_0 = -m\frac{U}{R}\left[1 - \exp\left(\frac{-2\pi + 2\pi\alpha}{Q}\right)\right]. \tag{4.22}$$

The ripple of i' is equal to $-i'_0$.

- From $\omega t = 0$, turning on K_1 makes u' equal to U. Current i', starting from an initial value equal to i'_0, increases according to the equation

$$i' = \frac{U}{R}\left\{1 - m - \left[1 - m\exp\left(\frac{-2\pi + 2\pi\alpha}{Q}\right)\right]\exp\left(\frac{-\omega t}{Q}\right)\right\}. \tag{4.23}$$

Current i' becomes zero for $\omega t = 2\pi\beta$, with

$$\beta = \frac{Q}{2\pi}\ln\left[\frac{1 - m\exp\left(\frac{-2\pi + 2\pi\alpha}{Q}\right)}{1 - m}\right]. \tag{4.24}$$

- For $2\pi\beta < \omega t < 2\pi\alpha$, i' is zero and u' equals U.

4.1.3.2 Characteristics

- *Average value of current i'*

The average value I' of current i' remains linked to the average value of voltage u' by the equation

$$RI' = U' - E = U' - mU. \tag{4.25}$$

But voltage U' is no longer solely dependent on the duty ratio α: because of ratio β, it is equally dependent on m and Q.

- Using the expressions of the current i' defined in the previous paragraph, the straightforward calculation of I' gives:
  - for the step-down chopper,

$$\frac{I'}{U/R} = \alpha - m\frac{Q}{2\pi}\ln\left[\frac{\exp(2\pi\alpha/Q) - 1 + m}{m}\right]; \tag{4.26}$$

- for the step-up chopper,

$$\frac{I'}{U/R} = (1-m)\frac{Q}{2\pi}\ln\left[\frac{1-m\exp((-2\pi+2\pi\alpha)/Q)}{1-m}\right] + m(\alpha-1). \tag{4.27}$$

It is difficult to make full use of these equations, which show that I', related to U/R, depends not only on m and on α as in continuous conduction but also on Q.

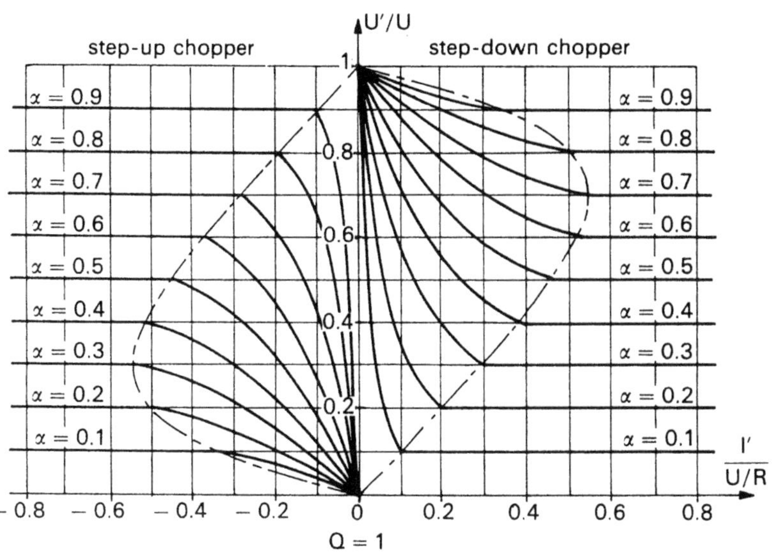

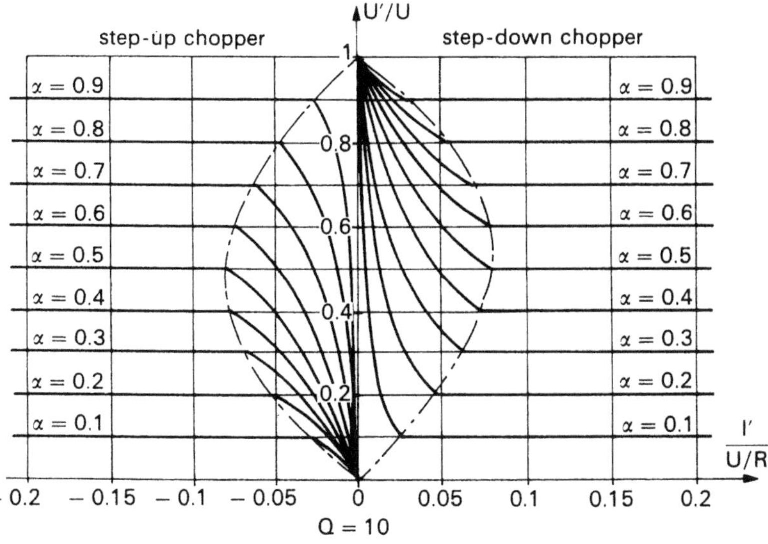

Fig. 4.16

- A system of parametric equations, taking m as a parameter, may conveniently be used to plot curves which show how the variations of U' are linked to those of I', for small values of I' corresponding to discontinuous conduction:

$$I' = I'(m); \quad U' = RI' + mU.$$

By replacing m by $(U'/U) - (RI'/U)$, implicit equations can be obtained.

Figure 4.16 provides two examples of families of curves $U'(I')$ plotted for different values of α, one for $Q = 1$, and the other for $Q = 10$.

The right-hand part applies to the step-down chopper ($I' > 0$) and the left-hand part to the step-up chopper ($I' < 0$). The dotted curves refer back to Fig. 4.8 and indicate the boundary between the two conduction modes.

For the step-down chopper, the decrease in I', going from the border of continuous conduction down to zero, makes U' increase from αU to U.

For the step-up chopper, the decrease in $|I'|$ from the border of continuous conduction down to zero, makes U' decrease from αU to zero.

In the case of the step-down chopper, if m is other than 1, it is necessary to take $\alpha = 0$ in order to bring I' down to zero. With the step-up chopper, if m is other than zero, it is necessary to take $\alpha = 1$ to reduce I' to zero.

Near to the axis at zero I', discontinuous conduction thus produces a zone in which operation differs completely from the normal control law ($U' = \alpha U$). Normally, by increasing L, it is possible to minimize the width of this zone and to operate with the minimal value of $|I'|$, which avoid entering it.

- *Average value of current i*

Current i which is delivered by the voltage source U is equal to current i' during the K_1 on-period; it is zero for the remainder of the operating cycle. This allows for its average value I to be calculated.

- For the step-down chopper,

$$I = \frac{1}{2\pi} \int_0^{2\pi\alpha} i' \mathrm{d}\omega t$$

gives, when taking Eq. (4.17) into account,

$$I = (1 - m)\frac{U}{R}\left\{\alpha - \frac{Q}{2\pi}\left[1 - \exp\left(\frac{-2\pi\alpha}{Q}\right)\right]\right\}. \tag{4.28}$$

- For the step-up chopper,

$$I = \frac{1}{2\pi} \int_0^{2\pi\beta} i' \mathrm{d}\omega t$$

gives, when taking Eqs. (4.23) and (4.24) into account,

$$I = \frac{U}{R}\frac{Q}{2\pi}\left\{-m + m\exp\left(\frac{-2\pi + 2\pi\alpha}{Q}\right) + (1-m)\ln\left[\frac{1 - m\exp((-2\pi + 2\pi\alpha)/Q)}{1-m}\right]\right\}. \tag{4.29}$$

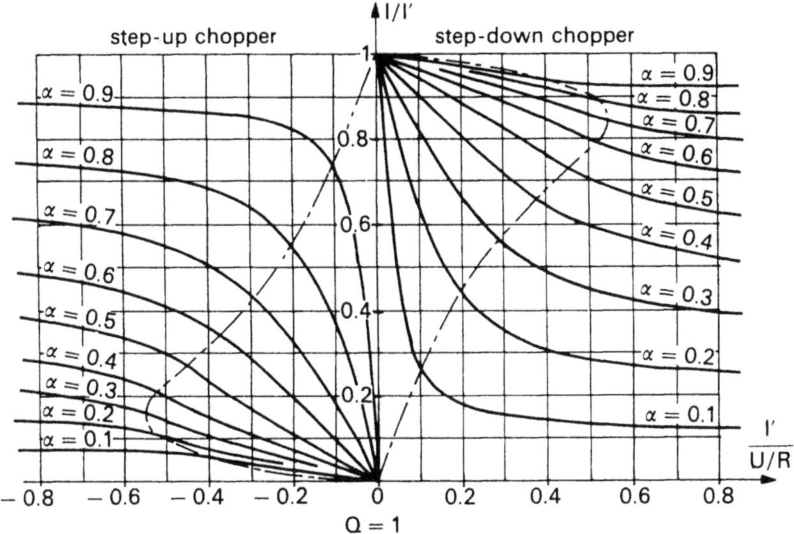

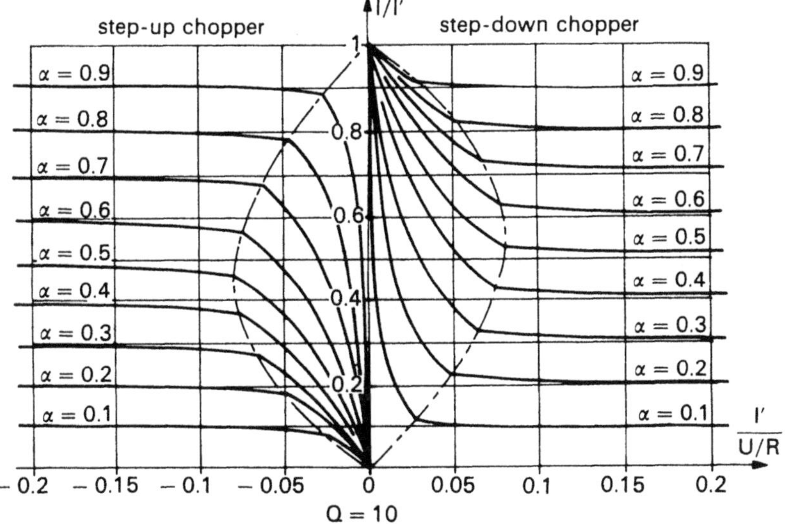

Fig. 4.17

For two values of Q, Fig. 4.17 shows the variations in the I/I' ratio, as a function of I' referred to U/R, for different values of α.

Once again, the right-hand part applies to the step-down chopper and the left-hand part to the step-up chopper; the dotted curves separate the two conduction modes.

When in continuous conduction (far from discontinuous conduction), the I/I' ratio is virtually equal to α. The plotted curves show how I/I' moves towards 1, whatever α may be, in cases where I' moves from a positive value towards zero. When I' is negative and moves towards zero, the I/I' ratio also moves towards zero, whatever α may be.

4.1.4 Discontinuous Conduction: Simplified Calculation

In most applications, the current "source" inductance L must be given a value high enough for the average value $|I'|$ of current i' – below which conduction becomes discontinuous – to be such that it makes $R|I'|$ negligible compared to U.

From then on, term Ri' can be ignored in the voltage equation, when analyzing the discontinuous conduction.

Voltage equation:

$$L\frac{\mathrm{d}i'}{\mathrm{d}t} + Ri' + E = u'$$

can thus be written as

$$L\frac{\mathrm{d}i'}{\mathrm{d}t} + E = u'. \tag{4.30}$$

As $L\mathrm{d}i'/\mathrm{d}t$ has an average value of zero, voltage u' has an average value equal to the EMF

$$U' = E = mU. \tag{4.31}$$

4.1.4.1 Equations

• *Step-down chopper* (Fig. 4.18)

From $\omega t = 0$ to $\omega t = 2\pi\alpha$, $L\mathrm{d}i'/\mathrm{d}t + mU = U$ gives

$$i' = \frac{U}{L\omega}(1 - m)\omega t\,. \tag{4.32}$$

From $\omega t = 2\pi\alpha$ to $\omega t = 2\pi\beta$, $L\mathrm{d}i'/\mathrm{d}t + mU = 0$ gives

$$i' = \frac{U}{L\omega}(2\pi\alpha - m\omega t)\,. \tag{4.32'}$$

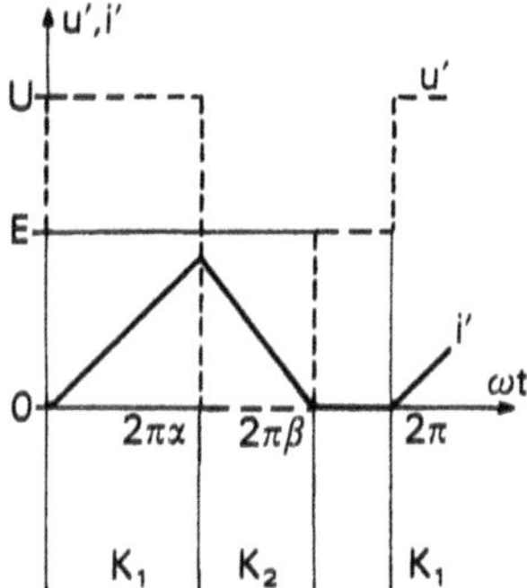

Fig. 4.18

Current i' falls to zero for $\omega t = 2\pi\beta$, with

$$\beta = \frac{\alpha}{m}.$$

The average value of current i' is equal to

$$I' = \frac{U}{L\omega}(1 - m)\pi\alpha\beta = \frac{U}{L\omega}\frac{1-m}{m}\pi\alpha^2. \tag{4.33}$$

The change to continuous conduction occurs for $\beta = 1$ or $m = \alpha$; the average value of current i' is then

$$I' = \frac{U}{L\omega}\pi(1 - m)m. \tag{4.34}$$

- *Step-up chopper* (Fig. 4.19)

For $2\pi\alpha < \omega t < 2\pi$, $\quad Ldi'/dt + mU = 0$ gives

$$i' = -\frac{mU}{L\omega}(\omega t - 2\pi\alpha). \tag{4.35}$$

For $0 < \omega t < 2\pi\beta$, $\quad Ldi'/dt + mU = U$ gives

$$i' = \frac{U}{L\omega}[(1 - m)\omega t - m2\pi(1 - \alpha)]. \tag{4.36}$$

Current i' falls to zero for $\omega t = 2\pi\beta$, with

$$\beta = \frac{m(1 - \alpha)}{1 - m}. \tag{4.37}$$

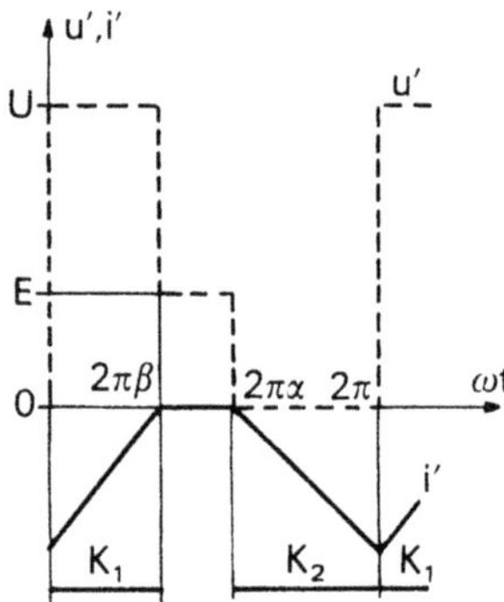

Fig. 4.19

The average value I' of current i' is

$$I' = -\frac{U}{L\omega}[m(1-\alpha)\pi(1+\beta-\alpha)]$$

$$I' = -\frac{U}{L\omega}\frac{m}{1-m}\pi(1-\alpha)^2. \tag{4.38}$$

The conduction ceases to be discontinuous for $\beta = \alpha$ or for $m = \alpha$; the corresponding value of I' is

$$I' = -\frac{U}{L\omega}\pi(1-m)m. \tag{4.39}$$

4.1.4.2 Characteristics

- The average value U' of voltage u' is referred to U and the average value I' of current i' to $U/L\omega$. Since U'/U equals m, Eqs. (4.33) and (4.38) show that the characteristics $U'(I')$ now depend on the single parameter α.

  We can thus obtain a single family of characteristics (Fig. 4.20). Equations (4.34) and (4.39) enable the zone in which discontinuous conduction takes place to be defined.

- Since R and the losses in the converter are ignored, the active power delivered (or received) by the voltage source U is equal to that received (or delivered) by voltage source E.

$$\frac{1}{2\pi}\int_0^{2\pi} Ei'\,\mathrm{d}\omega t = \frac{1}{2\pi}\int_0^{2\pi} Ui\,\mathrm{d}\omega t$$

$$EI' = UI.$$

Thus,

$$\frac{I}{I'} = \frac{E}{U} = \frac{U'}{U} = m. \tag{4.40}$$

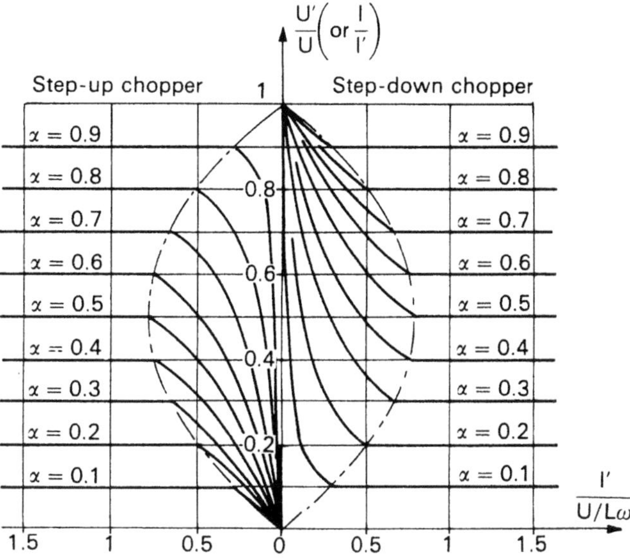

Fig. 4.20

The characteristics which give U'/U also provide the ratio between the average values of currents i and i'.

A comparison between the characteristics plotted for Q equal to 10 in Figs. 4.16 and 4.17 and those in Fig. 4.20 shows that when Q is equal to or above 10, R can be ignored in the analysis of discontinuous conduction.

4.1.5 Full-Bridge Choppers

Our analysis will be confined to the most frequently used type of full-bridge chopper, i.e. that which connects a current-reversible voltage source to a current- and voltage-reversible current source (Fig. 4.21);

A distinction must be made between the cases of sequential control and continuous control (see Sect. 3.3.1).

4.1.5.1 Sequential Control

In order to obtain an output voltage u' of a positive average value, the TC'_2 turn-on must be permanently controlled. In this case, points 0′ and B always have the same potential. The circuit acts as a chopper with two switches: switch K_1 (TC_1, D_1) which enables the two sources to be connected, switch K'_1 (TC'_1, D'_1) which enables the current source to be short-circuited.

- If only K_1 is controlled, the circuit acts as a step-down chopper.
- If only K'_1 is controlled, the circuit acts a step-up chopper.

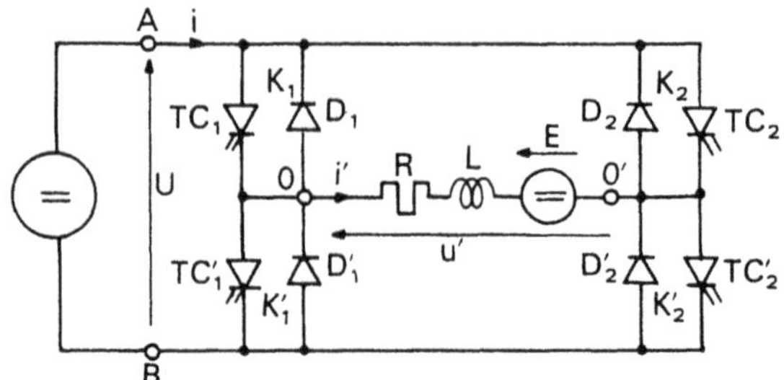

Fig. 4.21

- If K_1 and K'_1 are controlled complementarily, the circuit acts as a current-reversible chopper (see Fig. 4.20).

Similarly, to obtain a negative U', the turn-on of TC'_1 must be permanently controlled. Points 0 and B are connected whatever the polarity of current i'. Depending on whether K_2 alone, K'_2 alone or K_2 and K'_2 complementarily are controlled, the operational modes are those of the step-down chopper, the step-up chopper or the current-reversible chopper.

The whole of the analysis concerning the chopper with two switches is thus directly applicable.

4.1.5.2 Continuous Control

In continuous control, the control sequence of the switches is the same, whatever the polarities of the average values U' and I' of voltage u' and current i' may be:

- the turn-on of semiconductor devices TC_1 and TC'_2 are simultaneously controlled during a fractional part α of the switching cycle;
- *and* the turn-on of TC_2 and TC'_1 is under simultaneous control for the remainder of this cycle.

Since the controls of TC_1, and TC'_1, on the one hand, and of TC_2 and TC'_2, on the other, are complementary, current i' can flow whatever its polarity: *conduction is always continuous.*

Voltage u' across circuit R, L, E, is expressed by

$$u' = U, \quad \text{for } 0 < \omega t < 2\pi\alpha,$$

$$u' = -U, \quad \text{for } 2\pi\alpha < \omega t < 2\pi.$$

Thus, its average value is always given by

$$\frac{U'}{U} = 2\alpha - 1. \tag{4.41}$$

- *Current expressions*

– For $0 < \omega t < 2\pi\alpha$, K_1 and K_2 are closed: $i = i'$.

From $Ri' + L\mathrm{d}i'/\mathrm{d}t + E = U$, taking the periodicity of i' into account, it is possible to deduce that

$$i' = \frac{U}{R}\left[1 - m - 2\frac{\exp(2\pi/Q) - \exp(2\pi\alpha/Q)}{\exp(2\pi/Q) - 1}\exp\left(\frac{-\omega t}{Q}\right)\right]. \tag{4.42}$$

If current i' is positive, it flows via TC_1 and TC'_2; if it is negative, then it flows via D_1 and D'_2.

– For $2\pi\alpha < \omega t < 2\pi$, K_2 and K'_1 are closed, $i = -i'$.
From $Ri' + L\mathrm{d}i'/\mathrm{d}t + mU = -U$, it can be deduced that

$$i' = \frac{U}{R}\left\{-1 - m + 2\frac{\exp(2\pi/Q)\left[\exp(2\pi\alpha/Q) - 1\right]}{\exp(2\pi/Q) - 1}\exp\left(\frac{-\omega t}{Q}\right)\right\}. \tag{4.43}$$

If current i' is positive, it flows via D'_1 and D_2; if it is negative, then it flows via TC'_1 and TC_2.

Figure 4.22 gives, from top to bottom,

- the turn-on orders sent to the semiconductor devices,
- the waveform of voltage u',
- then, successively,
  a) for i' always positive,
  b) for i' either positive or negative,
  c) for i' always negative,
  the waveform of current i',
  the waveform of current i,
  and the diagram of semiconductor device conductions.

The left-hand curves apply for $\alpha > 0.5$, i.e. for a positive value of U'. The right-hand curves apply for $\alpha < 0.5$, i.e. for a negative value of U'.

It can be seen that, as in the case of a chopper with two switches,

- the complementary nature of the controls of TC_1 and TC'_1 and of TC_2 and TC'_2 is only used for low values of current I' (waveforms in Fig. 4.22b); in this case, it avoids conduction discontinuity;
- the commutations between semiconductor devices which are series-connected under voltage U are always of the controlled semiconductor device – diode type or vice versa. Necessary precautions can be taken to avoid short-circuiting the voltage source U.

- *Characteristics*

– *The average value of current i'* can be deduced from the two expressions of U':

$$U' = U(2\alpha - 1)$$

$$U' = RI' + mU.$$

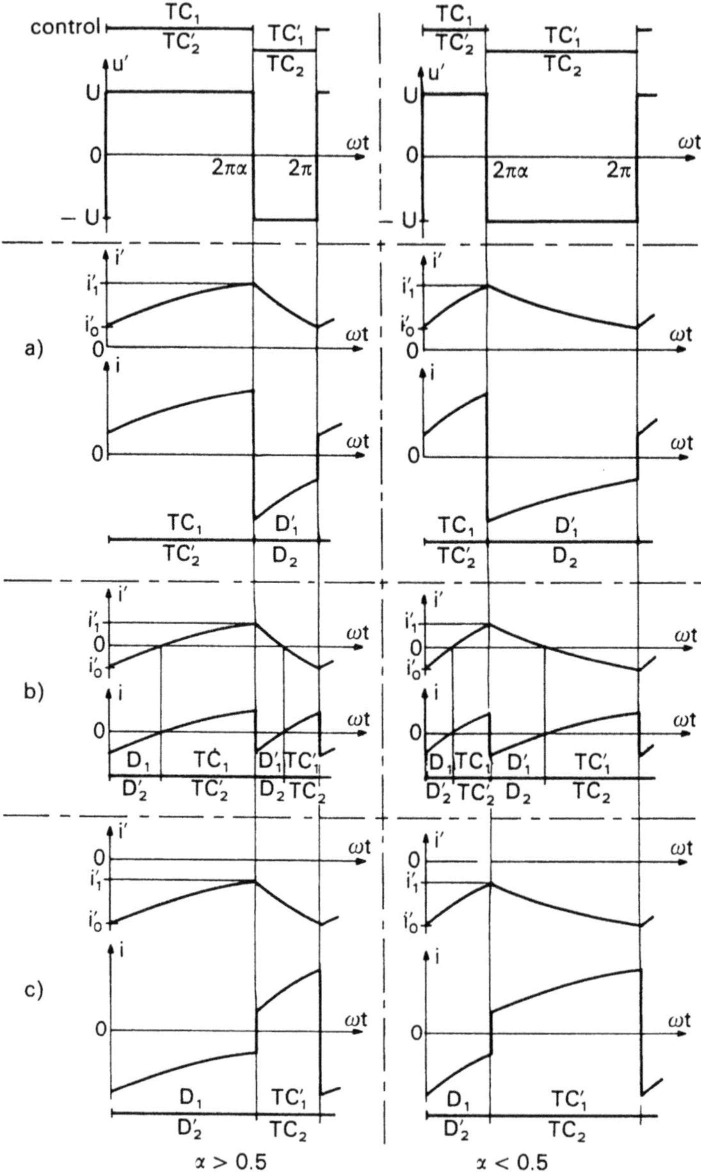

Fig. 4.22

Therefore

$$I' = \frac{U}{R}(2\alpha - 1 - m). \tag{4.44}$$

In the I', U' plane, the operating point is determined by the intersection of curve $U'(I')$ which characterizes the chopper. This is a horizontal line given by equation $U' = U(2\alpha - 1)$ according to the previous assumptions. If the impedance of voltage source U and the voltage drop in the conducting devices are taken into account, this line would have a slightly negative slope;

- of curve $U'(I')$ which characterizes the current source. This is a straight line given by equation $U' = RI' + mU$.

Figure 4.23 shows how α can be used to vary current I' for a positive or negative value of m. (Since voltage U' can be reversed, the chopper allows for the control energy exchange between voltage source U and a current source of which the EMF E can vary approximately from $-U$ to $+U$.)

The main interest of continuous control lies in that there is no problem to make I' and U' vary continuously.

- *Current i' ripple* is given by the minimum and maximum values of this current, obtained for $\omega t = 0$ and for $\omega t = 2\pi\alpha$, respectively:

$$i'_{\min} = i'_0 = \frac{U}{R}\left\{-1 - m + 2\frac{\exp(2\pi\alpha/Q) - 1}{\exp(2\pi/Q) - 1}\right\}$$

$$i'_{\max} = i'_1 = \frac{U}{R}\left\{-1 - m + 2\frac{\exp(2\pi/Q) - \exp((2\pi - 2\pi\alpha)/Q)}{\exp(2\pi/Q) - 1}\right\}.$$

This provides the peak-to-peak ripple in current i'

$$\Delta i' = \frac{2U}{R}\frac{[\exp(2\pi\alpha/Q) - 1][\exp((2\pi - 2\pi\alpha)/Q) - 1]}{\exp(2\pi/Q) - 1}. \tag{4.45}$$

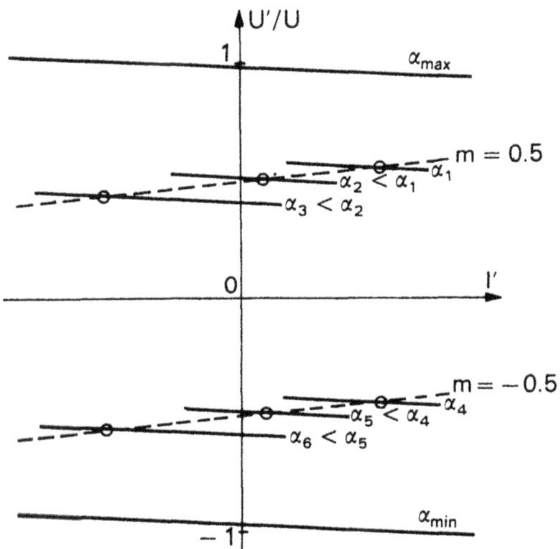

Fig. 4.23

For a given value of Q, the ripple is maximum for α equal to 0.5 and is then

$$\Delta i'_{max} = 2\frac{U}{R}\tanh\left(\frac{\pi}{2Q}\right)$$

or, for high value of Q,

$$\Delta i'_{max} \simeq \pi\frac{U}{L\omega}. \tag{4.46}$$

Comparing Eqs. (4.45) and (4.46) with Eqs. (4.11) and (4.12) shows that the current ripple produced by use of continuous control is *twice* that of a chopper with two switches or of a full-bridge chopper with sequential control. This important drawback must be considered when choosing the control mode.

- *The average value of current* i delivered by the voltage source can be calculated by

$$I = \frac{1}{2\pi}\left[\int_0^{2\pi\alpha} i'\,\mathrm{d}\omega t - \int_{2\pi\alpha}^{2\pi} i'\,\mathrm{d}\omega t\right]$$

giving

$$I = \frac{U}{R}\left\{1 - (2\alpha - 1)m - \frac{2Q}{\pi}\frac{\exp((2\pi - 2\pi\alpha)/Q) - 1}{\exp(2\pi/Q) - 1}[\exp(2\pi\alpha/Q) - 1]\right\}. \tag{4.47}$$

If current i' were perfectly smoothed by inductor L, this would give

$$I = \frac{U'}{U}I' = \frac{U}{R}(2\alpha - 1 - m)(2\alpha - 1). \tag{4.48}$$

By comparing Eqs. (4.47) and (4.48), on the one hand, and Eqs. (4.13) and (4.14), on the other, it can be seen that the difference $I - (U'I'/U)$ is four times greater than that provided by sequential control.

4.2 Correcting the Voltage Source

When the voltage source U has a noticeable inductance, this must be corrected by use of a filter.

- Using a highly simplified *example*, the adverse effects of this inductance will be illustrated. A step-down chopper (Fig. 4.24a) feeds a perfect current I' receptor; it is supplied by a voltage source U_S, the inductance of which is L_S.

- For $\omega t = 0$, TC_1 turn-on is controlled, while diode D_2 lets current I' flow. Due to L_S, current i cannot change suddenly from zero to I', even if voltage v_{T_1} across device TC_1 instantaneously becomes zero. During the increase in i, diode D_2 continues to conduct.

  The rise in i is governed by the equation

$$U_S - L_S \, di/dt = 0$$

which gives a rise time t_{on} such that

$$\omega t_{on} = L_S \omega I' / U_S .$$

- If the turn-off order given to TC_1 for $\omega t = 2\pi\alpha$ effectively opens this switch, it brings the diode D_2 into conduction. The latter's current cannot change instantaneously from zero to I' since, on account of L_S, current i cannot suddenly fall to zero. During the decrease of i, the diode is conducting and u' is zero.

  Since the decrease in i is governed by the equation

$$U_S - L_S \, di/dt = v_{T_1} ,$$

v_{T_1} must be higher than U_S for current i to be able to decrease.

If v_{T_1} remained constant and equal to $2U_S$, the fall time of current i would be equal to its rise time.

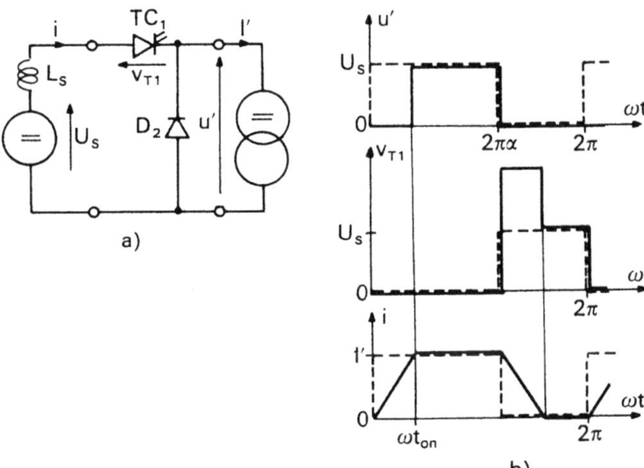

Fig. 4.24

Figure 4.24b shows, in simplified form, the waveform modifications due to inductance L_S. The dashed lines correspond to a zero inductance and the full lines to a large inductance, at the time scale of the operating frequency.

The voltage source inductance reduces the possible operating frequency, as

- the rise and fall times of current i must remain small with regard to the cycle, if the aim is to reach maximum values of α near to 1;
- there are important losses in device TC_1 at each turn-off; the number of commutations per second must thus be reduced.

- Generally, a *filter* has to be placed between the imperfect voltage source and the chopper (Fig. 4.25). This filter plays a double role:
  - its capacitor enables it to restore the properties of a voltage source at the chopper input;
  - its inductor enables it to smooth the current which is delivered by the voltage source.

- For most of the present analysis of the input filter, we will use the circuit configuration shown in Fig. 4.26, i.e. a *chopper with two switches.* In this figure are shown the notations used.

The internal resistance of voltage source U_S is not taken into account . Its internal inductance is added to that of the filter to form L_S.

The switches and the current source I' are assumed to be perfect. Depending on the sign of I', the chopper can operate either as a step-down chopper ($I' > 0$) or a step-up chopper ($I' < 0$).

Two parameters are needed to characterize this system. We take

$$k_i = \sqrt{\frac{L_S}{C}}\frac{I'}{U_S}$$

$$k_f = \frac{\omega_f}{\omega} = \frac{1}{\omega\sqrt{L_S C}} \tag{4.49}$$

The first parameter relates the filter characteristic impedance to the normalizing impedance U_S/I' which characterizes the supply and the load.

The second parameter relates the filter resonance angular frequency to the switching angular frequency.

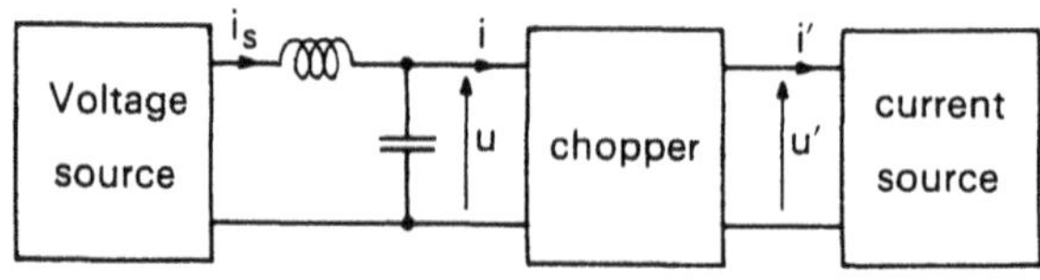

Fig. 4.25

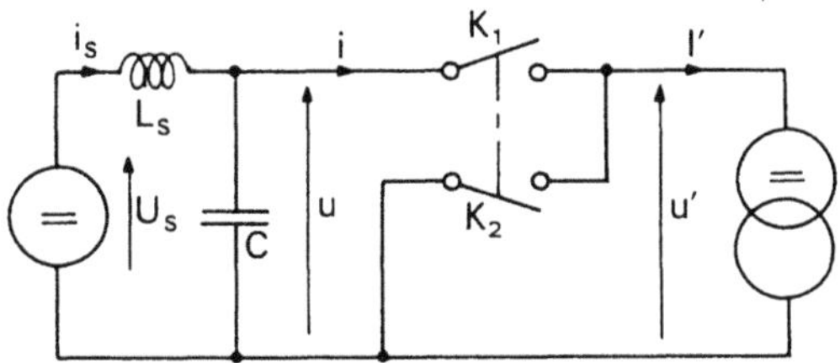

Fig. 4.26

After an analysis of the operation and the definition of the characteristics in the case of the chopper with two switches, we will show how the results obtained can be extended to the case of the full-bridge chopper.

4.2.1 Operation: Equations

Current i_S through inductor L_S and voltage u across capacitor C are solutions of the following system of equations:

$$L_S \frac{di_S}{dt} + u = U_S \tag{4.50a}$$

$$C \frac{du}{dt} = i_S - i \tag{4.50b}$$

with i equal to I' for $0 < \omega t < 2\pi\alpha$ and equal to zero for $2\pi\alpha < \omega t < 2\pi$.

In order to find the expressions of i_S and u in steady state,[1] use is made of the fact that neither i_S nor u is affected by any discontinuity.

For $0 < \omega t < 2\pi\alpha$, solving the system (4.50) gives

$$u = U_S - (U_S - u_0)\cos k_f\omega t + L_S k_f \omega (i_{S_0} - I')\sin k_f \omega t$$

$$i_S = I' + \frac{U_S - u_0}{L_S k_f \omega} \sin k_f \omega t + (i_{S_0} - I')\cos k_f \omega t$$

if u_0 and i_{S_0} denote the values of u and i_S for $\omega t = 0$.

For $2\pi\alpha < \omega t < 2\pi$, solving the system (4.50) gives

$$u = U_S - (U_S - u_1)\cos[k_f(\omega t - 2\pi\alpha)] + L_S\, k_f \omega i_{S_1} \sin[k_f(\omega t - 2\pi\alpha)]$$

$$i_S = \frac{U_S - u_1}{L_S k_f \omega} \sin[k_f(\omega t - 2\pi\alpha)] + i_{S_1} \cos[(k_f(\omega t - 2\pi\alpha)]$$

by denoting as u_1 and i_{S_1} the values of u and i_S for $\omega t = 2\pi\alpha$.

[1] It may seem paradoxical to analyze the steady state behaviour of a purely oscillating system. However, the current source is, in fact, an RLE circuit, which thus is damping the system and makes it tend to operate periodically.

The continuity of u and i_S for $\omega t = 2\pi\alpha$ gives the following equations between u_1, i_{S_1} and u_0, i_{S_0}:

$$u_1 = U_S - (U_S - u_0)\cos 2\pi k_f \alpha + L_S k_f \omega (i_{S_0} - I')\sin 2\pi k_f \alpha$$

$$i_{S_1} = I' + \frac{U_S - u_0}{L_S k_f \omega} \sin 2\pi k_f \alpha + (i_{S_0} - I')\cos 2\pi k_f \alpha .$$

By substituting these values in the expressions of u and i_S for interval $(2\pi\alpha, 2\pi)$, and by expressing that u and i_S return to values u_0 and i_{S_0} for $\omega t = 2\pi$, we can obtain

$$u_0 = U_S\left[1 + k_i \frac{\sin[\pi k_f(1-\alpha)]\sin(\pi k_f \alpha)}{\sin(\pi k_f)}\right] \tag{4.51a}$$

$$i_{S_0} = I'\left[1 - \frac{\sin[\pi k_f(1-\alpha)]\cos(\pi k_f \alpha)}{\sin(\pi k_f)}\right]. \tag{4.51b}$$

From u_0 and i_{S_0}, u_1 and i_{S_1} can be deduced. If u_0, i_{S_0}, u_1, i_{S_1} are then replaced in the expressions of u and i during both periods, we obtain

for $0 < \omega t < 2\pi\alpha$:

$$u = U_S\left[1 - k_i \frac{\sin[\pi k_f(1-\alpha)]}{\sin(\pi k_f)} \sin[k_f(\omega t - \pi\alpha)]\right] \tag{4.52a}$$

$$i_S = I'\left[1 - \frac{\sin[\pi k_f(1-\alpha)]}{\sin(\pi k_f)} \cos[k_f(\omega t - \pi\alpha)]\right] \tag{4.52b}$$

for $2\pi\alpha < \omega t < 2\pi$:

$$u = U_S\left[1 + k_i \frac{\sin(\pi k_f \alpha)}{\sin(\pi k_f)} \sin[k_f(\omega t - \pi - \pi\alpha)]\right] \tag{4.53a}$$

$$i_S = I' \frac{\sin(\pi k_f \alpha)}{\sin(\pi k_f)} \cos[k_f(\omega t - \pi - \pi\alpha)] . \tag{4.53b}$$

4.2.2 Ripple in Voltage u and Current i_S

As shown by Eqs. (4.52) and (4.53), the waveform of current i_S depends only on α and k_f. Similarly, if the variations of voltage u on either side of U_S are proportional to k_i, the form of these variations only depends on k_f and on α.

The behaviour of the filter is thus essentially dependent on parameter k_f. We will examine the effect of k_f for the step-down chopper (positive I' and k_i) and then for the step-up chopper (negative I' and k_i). We will limit our analysis to values of k_f between 0 and 1, since, as will be seen, this parameter must be given a value much lower than 1 in order to obtain acceptable ripples.

4.2.2.1 Step-down Chopper

- *k_f less than 0.5*

– *Voltage u.* Eqs. (4.52a) and (4.53a) show that, if k_f is less than 0.5, voltage u reaches its maximum for $\omega t = 0$ and its minimum for $\omega t = 2\pi\alpha$, whatever the value of α. This gives

$$u_{\min} = U_S\left[1 - k_i \frac{\sin[\pi k_f(1-\alpha)]\sin(\pi k_f\alpha)}{\sin(\pi k_f)}\right]$$

$$u_{\max} = U_S\left[1 + k_i \frac{\sin[\pi k_f(1-\alpha)]\sin(\pi k_f\alpha)}{\sin(\pi k_f)}\right].$$

These two values are symmetrical in relation to the average value U_S of voltage u.

The peak-to-peak ripple in voltage u is given by

$$\Delta u = u_{\max} - u_{\min} = 2k_i U_S \frac{\sin[\pi k_f(1-\alpha)]\sin(\pi k_f\alpha)}{\sin(\pi k_f)}. \tag{4.54}$$

The derivative of Δu in relation to α indicates that the ripple is maximum for α equal to 0.5. Its value is then

$$\Delta u_{\max} = k_i U_S \tan\left(\frac{\pi}{2} k_f\right). \tag{4.55}$$

For small values of k_f, the following approximation can be used:

$$\Delta u_{\max} \simeq \frac{\pi}{2} k_i k_f U_S. \tag{4.55'}$$

– *Current i_S.* Eqs. (4.52b) and (4.53b) show that current i_S is minimum for $\omega t = \pi\alpha$, i.e. in the middle of interval $0, 2\pi\alpha$. It is at its maximum for $\omega t = \pi(1+\alpha)$, i.e. in the middle of interval $(2\pi\alpha, 2\pi)$.

Using the peak values of current i_S, i.e.

$$i_{S_{\min}} = I'\left[1 - \frac{\sin[\pi k_f(1-\alpha)]}{\sin(\pi k_f)}\right]$$

$$i_{S_{\max}} = I' \frac{\sin(\pi k_f\alpha)}{\sin(\pi k_f)}$$

its ripple can be deduced:

$$\Delta i_S = i_{S_{\max}} - i_{S_{\min}} = I'\left[\frac{\sin(\pi k_f\alpha) + \sin[\pi k_f(1-\alpha)]}{\sin(\pi k_f)} - 1\right]. \tag{4.56}$$

For a given k_f, this ripple is also maximum for α equal to 0.5. It is then given by

$$\Delta i_{S_{max}} = I' \frac{1 - \cos\left(\frac{\pi}{2} k_f\right)}{\cos\left(\frac{\pi}{2} k_f\right)}. \tag{4.57}$$

If k_f is small, the following can be used:

$$\Delta i_{S_{max}} \simeq \frac{\pi^2}{8} k_f^2 I'. \tag{4.57'}$$

- Figure 4.27 indicates the waveforms of voltage u and current i_S, for $k_i = 1$ and $k_f = 0.25$, for α equal to 0.25, 0.5 and 0.75, respectively.

There is only a small ripple in the current i_S. The very high values of the voltage u ripple come from taking k_i equal to 1.

- k_f *above 0.5*

- *Voltage u.* When k_f is between 0.5 and 1, the maximum and minimum of u do not always occur for $\omega t = 0$ and for $\omega t = 2\pi\alpha$. It depends on α.
  - If $1 - 1/2k_f < \alpha < 1/2k_f$,
    in other words, if $k_f\alpha$ and $k_f(1 - \alpha)$ are both less than 0.5, the peak values of u remain at $\omega t = 0$ and $\omega t = 2\pi\alpha$. The ripple Δu remains that given by Eq. (4.54).
  - If $\alpha < 1 - 1/2k_f$,
    the peak values of u are reached within period $(2\pi\alpha, 2\pi)$ and the value of ripple is

$$\Delta u = 2k_i U_S \frac{\sin(\pi k_f \alpha)}{\sin(\pi k_f)}.$$

  - If $\alpha > \dfrac{1}{2k_f}$,

the peak values lie within period $(0, 2\pi\alpha)$ and the value of ripple Δu is

$$\Delta u = 2k_i U_S \frac{\sin[\pi k_f(1 - \alpha)]}{\sin(\pi k_f)}.$$

In Figure 4.28, the waveforms of u and of i_S are shown for $k_i = 1$ and $k_f = 0.75$, for α equal to 0.25, 0.5 and 0.75, respectively. These waveforms correspond to the three above-mentioned cases: the first for $\alpha = 0.5$, the second for $\alpha = 0.25$, and the third for $\alpha = 0.75$.

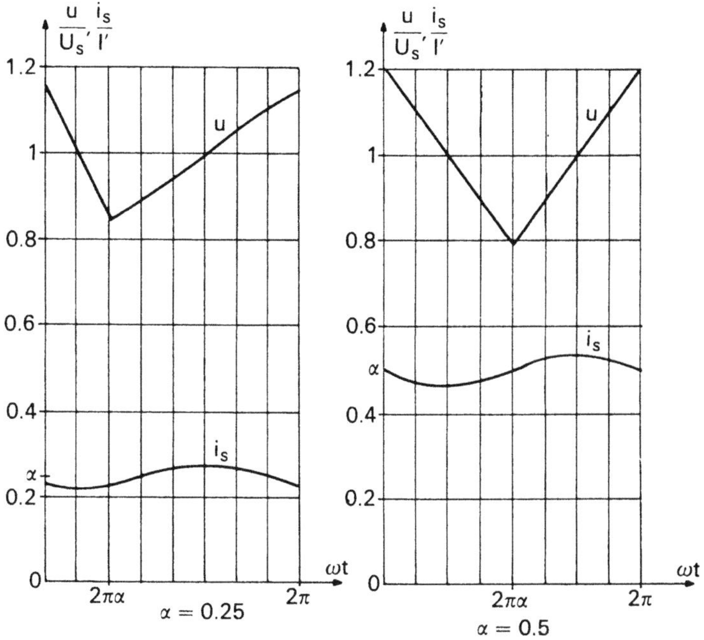

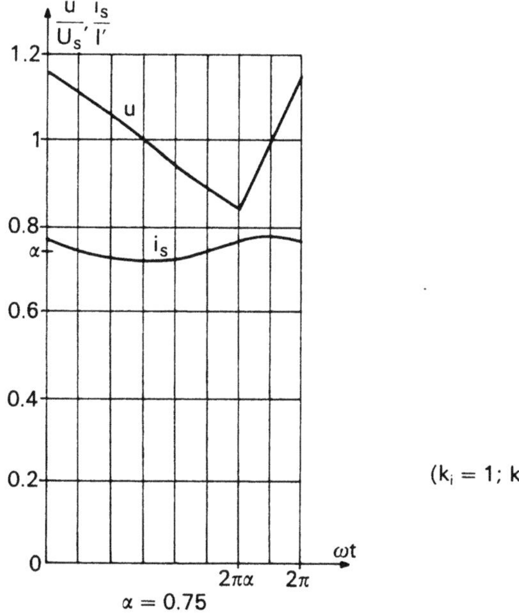

($k_i = 1$; $k_f = 0.25$)

Fig. 4.27

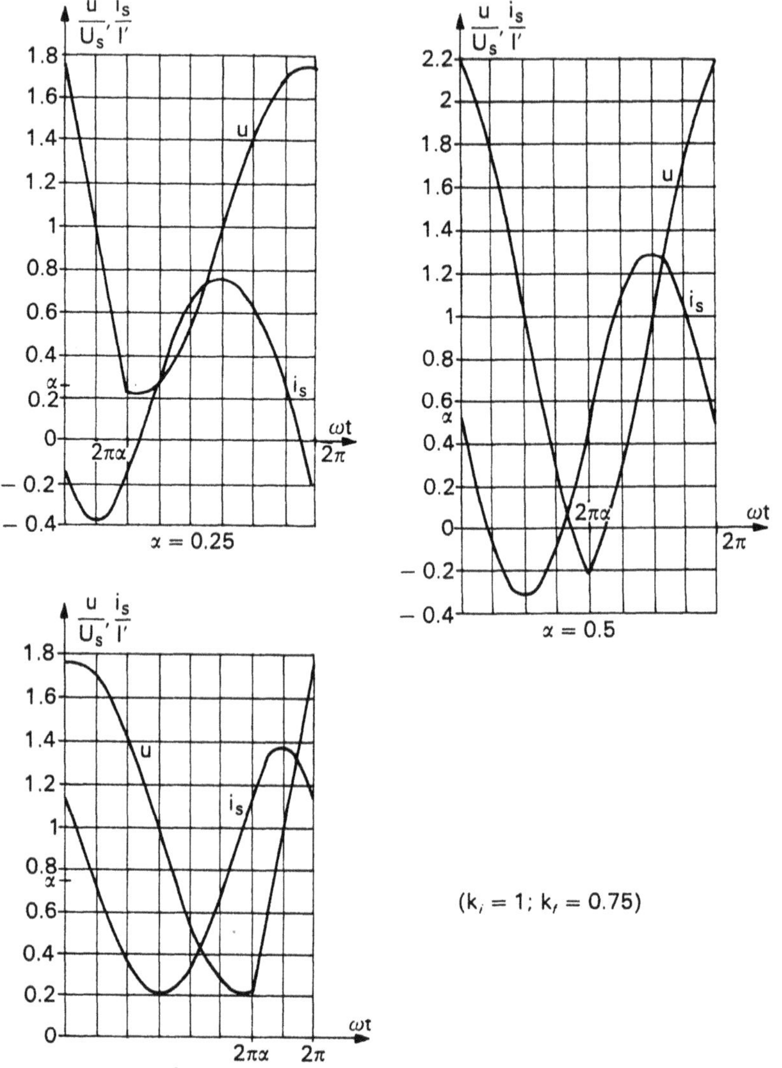

Fig. 4.28

However, for given k_i and k_f, the ripple is always strongest for α equal to 0.5. If k_f remains less than 1, α equal to 0.5 is to be found between $1 - k_f/2$ and $k_f/2$: Eqs. (4.55) and (4.55′) can still be used.

- *Current* i_S. Equations (4.52b) and (4.53b) show that current i_S reaches its minimum value for $\omega t = \pi\alpha$ and its maximum value for $\omega t = (1 + \alpha)\pi$, even if k_f is above 0.5. The equations defined for k_f below 0.5 can still be used.

It should be noted that the minimum value of the current i_S becomes negative when α becomes less than $2 - (1/k_f)$.

In fact, voltage U_S is often obtained from a rectifier and current i_S cannot be reversed. A value of k_f less than 0.5 must therefore be chosen in order to ensure that the current delivered by the voltage source is unidirectional, whatever α may be.

4.2.2.2 Step-up Chopper

Current I' and parameter k_i are negative in the step-up chopper operating mode. Equations (4.52) and (4.53) show that

- the direction of current i_S
- and the variations of u around its mean values U_S

are inverted.

Figure 4.29 shows the correspondence between waveforms, for the same values of α, of k_f and of $|I'|$.

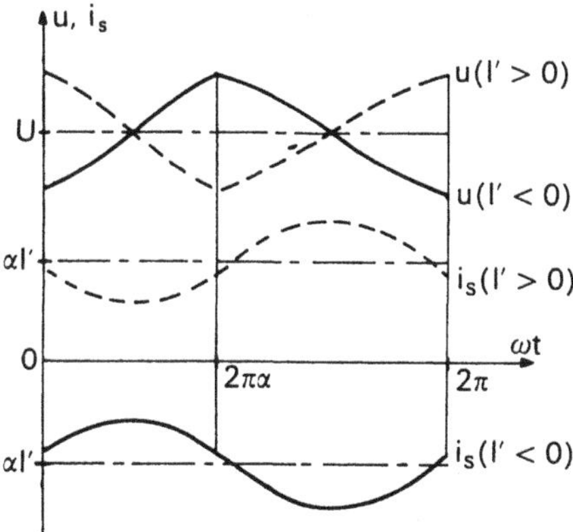

Fig. 4.29

In the equations defined for the step-down chopper, u_{max} must simply be replaced by u_{min}, u_{min} by u_{max}, i_{Smax} by i_{Smin}, i_{Smin} by i_{Smax}.

The expression of ripples Δu and Δi_S can again be used, but absolute values of I' and k_i must be taken.

4.2.3 Characteristics: Choice of C and L_S

For different values of k_f less than 0.5, Fig. 4.30 shows the variation of voltage ripple Δu as a function of α. Δu is related to $|k_i| U_S$.

Under the same conditions, Fig. 4.31 shows the variation of Δi_S. This ripple is related to $|I'|$.

In the majority of cases, the maximum current ripple is limited to a value less than 10 or 20%. The corresponding value of k_f then enables the simplified Eqs.

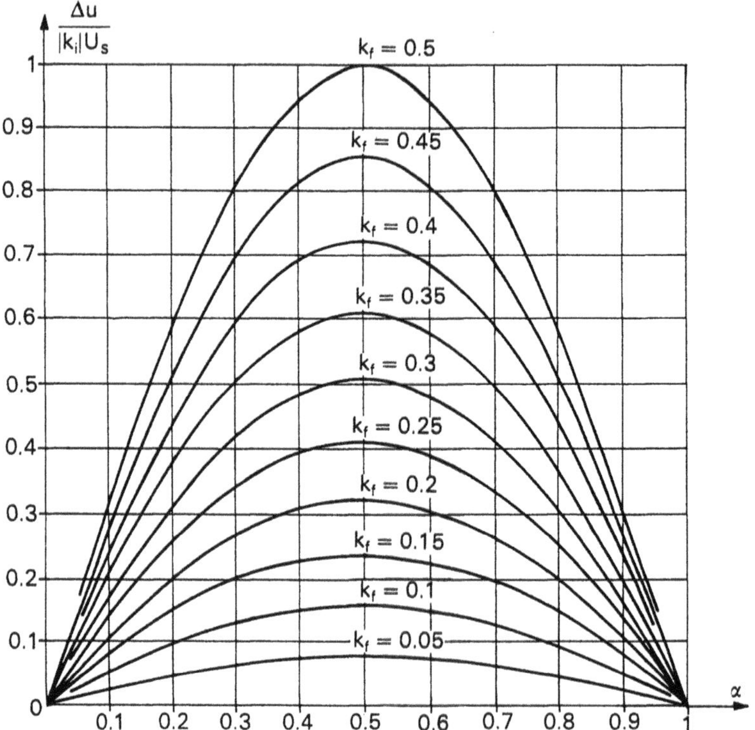

Fig. 4.30

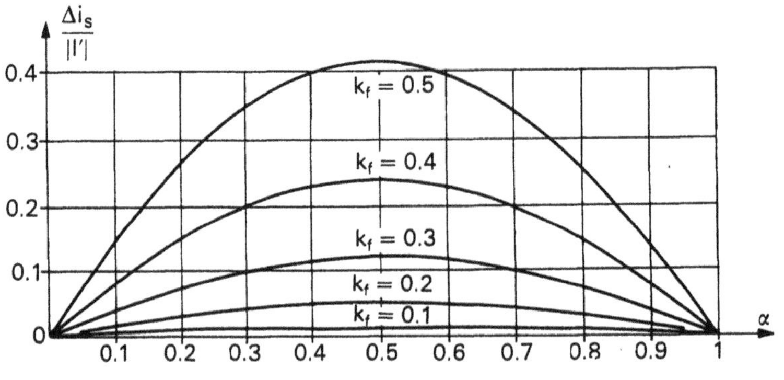

Fig. 4.31

(4.55′) and (4.57′) to be used:

$$\Delta u_{\max} \simeq \frac{\pi}{2}\, k_\mathrm{i} k_\mathrm{f}\, U_\mathrm{S} = \frac{\pi}{2}\sqrt{\frac{L_\mathrm{S}}{C}}\,\frac{I'}{U_\mathrm{S}}\,\frac{1}{\omega\sqrt{L_\mathrm{S} C}}\,U_\mathrm{S} = \frac{\pi}{2\omega}\,\frac{I'}{C}$$

$$\Delta i_{\mathrm{S}_{\max}} \simeq \frac{\pi^2}{8}\, k_\mathrm{f}^2\, I' = \frac{\pi^2}{8\omega^2}\,\frac{I'}{L_\mathrm{S} C}$$

C and L_S can thus be obtained:

$$C \geqslant \frac{\pi}{2\omega} \frac{I'}{\Delta u_{max}}$$

$$L_S \geqslant \frac{\pi}{4\omega} \frac{\Delta u_{max}}{\Delta i_{S_{max}}}. \tag{4.58}$$

4.2.4 Full-Bridge Choppers

- In the case of *sequential control* of the full-bridge chopper, current i at the chopper input (see Fig. 3.13) is once more equal to
  I' or $-I'$ during a part of the switching cycle,
  zero during the remainder of this cycle,
  if I' denotes the current in the supposed perfect current source.
  The behaviour of the filter is identical to that analyzed in the case of the chopper with two switches. All the equations and characteristics laid down for the latter can be of direct use.

- In the case of *continuous control*, the simultaneous turn-on of K_1 and K'_2, or of K_2 and K'_1 is added to the complementary control of switches K_1 and K'_1, on one hand, K_2 and K'_2, on the other. This provides the equivalent configuration in Fig. 4.32a.

- Current i_S and voltage u remain the solutions of the system of Eqs. (4.50):

$$L_S \frac{di_S}{dt} + u = U_S$$

$$C \frac{du}{dt} = i_S - i .$$

But current i, which was equal to

I', for $0 < \omega t < 2\pi\alpha$,

0, for $2\pi\alpha < \omega t < 2\pi$

is now equal to

I', for $0 < \omega t < 2\pi\alpha$ (K_1 and K'_2 closed),

$-I'$, for $2\pi\alpha < \omega t < 2\pi$ (K_2 and K'_1 closed).

- It can be noted that current i (Fig. 4.32b) is the sum of
  - a current equal to $2I'$ for $0 < \omega t < 2\pi\alpha$, and zero for the remainder of the cycle
  - and a constant current equal to $-I'$.

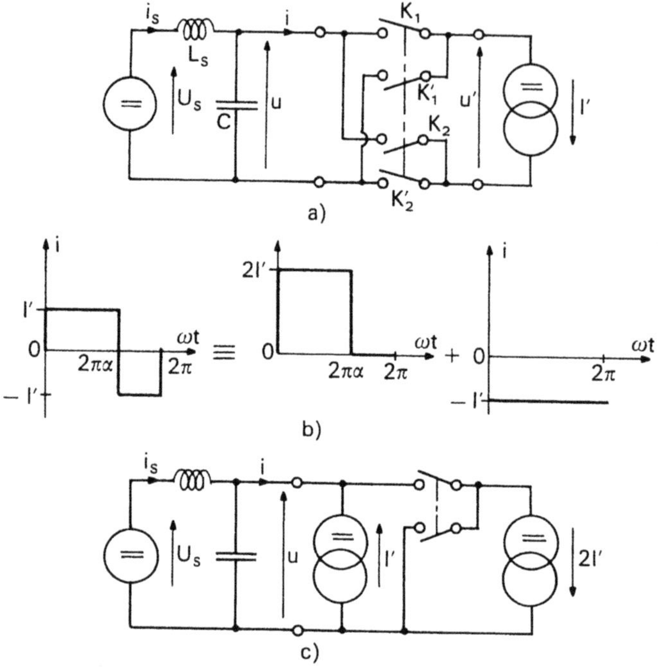

Fig. 4.32

As concerns the filter, the resulting configuration is that of Fig. 4.32c with a two-switch chopper.

- The previous remark means that there is no need to repeat the analysis of the filter in the case of the full-bridge chopper.

  The constant current $-I'$ produces no ripple on either u or on i_S.

Since the current is doubled at the output of the equivalent double-switch chopper, for the same values of I', k_i, k_f, and α, the *ripple Δu on voltage u and the ripple Δi_S on current i_S are doubled.*

Providing that this doubling is taken into account, the equation defined for the chopper with two switches can be used.

In particular, for k_f less than 0.5, the ripples are given by

$$\Delta u = 4k_i U_S \frac{\sin[\pi k_f(1-\alpha)]\sin(\pi k_f \alpha)}{\sin(\pi k_f)}$$

$$\Delta i_S = 2I'\left[\frac{\sin[\pi k_f(1-\alpha)] + \sin(\pi k_f \alpha)}{\sin(\pi k_f)} - 1\right]. \tag{4.59}$$

Their maximum values, obtained for $\alpha = 0.5$, are

$$\Delta u_{\max} = 2k_i U_S \tan(\pi k_f/2)$$
$$\Delta i_{S_{\max}} = 2I' \frac{1 - \cos(\pi k_f/2)}{\cos(\pi k_f/2)} \tag{4.60}$$

or, if k_f is low,

$$\Delta u_{\max} \simeq \pi k_i k_f U_S$$
$$\Delta i_{S_{\max}} \simeq \frac{\pi^2}{4} k_f^2 I'. \tag{4.61}$$

The characteristics shown in Figs. 4.30 and 4.31 can be used, providing that the values noted for Δu and Δi_S are multiplied by two.

Equations (4.58) become:

$$C \geqslant \frac{\pi}{\omega} \frac{I'}{\Delta u_{\max}}$$
$$L_S \geqslant \frac{\pi}{4\omega} \frac{\Delta u_{\max}}{\Delta i_{S_{\max}}}. \tag{4.62}$$

In order to obtain the same maximum ripple on voltage u and current i_S, *capacitance C must be doubled.*

4.2.5 The "First-Harmonic" Method

Figure 4.33 shows that filter L_S, C is inserted between a DC voltage source (internal impedance assumed to be zero) and a switched-current source (internal impedance assumed to be infinite).

Since $U_S = L_S \, \mathrm{d}i_S/\mathrm{d}t + u$

$$i = i_S + C \frac{\mathrm{d}u}{\mathrm{d}t},$$

the average values are linked by

$$U = U_S; \quad I_S = I$$

Amplitudes $U_1\sqrt{2}$ and $I_{S_1}\sqrt{2}$ of terms with angular frequency ω of voltage u and current i_S are linked to amplitude $I'_1\sqrt{2}$ of term with the same angular

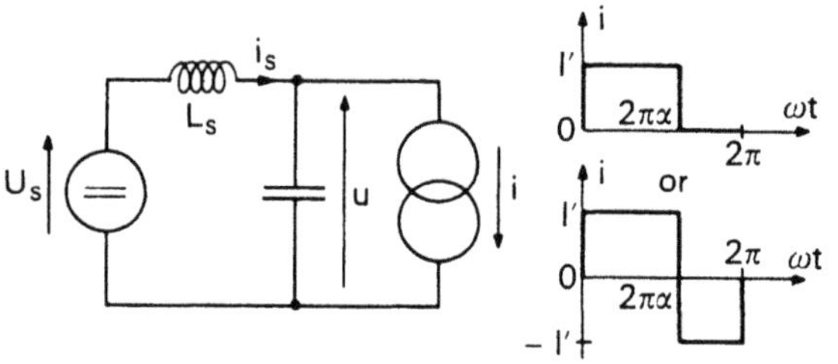

Fig. 4.33

frequency of current i by

$$I_{S_1} = I_1 \frac{1/C\omega}{|L_S\omega - 1/C\omega|} = \frac{I_1}{|L_S C\omega^2 - 1|}$$

$$U_1 = L_S\omega I_{S_1} = \frac{L_S\omega I_1}{|L_S C\omega^2 - 1|}.$$

If only the terms with angular frequency ω are taken into account, the peak-to-peak variations of voltage u and current i_S are given by

$$\Delta u = 2U_1\sqrt{2}; \quad \Delta i_S = 2I_{S_1}\sqrt{2}.$$

- For the *two-switch choppers* and the full-bridge chopper with sequential control,

$$I = \alpha I'; \quad I_1\sqrt{2} = \frac{2}{\pi} I' \sin \pi\alpha.$$

This gives the ripple values, in cases where $L_S C\omega^2$ is greater than 1 ($k_f < 1$):

$$\Delta u = \frac{4}{\pi} \frac{L_S\omega}{L_S C\omega^2 - 1} I' \sin \pi\alpha$$

$$\Delta i_S = \frac{4}{\pi} \frac{1}{L_S C\omega^2 - 1} I' \sin \pi\alpha. \tag{4.63}$$

When α varies, the ripples reach their maximum for $\alpha = 0.5$, and have the following values:

$$\Delta u_{max} = \frac{4}{\pi} I' \frac{L_S\omega}{L_S C\omega^2 - 1}; \quad \Delta i_{S_{max}} = \frac{4}{\pi} \frac{I'}{L_S C\omega^2 - 1}. \tag{4.64}$$

If $LC\omega^2$ is much greater than 1,

$$\Delta u_{max} \simeq \frac{4}{\pi} \frac{I'}{C\omega}; \quad \Delta i_{S_{max}} \simeq \frac{4}{\pi} \frac{I'}{L_S C\omega^2}. \tag{4.65}$$

- To compare these results with those of the more precise analysis, parameters k_f and k_i can be included in the latter equations, which then become

$$\Delta u_{max} \simeq \frac{4}{\pi} k_i k_f U_S = 1.27\, k_i k_f U_S$$

$$\Delta i_{S_{max}} \simeq \frac{4}{\pi} k_f^2 I' = 1.27\, k_f^2 I'.$$

The values given by Eqs. (4.55′) and (4.57′) were

$$\Delta u_{max} \simeq 1.57\, k_i k_f U_S ; \quad \Delta i_{S_{max}} \simeq 1.23\, k_f^2 I'.$$

As could be predicted from the waveforms in Figs 4.27 and 4.28, the "first-harmonic" approximation is more accurate for current i_S than for voltage u.

- In the case of *full-bridge chopper with continuous control*, the average value and the component with angular frequency ω of current i are given by

$$I = I'(2\alpha - 1) ; \quad I_1\sqrt{2} = \frac{4}{\pi} I' \sin \pi\alpha.$$

Ripples Δu and Δi_S, given by Eqs. (4.64) and (4.65), must thus be *multiplied by two.*

4.3 Multiphase Choppers

The first two sections of this chapter have indicated that the ripple in the output current i of a chopper, as well as the ripple in its input voltage u and the ripple in current i_S which it takes from the supply, are inversely proportional to the switching frequency.

If the amplitude of these ripples is imposed, any increase in the switching frequency leads to a proportional decrease in the filter components. These components usually account for a considerable part of the chopper size and of its total cost. The highest possible frequency should therefore be chosen. However, limits are imposed by the duration of semiconductor device commutations and by the commutation losses, as noted in Chap. 2

Nevertheless, *the frequency "seen" by the generator and load* – placed on either side of the chopper – *can be multiplied* without increasing the switching frequency. This can be achieved by staggering several single-phase choppers.

We shall consider the most common case which consists in staggering n choppers, each with two switches, connected between an input voltage source and an output current source.

4.3.1 Switch Operation, Output Waveforms

Figure 4.34 shows n paralleled choppers, with their "switches" K_{11} and K_{21}, K_{12} and $K_{22}, \ldots, K_{1j}$ and $K_{2j}, \ldots, K_{1n}$ and K_{2n}. Each elementary chopper must be provided with a smoothing inductor L' to prevent the different chopper switches short-circuiting each other.

In order to obtain a frequency multiplication by n with regard to the input and output, the n choppers must operate at the same frequency, at the same duty ratio α, and their control signals must be shifted between them by $(1/n)^{\text{th}}$ of a cycle.

In this analysis of output current i' and currents $i'_1, i'_2, \ldots, i'_j \ldots, i'_n$ supplied by the choppers, it is assumed that

- the voltage source and the switches are perfect,
- the resistance of inductors L' is negligible,
- the current receptor is formed by the series connection of a resistance R, an inductance L and an EMF E.
- the conduction of each chopper is continuous.

The last assumption means that, for some operating conditions, the analysis is only valid if choppers are current-reversible.

4.3.1.1 Output Current Expression, Ripple

- Switch K_{11} is turned on for $0 < \omega t < 2\pi\alpha$, K_{12} for $2\pi/n < \omega t < 2\pi\alpha + 2\pi/n, \ldots$ Cycle 2π can be divided into $2n$ intervals.

$$\text{If } \frac{l}{n} \leqslant \alpha \leqslant \frac{l+1}{n}, \text{ with } 0 \leqslant l \leqslant n-1,$$

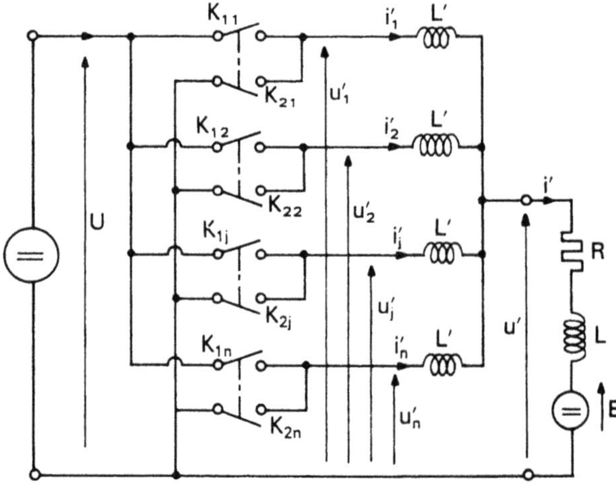

Fig. 4.34

there are either $l+1$ or l switches of the series $K_{11}, K_{12}, \ldots, K_{1n}$ closed. For example, Fig. 4.35 corresponds to $n = 6$, $\alpha = 0.55$ and, thus, to $l = 3$.

Current i' and currents $i'_1, i'_2, \ldots, i'_j, \ldots, i'_n$ are linked by

$$i' = \sum_1^n i'_j \tag{4.66}$$

with

$$u'_j - L'\frac{di'_j}{dt} = Ri' + L\frac{di'}{dt} + E \qquad (1 \leqslant j \leqslant n) \tag{4.67}$$

where voltage u'_j equals U when switch K_{1j} is on and zero when this switch is off, since, in that case, K_{2j} is on.

- By considering an nth part of the switching cycle, e.g. interval $(0, 2\pi/n)$, the expressions of current i' can be determined, on account of the latter's periodicity.

Let

$$\alpha' = \alpha - \frac{l}{n} \tag{4.68}$$

Summing the n Eqs. (4.67) and then dividing the sum by n gives:

– for $0 < \omega t < 2\pi\alpha'$,

$$\frac{1}{n}\left(\sum_1^n u'_j - L'\sum_1^n \frac{di'_j}{dt}\right) = Ri' + L\frac{di'}{dt} + E$$

$$\frac{1}{n}\left[(l+1)U - L'\frac{di'}{dt}\right] = Ri' + L\frac{di'}{dt} + E$$

Thus

$$\frac{l+1}{n}U = Ri' + \left(L + \frac{L'}{n}\right)\frac{di'}{dt} + E \tag{4.69a}$$

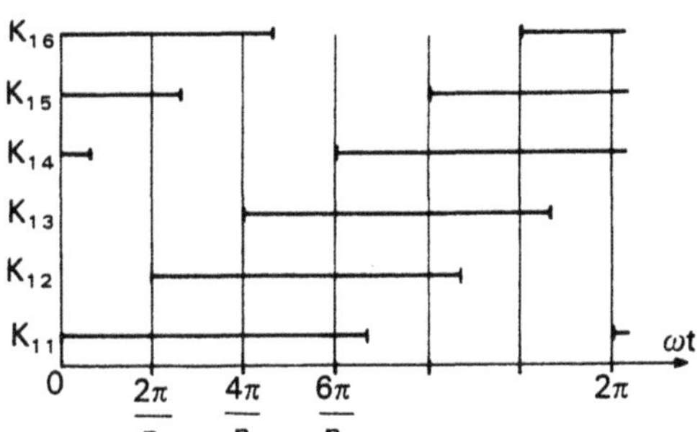

Fig. 4.35

– for $2\pi\alpha' < \omega t < 2\pi/n$,

$$\frac{l}{n}U = Ri' + \left(L + \frac{L'}{n}\right)\frac{di'}{dt} + E. \tag{4.69b}$$

Taking into account that, for $\omega t = 2\pi/n$, current i' takes again the same value as for $t = 0$, we can obtain:

– for $0 < \omega t < 2\pi\alpha'$,

$$i' = \frac{U}{R}\left[\frac{l}{n} - m + \frac{1}{n} - \frac{1}{n}\frac{\exp(2\pi/nQ') - \exp(2\pi\alpha'/Q')}{\exp(2\pi/nQ') - 1}\exp\left(\frac{-\omega t}{Q'}\right)\right] \tag{4.70a}$$

– for $2\pi\alpha' < \omega t < 2\pi/n$,

$$i' = \frac{U}{R}\left[\frac{l}{n} - m + \frac{1}{n}\frac{\exp(2\pi\alpha'/Q') - 1}{\exp(2\pi/nQ') - 1}\exp\left(\frac{2\pi}{nQ'}\right)\exp\left(\frac{-\omega t}{Q'}\right)\right] \tag{4.70b}$$

with

$$Q' = \frac{1}{R}\left(L + \frac{L'}{n}\right)\omega.$$

- *The peak-to-peak ripple* $\Delta i'$ in current i' is

$$\Delta i' = i'(2\pi\alpha') - i'(0)$$

$$\Delta i' = \frac{U}{nR}\frac{[\exp(2\pi/Q')(1/n - \alpha') - 1]\,[\exp(2\pi\alpha'/Q') - 1]}{\exp(2\pi/nQ') - 1}. \tag{4.71}$$

This ripple is maximum for α' equal to $1/2n$, and is then.

$$\Delta i'_{max} = \frac{U}{nR}\tanh\left(\frac{\pi}{2nQ'}\right). \tag{4.72}$$

For high values of nQ' (in practice $nQ' \geqslant 4$), this gives, as a close approximation,

$$\Delta i'_{max} = \frac{U}{nR}\frac{\pi}{2nQ'} = \frac{1}{n^2}\frac{\pi}{2}\frac{U}{(L + L'/n)\,\omega}. \tag{4.72'}$$

Figure 4.36 gives the waveforms of voltage $u'_1, u'_2, \ldots, u'_n$ and $(1/n)\sum_1^n u'_j$, of current i', for $n = 4$ and $\alpha = 1/3$. The average value of the five voltages is αU, and that of the current i' is $(\alpha U - E)/R$.

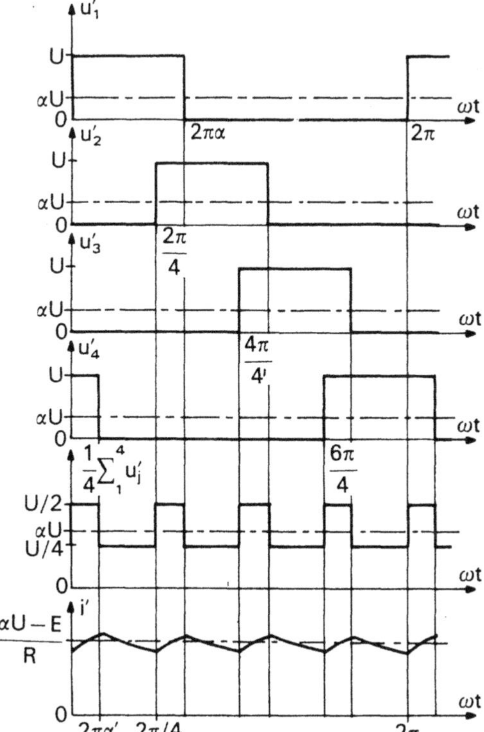

Fig. 4.36

4.3.1.2 Approximate Calculation of the Output Current Ripple

- Resistance R of the current source i' must be taken into account in determining the average value I' of the latter, using the intersection of curve $U'(I')$ of the chopper and curve $U'(I')$ of the source (see Figs. 4.3 and 4.23).

  Once more, this gives

$$U' = \alpha U \qquad \text{and} \qquad U' = RI' + E$$

and thus,

$$I' = \frac{\alpha U - E}{R} = \frac{U}{R}(\alpha - m). \tag{4.73}$$

- However, in analysing ripple $\Delta i'$ in current i', R can be ignored and the current source appears as a LE load, if Q' has a high value. If the resistance is ignored, E is equal to αU and the currents must be related to $U/(L + L'/n)\omega$.

  Equations (4.69a) and (4.69b) thus give:

  - for $0 < \omega t < 2\pi\alpha'$,

$$i' = i'_0 + \frac{U}{(L + L'/n)\,\omega}\left(\frac{l+1}{n} - \alpha\right)\omega t;$$

• for $2\pi\alpha' < \omega t < 2\pi/n$,

$$i = i'_0 + \frac{U}{(L + L'/n)\,\omega}\left[\frac{2\pi\alpha'}{n} + \left(\frac{l}{n} - \alpha\right)\omega t\right]$$

where i'_0 is the value of i' for $\omega t = 0$.

Since $l/n < \alpha < (l+1)/n$, current i' increases linearly during the first period and decreases linearly by an equal amount during the second.

The peak-to-peak ripple in current i' is equal to

$$\Delta i' = \frac{U}{(L + L'/n)\,\omega}\,2\pi\alpha'\left(\frac{1}{n} - \alpha'\right). \tag{4.74}$$

This ripple is maximum for $\alpha' = 1/2n$ and then has the value

$$\Delta i'_{\max} = \frac{1}{n^2}\,\frac{\pi}{2}\,\frac{U}{(L + L'/n)\,\omega}$$

which is identical to Eq. (4.72').

4.3.1.3 Current in the Elementary Choppers

The average value I'_1 of the current in each of the n choppers is

$$I'_1 = \frac{I'}{n} = \frac{1}{n}\frac{U}{R}(\alpha - m). \tag{4.75}$$

- In order to follow the variations of these currents on either side of their mean value, resistance R can be ignored.

  Equation (4.67) becomes

$$u'_j - L'\frac{\mathrm{d}i'_j}{\mathrm{d}t} = L\frac{\mathrm{d}i'}{\mathrm{d}t} + \alpha U, \qquad \text{with} \qquad 1 \leqslant j \leqslant n.$$

Equation (4.69a), which is valid during periods where only $l + 1$ switches of the series $K_{11}, K_{12}, \ldots, K_{1n}$ are closed, becomes

$$\frac{l+1}{n}U = \left(L + \frac{L'}{n}\right)\frac{\mathrm{d}i'}{\mathrm{d}t} + \alpha U.$$

Equation (4.69b), which can be used during periods where only l switches are closed, can be written as

$$\frac{l}{n}U = \left(L + \frac{L'}{n}\right)\frac{\mathrm{d}i'}{\mathrm{d}t} + \alpha U.$$

Values of $\mathrm{d}i'/\mathrm{d}t$ can be deduced from these equations. By substituting them into Eq. (4.67) where R is neglected, this gives:

- for $0 + k\frac{2\pi}{n} < \omega t < 2\pi\alpha' + k\frac{2\pi}{n}$,

$$L'\frac{\mathrm{d}i'_j}{\mathrm{d}t} = u'_j - \alpha U - \frac{L}{L + L'/n}U\left(\frac{l+1}{n} - \alpha\right); \tag{4.76a}$$

- for $2\pi\alpha' + k\frac{2\pi}{n} < \omega t < (k+1)\frac{2\pi}{n}$,

$$L'\frac{\mathrm{d}i'_j}{\mathrm{d}t} = u'_j - \alpha U - \frac{L}{L + L'/n}U\left(\frac{l}{n} - \alpha\right); \tag{4.76b}$$

with k successively equal to 1, 2, . . . , $n - 1$.

- In order to follow the variations of current i'_1 of the first chopper, for example, throughout a cycle, $2n$ intervals must be considered (see Fig. 4.36):
  - the $(2l + 1)$ intervals during which K_{11} is closed and in which u'_1 is equal to $+U$. Among these intervals, there are
    - $l + 1$ intervals with a width of $2\pi\alpha'$, where there are $l + 1$ switches of series K_{11} on:

$$L'\frac{\mathrm{d}i'_1}{\mathrm{d}t} = U(1 - \alpha) - \frac{L}{L + L'/n}U\left(\frac{l+1}{n} - \alpha\right);$$

    - l intervals with a width of $(2\pi/n) - 2\pi\alpha'$, where only l switches of this series are on:

$$L'\frac{\mathrm{d}i'_1}{\mathrm{d}t} = U(1 - \alpha) - \frac{L}{L + L'/n}U\left(\frac{l}{n} - \alpha\right);$$

  - the $(2n - 2l - 1)$ intervals during which K_{11} is open and where u'_1 is zero. Among these intervals, there are
    - $n - l - 1$ intervals with a width of $2\pi\alpha'$, where the variation in i'_1 is given by

$$L'\frac{\mathrm{d}i'_1}{\mathrm{d}t} = -\alpha U - \frac{L}{L + L'/n}U\left(\frac{l+1}{n} - \alpha\right);$$

    - $n - l$ intervals with a width of $(2\pi/n) - 2\pi\alpha'$, where

$$L'\frac{\mathrm{d}i'_1}{\mathrm{d}t} = -\alpha U - \frac{L}{L + L'/n}U\left(\frac{l}{n} - \alpha\right).$$

In each cycle, the waveform of current i'_1 is composed of $2n$ straight line segments, since the derivative of current i'_1 is constant during each of $2n$ intervals. This derivative is positive during the $2l + 1$ first intervals and negative during the last $2n - 2l - 1$ ones. The waveform in Fig. 4.37, like that in Fig. 4.35, corresponds to $n = 6$, $\alpha = 0.55$ and, thus, $l = 3$.

- Current i'_1 ripple can be obtained by simply adding together its increases during the $2l + 1$ intervals when K_{11} is closed:

$$\Delta i'_1 = (l+1)\frac{1}{L'\omega}\left[U(1-\alpha) - \frac{L}{L+L'/n}U\left(\frac{l+1}{n} - \alpha\right)\right]2\pi\alpha'$$

$$+ l\frac{1}{L'\omega}\left[U(1-\alpha) - \frac{L}{L+L'/n}U\left(\frac{l}{n} - \alpha\right)\right]\left(\frac{2\pi}{n} - 2\pi\alpha'\right).$$

Taking Eq. (4.68) into account, the following is obtained, after several simplifications:

$$\Delta i'_1 = \frac{U}{L'\omega}\left[2\pi\alpha(1-\alpha) - \frac{L}{L+L'/n}2\pi\alpha'\left(\frac{1}{n} - \alpha'\right)\right]. \tag{4.77}$$

This ripple is maximum for $\alpha = 0.5$ and has the following values:

- if n is even,

$$\Delta i'_{1\max} = \frac{\pi}{2}\frac{U}{L'\omega}; \tag{4.78a}$$

- if n is odd,

$$\Delta i'_{1\max} = \frac{\pi}{2}\frac{U}{L'\omega}\left(1 - \frac{1}{n^2}\frac{L}{L+L'/n}\right) \tag{4.78b}$$

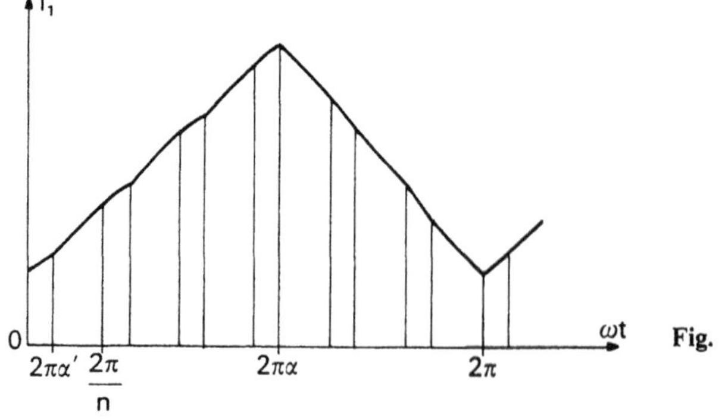

Fig. 4.37

4.3.2 Input-Filter Behaviour

As in the case of the single-phase chopper, the current ripple at the output of each paralleled chopper is ignored, in order to analyze the input filter.

In these conditions, on the filter side, chopper j takes a current equal to I'/n when switch K_{1j} is on and no current when it is off. The total current i (Fig. 4.38) at the filter output is a square-wave current with a frequency equal to n times that of the switching frequency:

- i equals $I'(l+1)/n$ during intervals $2\pi\alpha'$ wide when $l+1$ switches K_{1j} are on;
- i equals $I'l/n$ during intervals $2\pi/n - 2\pi\alpha'$ wide when only l switches K_{1j} are on.

Apart from the $I'l/n$ component, the filter output current is the same as with a single chopper:

- absorbing a chopped current equal to I'/n or to zero,
- with a switching angular frequency equal to $n\omega$.

The filter–chopper system can be analyzed by using a configuration similar to that shown in Fig. 4.38

If k_i and k_f are once more used to denote the ratios

$$k_i = \sqrt{\frac{L_S}{C}}\frac{I'}{U_S}; \quad k_f = \frac{1}{\omega\sqrt{L_S C}};$$

by analogy with Eqs. (4.52) and (4.53), this yields:

– from $\omega t = 0$ to $\omega t = 2\pi\alpha'$,

$$u = U_S\left[1 - \frac{k_i}{n}\frac{\sin[\pi k_f(1/n - \alpha')]}{\sin(\pi k_f/n)}\sin[k_f(\omega t - \pi\alpha')]\right]$$

$$i_S = \frac{I'}{n}\left[l + 1 - \frac{\sin[\pi k_f(1/n - \alpha')]}{\sin(\pi k_f/n)}\cos[k_f(\omega t - \pi\alpha')]\right];$$

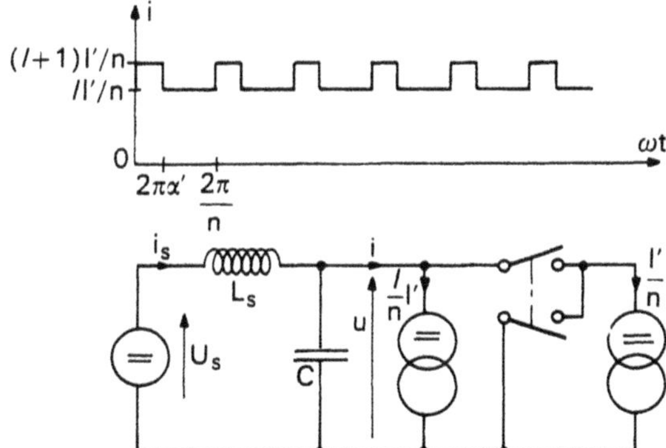

Fig. 4.38

– from $\omega t = 2\pi\alpha'$ to $\omega t = 2\pi/n$,

$$u = U_S\left[1 + \frac{k_i}{n}\frac{\sin(\pi k_f \alpha')}{\sin(\pi k_f/n)}\sin\left(k_f\left[\omega t - \pi\left(\frac{1}{n} + \alpha'\right)\right]\right)\right]$$

$$i_S = \frac{I'}{n}\left[l + \frac{\sin(\pi k_f \alpha')}{\sin(\pi k_f/n)}\cos\left(k_f\left[\omega t - \pi\left(\frac{1}{n} + \alpha'\right)\right]\right)\right].$$

For k_f/n less than 0.5, the ripple in voltage u and that in current i_S are given by

$$\Delta u = 2\frac{k_i}{n}U_S\frac{\sin[\pi k_f(1/n - \alpha')]}{\sin(\pi k_f/n)}\sin(\pi k_f \alpha') \tag{4.79}$$

$$\Delta i_S = \frac{I'}{n}\left[\frac{\sin(\pi k_f \alpha') + \sin[\pi k_f(1/n - \alpha')]}{\sin(\pi k_f/n)} - 1\right]. \tag{4.80}$$

For values of α which make them maximum, these ripples are

$$\Delta u_{max} = \frac{k_i}{n}U_S\tan\left(\frac{\pi}{2}\frac{k_f}{n}\right)$$

$$\Delta i_{S_{max}} = \frac{I'}{n}\frac{1 - \cos(\pi k_f/2n)}{\cos(\pi k_f/2n)}.$$

For the small values of k_f/n which are usually used, the following can be used:

$$\Delta u_{max} \simeq \frac{1}{n^2}k_i k_f \frac{\pi}{2}U_S \tag{4.81}$$

$$\Delta i_{S_{max}} \simeq \frac{1}{n^3}k_f^2\frac{\pi^2}{8}I'. \tag{4.82}$$

4.3.3 Characteristics

The characteristics showing the variations, as a function of α and for different values of n, of

the current ripple in the current receptor,
the current ripple in the choppers,
the voltage ripple at the filter output,
and the current ripple in the voltage source,

enable the effects of the increase of n to be evaluated.

All these characteristics are symmetrical in relation to $\alpha = 0.5$ and their representation will therefore be limited to α taken between 0 and 0.5.

4.3.3.1 Output-Current Ripple $\Delta i'$

– The continuous lines in Fig. 4.39 show the curves which give $\Delta i'$ as a function of α, for $n = 1$, $n = 2$ and $n = 3$, for $Q' = 1$ and then for $Q' = 3$. The ripple is calculated on the basis of Eq. (4.71).

$\Delta i'$ is multiplied by $2/\pi$ and related to $U/\mathscr{L}\omega$, the inductance $\mathscr{L}$ being

$$\mathscr{L} = L + \frac{L'}{n} \tag{4.83}$$

$\mathscr{L}$ is the *current i′ total smoothing inductance.* Indeed, the total energy stored in inductor L, through which flows a current whose average value equals I', and in n inductors L', through which flows a current whose average value is I'/n, is

$$LI'^2 + nL'\left(\frac{I'}{n}\right)^2 = \left(L + \frac{L'}{n}\right)I'^2 = \mathscr{L}I'^2.$$

$\mathscr{L}$ plays the same role as L in the single chopper

– The same figure shows, in broken lines, the variations in $\Delta i'$ obtained by ignoring resistance R, i.e. by using Eq. (4.74).

It can be seen that the gap between the two types of curve decreases as Q' (equal to $\mathscr{L}\omega/R$) becomes higher and as the number of staggered choopers increases.

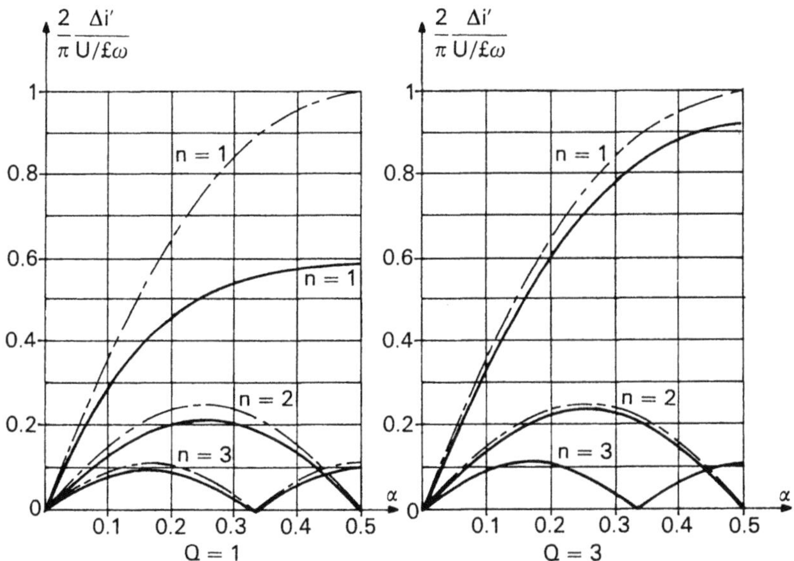

Fig. 4.39

As soon as Q' becomes sufficiently high to ensure an effective smoothing of current i', it is possible to calculate the ripple in the latter, taking only the inductance into account.

- As in Eq. (4.72′) the curves in Fig. 4.39 indicate that the maximum ripple in i' is inversely proportional to the square of the number of staggered choppers.

4.3.3.2 Current Ripple $\Delta i_1'$ in the Choppers

- Using Eq. (4.77) – obtained by assuming a negligible R – the curves of Fig. 4.40 have been plotted. They give current ripple $\Delta i_1'$ in a chopper, multiplied by $2/\pi$ and related to $U/\mathscr{L}\omega$, for two values of the ratio $(L'/n)/L$.
- For a given total smoothing inductance $\mathscr{L}$, thus for a given current ripple $\Delta i'$, current ripple $\Delta i_1'$ in each chopper decreases as the part of L'/n of $\mathscr{L}$ in each chopper increases. For n equal to 2 for example, $\Delta i_1'$ is only less than $\Delta i'$ (and thus $\Delta i_1'/(I'/2)$ less than $2\Delta i'/I'$), if L'/n is greater than L.
- To reduce current ripple in the choppers and especially the current which has to be switched when "switches" K_{11}, K_{12}, ... are turned off, part L'/n of the total inductance $\mathscr{L}$ must be increased.

  Whatever the case, the *relative ripple* $\Delta i_1'/(I'/n)$ is greater than $\Delta i'/I$ from the moment that n is greater than 1. If there are problems of discontinuous conduction, they must be examined at the level of the elementary chopper.

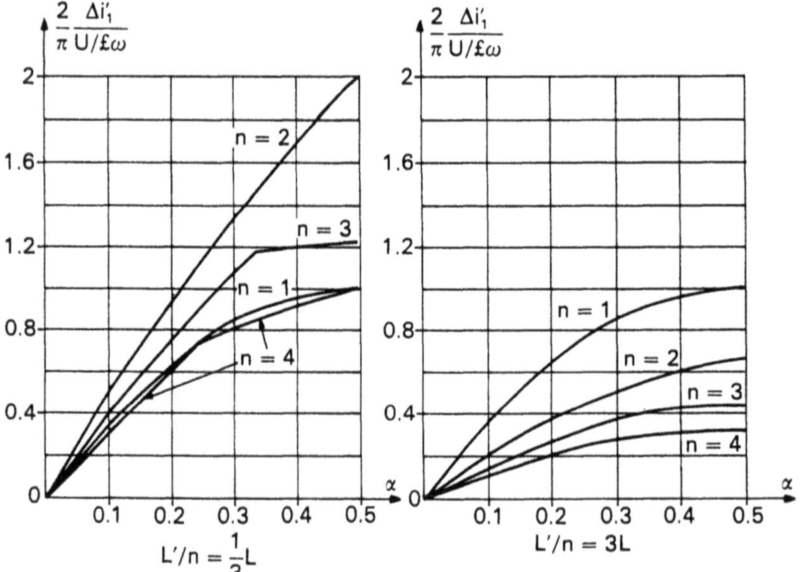

Fig. 4.40

4.3.3.3 Output Voltage Ripple Δu in the Filter

For two values of k_f and the lowest values of n, Fig. 4.41 provides variations of voltage ripple Δu in voltage u across the capacitor. This ripple is calculated using Eq. (4.79) and related to $k_i U_S$.

It can be observed that the maximum ripple of u is divided by n^2 or more, for a given input filter.

4.3.3.4 Supply Current Ripple Δi_S

For two values of k_f and the lowest values of n, Fig. 4.42 shows variations of ripple Δi_S in current i_S delivered by voltage supply U_S. This ripple is related to the average current I' through the current source.

These curves show that the maximum ripple in i_S is divided by n^3 or more, for a given filter.

4.3.4 Remarks about the Filter Common to Several Separate Choppers

In the case of n choppers, supplying n separate but identical loads (Fig. 4.43) and using the same duty ratio, it is possible to obtain – with regard to the common

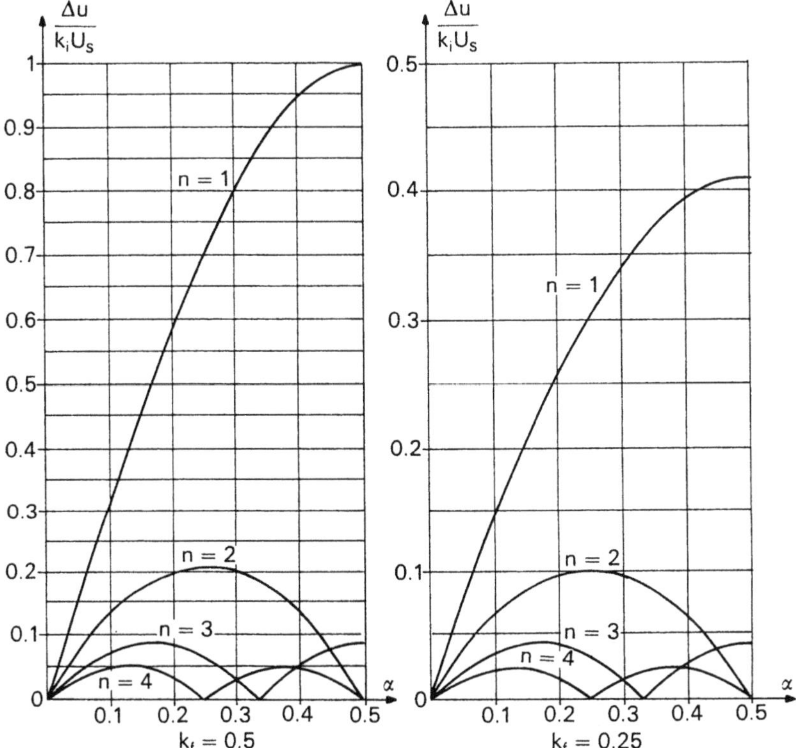

Fig. 4.41

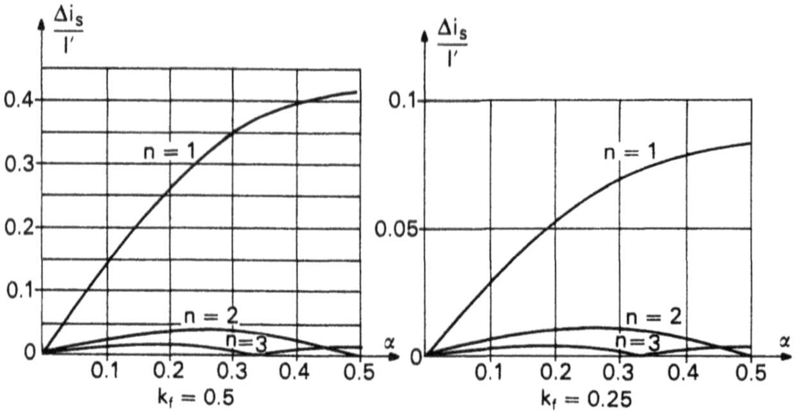

Fig. 4.42

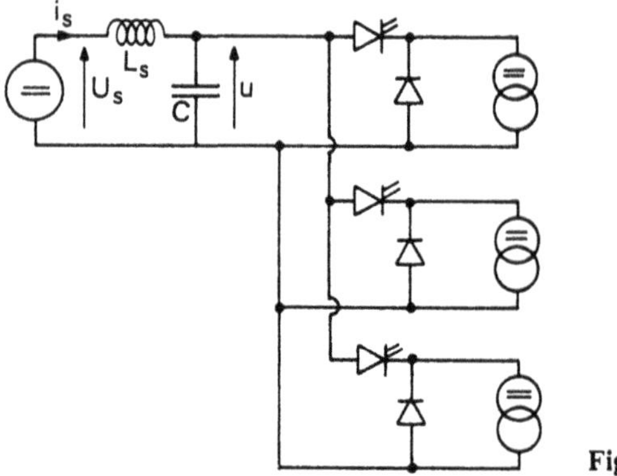

Fig. 4.43

input filter – the same effect as with multiphase chopper. To achieve this, the chopper controls must be shifted between themselves by an nth of a cycle.

This solution is often used in electric traction drives where the supply of all the motors is divided into several identical subgroups.

4.4 Notes on Choppers in Traction Applications

- Choppers are, like rectifiers, widely used in the field of DC motor variable-speed drives. In the case of motors *with separate excitation*,
  - the speed is proportional to the armature EMF,
  - the torque is proportional to the current through the armature.

  The torque can be reversed without changing the connections, by reversing the polarity of the difference between the voltage applied to the armature and its internal EMF.

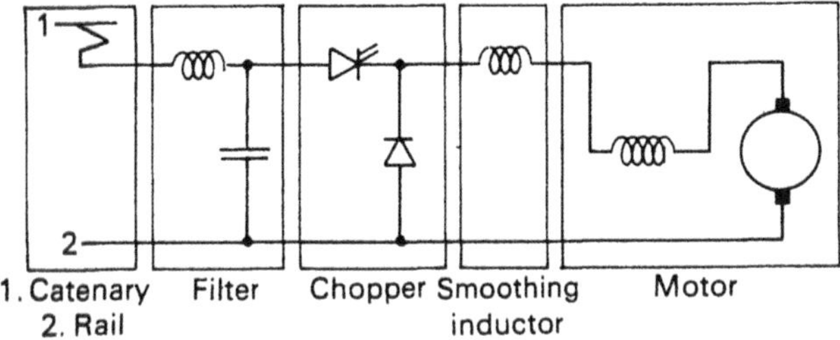

Fig. 4.44

In the analysis of rectifiers (Vol. 1 Chap. 9 § 3 and 4), we emphasized the problems concerning fast torque reversal by fast current reversal.

In the case of choppers, all the observations about the continuous nature of the characteristics $U'(I')$ when the axis $I' = 0$ is crossed, are of major interest in the speed variation of separately excited motors.

- Traction motors are usually *series motors* (If motors with separate excitation are used, they can be given series-type characteristics by making the inductor current servo-controlled by the current in the armature).

  For vehicles powered by a battery or by a DC catenary or by an AC catenary (in cases where the traction system includes a diode rectifier), a chopper frequently drives the motors.

  By using a chopper, it is possible to achieve
  - motoring operation,
  - rheostatic braking operation,
  - regenerative braking operation.

  The change from one mode to another results from a change in connections.

- Current ripple in the motor can be limited by series-connecting a smoothing inductor with the latter.

  An input filter, comprising essentially an inductor and a capacitor, must be used to reduce any disturbances caused by the chopper and to ensure that the latter is fed by the equivalent of a voltage source.

  Figure 4.44 represents schematically the system for operating in motoring mode, with the converter connected as a step-down chopper.

4.4.1 Traction Operation

4.4.1.1 Determination of the Motor Current

In order to analyze current i' in the motor, the switches and the voltage source U are assumed to be perfect (Fig. 4.45)

- R_a, R_f, R_S are used to denote the resistances of the armature, the field winding and the smoothing inductor, and L_a, L_f, L_S the corresponding inductances.

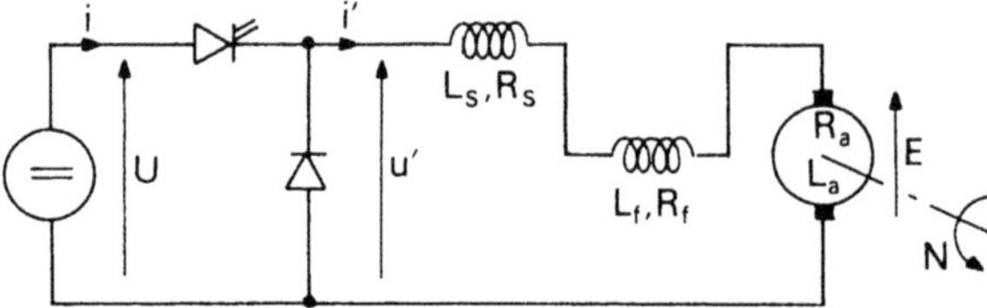

Fig. 4.45

This gives:

$$u' = (R_a + R_f + R_S)i' + (L_a + L_f + L_S)\frac{di'}{dt} + E$$

with

$$u' = U, \text{ for } 0 < \omega t < 2\pi\alpha$$

$$u' = 0, \text{ for } 2\pi\alpha < \omega t < 2\pi.$$

The electromotive force E, developed by the motor, can be deduced from its magnetic characteristic $E(i')$, taken at nominal speed N_{nom}. To simplify this characteristic, the hysteresis has been neglected in Fig. 4.46.

If the characteristic is linearised around the mean value I' of the current, it is possible to take the following value for E at speed N:

$$E = E_{10}\frac{N}{N_{nom}} + k_1' N i'.$$

When substituted in the voltage equation, this yields

$$u' = (R_a + R_f + R_S + k_1' N)i' + (L_a + L_f + L_S)\frac{di'}{dt} + E_{10}\frac{N}{N_{nom}}.$$

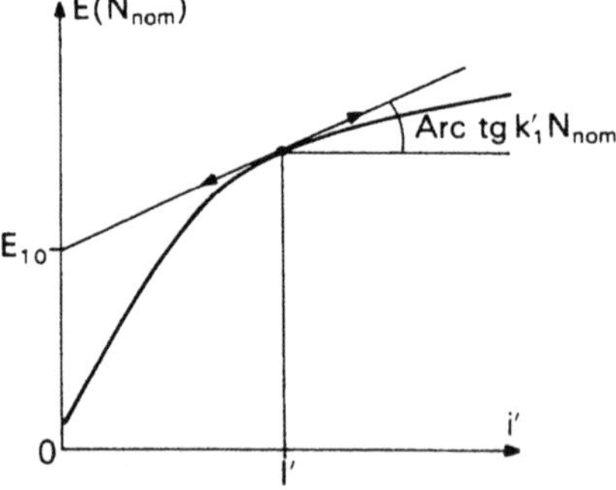

Fig. 4.46

If it is assumed that the rotation speed of the motor remains constant over a switching period, the motor with its smoothing inductor appears as a RLE load. The parameters of the latter depend on the rotation speed N and on the mean armature current I'.

The result noted in the first part of this chapter can thus be used, providing that

$$R = R_a + R_f + R_S + k_1' N$$

$$L = L_a + L_f + L_S$$

$$Q = \frac{(L_a + L_f + L_S)\,\omega}{R_a + R_f + R_S + k_1' N}$$

$$m = \frac{1}{U} E_{10} \frac{N}{N_{nom}}.$$

- The mean values are linked by the intersection of the straight line and the curve of equations

$$U' = \alpha U$$

$$U' = I' R(N, I') + m(N, I').$$

Current ripple in the motor is given by Eq. (4.11) and its maximum value by Eq. (4.12):

$$\Delta i' = \frac{U}{R} \frac{[\exp(2\pi\alpha/Q) - 1]\,[\exp((2\pi - 2\pi\alpha)/Q) - 1]}{\exp(2\pi/Q) - 1}$$

$$\Delta i'_{max} = \frac{U}{R} \tanh\left(\frac{\pi}{2Q}\right).$$

Because of R and Q, these values depend on the mean operating point.

However, for sufficiently high values of Q (in practice, $Q \geqslant 4$), the previous expressions of $\Delta i'$ can be approached by their series expansion limited to the first order. It can thus be noted that the operating point has no further influence on the current ripple.

$$\Delta i' \simeq 2\pi\alpha(1 - \alpha) \frac{U}{(L_a + L_f + L_S)\,\omega}$$

$$\Delta i'_{max} \simeq \frac{\pi}{2} \frac{U}{(L_a + L_f + L_S)\,\omega}. \tag{4.84}$$

The latter equation enables the value given to smoothing inductance to be determined, if the following are known:

- the switching frequency,

- the armature and field winding inductances,
- the maximum tolerable current ripple.

This ripple is usually fixed by the satisfactory commutation of the machine and by the torque pulsation which can be accepted for the relevant application.

4.4.1.2 Choice of Chopper Frequency: Remarks

In Chap. 1, we presented the various time-scales to be considered in analysis of choppers as well as the advantages of frequency increase. This latter point has been largely emphasized in the present chapter.

- When the controlled on/off switch used in the chopper is a thyristor equipped with a turn-off auxiliary circuit, a minimum turn-on time $t_{on, min}$ must be allowed. This may last tens or hundreds of microseconds as it will be seen in Chap. 5.

Consequently, the duration of this interval referred to the switching cycle must be taken into account, unless extremely low frequencies are used. The result is that, at a given switching frequency, the duty ratio α and the mean output voltage cannot be brought below the following values:

$$\alpha_{min} = f \cdot t_{on, min}$$

$$U'_{min} = U \cdot f \cdot t_{on, min} .$$

However, for traction applications, one of the roles of the chopper is to limit the current absorbed by the motor at starting. If I'_{max} is used to denote the maximum acceptable mean current, the following is found

$$(R_a + R_f + R_S) I'_{max} = U'_{min} = U \cdot f \cdot t_{on, min} .$$

The result is that the frequency, at least when run-up begins, must be less than a maximum given by

$$f_{max} = \frac{(R_a + R_f + R_S) I'_{max}}{U t_{on, min}} .$$

The input filter frequency must remain less than half that frequency, as shown in Sect. 4.2.1.

- As the motor speed increases, its EMF increases also and the value of U' and thus of the duty ratio α can equally be increased.

  This increase in α can be obtained in two ways:
  - either by maintaining the frequency at a constant level and increasing turn-on time of the controlled thyristor,
  - or by maintaining the thyristor turn-on time at a constant level and increasing the frequency.

For a given value of U'/U, the value of α is the same for either procedure. But, when f is increased, the second gives

- a lower ripple in current i', for a given smoothing inductance,
- a lower ripple in voltage u and current i_S, for a given input filter.

The second procedure increases the number of commutations and thus the commutation losses. However it usually provides more advantages.

- Nevertheless, in the case of catenary supply, variable frequency operation has, in most cases, to be excluded. The chopper produces interference at the switching frequency and at multiples of this frequency. If a chopper operates at frequency equal to or near to that of the control or signalisation signals transmitted by the catenary, these signals can undergo considerable disturbance.

  However, by use of two or three fixed frequencies which are generally multiples of each other, it is possible to closely approximate the advantages which occur with variable frequency. As soon as the duty ratio reaches a sufficient value, the frequency is changed.

4.4.2 Rheostatic Braking

During the rheostatic braking, the DC machine operates as a generator supplying a resistance Rh (Fig. 4.47). The controlled on/off switch K enables the apparent value of the resistance seen from the generator to be varied. The field winding connections with the armature must be reversed permitting the voltage across the machine to build-up using its residual magnetism.

In analyzing this braking, it will be assumed that controlled switch K is perfect and that speed N does not vary during the switching cycle. In the case of the magnetic characteristic, the residual flux but not the hysteresis will be taken into account.

4.4.2.1 Current Build-up: Operational Boundaries

– If K is permanently on ($\alpha = 1$), the EMF and current are linked by

$$(R_a + R_f + R_S)\,i' + (L_a + L_f + L_S)\frac{di'}{dt} = E;$$

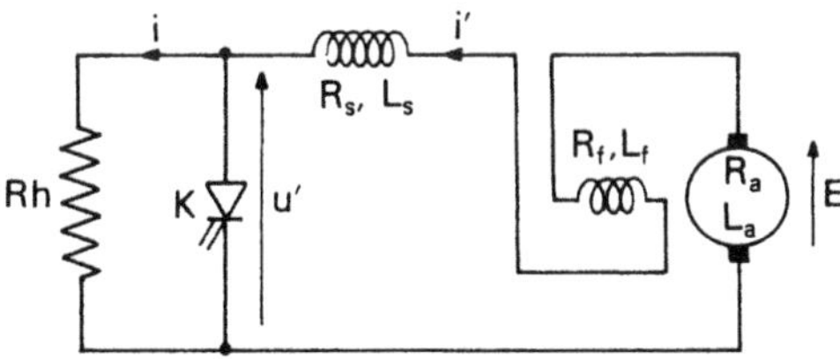

Fig. 4.47

thus,

$$I' = \frac{E}{R_a + R_f + R_S}.$$

– If K is permanently off ($\alpha = 0$), from

$$(R_a + R_f + R_S + \mathrm{Rh})\, i' + (L_a + L_f + L_S)\frac{\mathrm{d}i'}{\mathrm{d}t} = E$$

the following can be deduced:

$$I' = \frac{E}{R_a + R_f + R_S + \mathrm{Rh}}.$$

In Fig. 4.48, the magnetic characteristic of the machine at a rotation speed N has been plotted. The intersection of this with the straight line of slope $R_a + R_f + R_S$ and $R_a + R_f + R_S + \mathrm{Rh}$ gives the two boundary values of the current at this speed:

I'_{max} for K always on (point A),
I'_{min} for K always off (point B).

The current can only build up to a noticeable value if the slope of the straight line corresponding to $R_a + R_f + R_S$ is less than the initial slope of magnetic characteristic $k'_0 N$:

$$k'_0 N > R_a + R_f + R_S$$

$$N > \frac{1}{k'_0}(R_a + R_f + R_S). \tag{4.85}$$

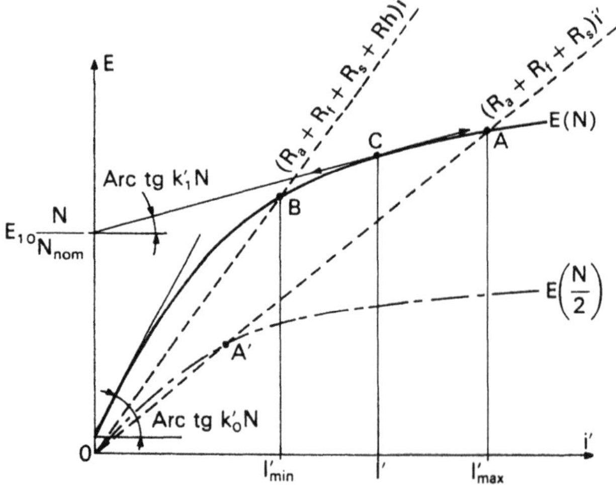

Fig. 4.48

Below a certain speed, the voltage build-up of the series motor is impossible.

Moreover, the closer to this minimal speed, the lower I'_{max} will be. (In Fig. 4.48 the curve $E(i')$ corresponding to $N/2$ has been plotted. The point corresponding to the maximum current passes from A to A'.)

4.4.2.2 Controlling the Mean Current

- The mean value of the current can be controlled by acting on the duty ratio α of the chopper.

  By linearizing the magnetic characteristic around the point corresponding to I' (point C in Fig. 4.48), the following value can be taken for E:

$$E = E_{10} \frac{N}{N_{nom}} + k'_1 N i'.$$

Hence, the equation giving current i' is

$$E_{10} \frac{N}{N_{nom}} - (L_a + L_f + L_S) \frac{di'}{dt} - (R_a + R_f + R_S - k'_1 N) i' = u'; \tag{4.86}$$

with

$u' = 0$, when K is on,
$u' = \mathrm{Rh}\, i'$, when K is off.

The solution of this equation gives:

- for $0 < \omega t < 2\pi\alpha$, (K on):

$$i' = i'_0 \exp\left(\frac{-\omega t}{Q}\right) + \frac{E_{10} N/N_{nom}}{R_a + R_f + R_S - k'_1 N}\left[1 - \exp\left(\frac{-\omega t}{Q}\right)\right]$$

with

$$Q = \frac{(L_a + L_f + L_S)\omega}{R_a + R_f + R_S - k'_1 N}.$$

with i'_0 denoting the value of i' for $\omega t = 0$

- for $2\pi\alpha < \omega t < 2\pi$, (K off):

$$i' = i'_1 \exp\left(-\frac{\omega t - 2\pi\alpha}{Q'}\right) + \frac{E_{10} N/N_{nom}}{R_a + R_f + R_S + \mathrm{Rh} - k'_1 N} \times\left[1 - \exp\left(-\frac{\omega t - 2\pi\alpha}{Q'}\right)\right]$$

with

$$Q' = \frac{(L_a + L_f + L_S)\omega}{R_a + R_f + R_S + \mathrm{Rh} - k'_1 N}.$$

Value i'_1 of current i' at the beginning of this period is linked to i'_0 by

$$i'_1 = i'_0 \exp\left(\frac{-2\pi\alpha}{Q}\right) + \frac{E_{10} N/N_{nom}}{R_a + R_f + R_S - k'_1 N}\left[1 - \exp\left(\frac{-2\pi\alpha}{Q}\right)\right].$$

Figure 4.49 shows the waveforms of current i' and voltage u' (equal to Rhi' when K is off, zero when K is on).

- Rheostatic braking with a given mean current I' is only possible between two speeds:
  - the lowest is obtained with K permanently on. Voltage u' is then zero and Eq. (4.86) gives

$$E_{10}\frac{N}{N_{nom}} - (R_a + R_f + R_S - k'_1 N) I' = 0$$

  - the highest is obtained with K permanently off. Voltage u' is then always equal to Rhi' and Eq. (4.86) gives

$$E_{10}\frac{N}{N_{nom}} - (R_a + R_f + R_S - k'_1 N) I' = \mathrm{Rh} I'$$

Hence the following range of possible speeds:

$$\frac{(R_a + R_f + R_S) I'}{E_{10}/N_{nom} + k'_1 I'} < N < \frac{(R_a + R_f + R_S + \mathrm{Rh}) I'}{E_{10}/N_{nom} + k'_1 I'}. \tag{4.87}$$

- *Remark*: Equation (4.87) shows that Rh must be increased in order to extend the higher limit of the possible braking speeds, for a given value of I'. But this

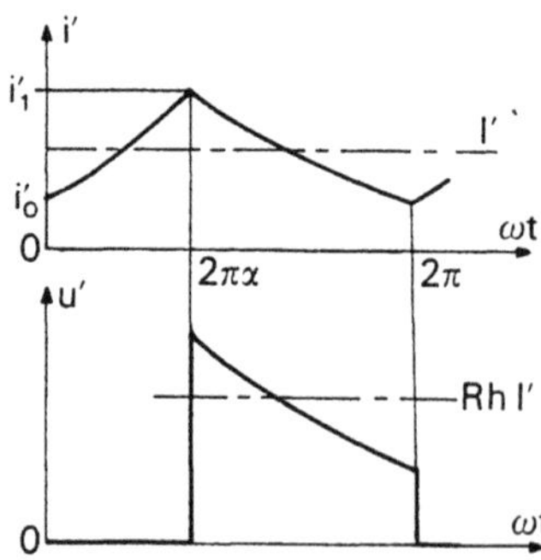

Fig. 4.49

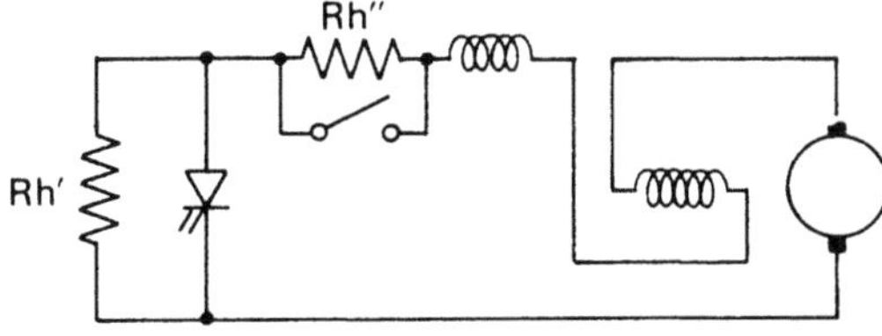

Fig. 4.50

increases the voltages which appear across the machine and the switch, when the latter is turned off. Therefore, the switch is sometimes made to act only on part Rh' of the braking resistance, during high-speed braking (Fig. 4.50). At low speeds, Rh'' is short-circuited.

4.4.3 Regenerative Braking

Braking with energy recovery is achieved by connecting the semiconductor devices in such a way as to obtain a step-up chopper placed between the power supply and the machine (Fig. 4.51). As in the case of rheostatic braking, the connections between armature and field winding must be reversed in relation with motoring operation.

It is clear that operation in this mode is only possible if

- the condition of voltage build-up in generator operation is satisfied by the machine Eq. (4.85);
- the power supply is able to absorb energy. (In the case of a system fed by a catenary, this is only possible insofar as other traction systems simultaneously absorb power at the catenary.)

We are going to examine *the conditions needed for a stable operating mode* to be established, by assuming (Fig. 4.52) that the supply and the switches are perfect and that the rotation speed remains constant during the switching cycle. Our analysis will be limited to operation in continuous conduction.

- Using for voltages and currents the sign conventions denoted in Fig. 4.52 and by linearizing the magnetic characteristic around the point corresponding to the mean current I' delivered, the following equation can be written:

$$u' + (R_a + R_f + R_S - k_1' N)i' + (L_a + L_f + L_S)\frac{di'}{dt} = E_{10}\frac{N}{N_{nom}} \quad (4.88)$$

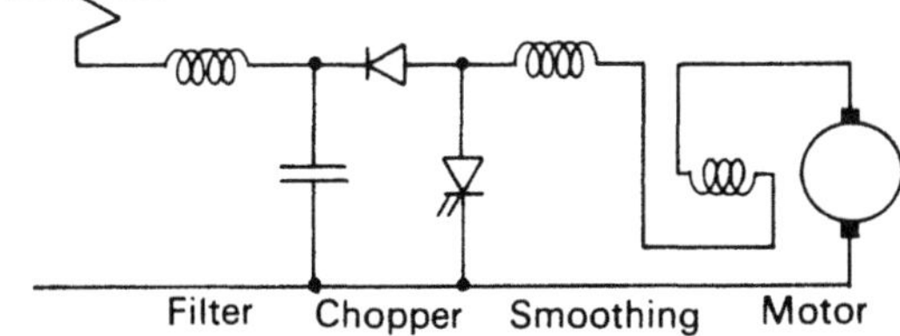

Fig. 4.51

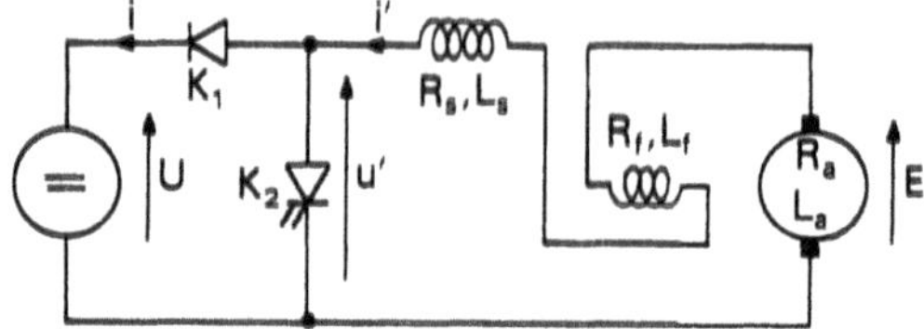

Fig. 4.52

with

$u' = 0$, when K_2 is on and K_1 off,
$u' = U$, when K_2 is off and K_1 on.

If i'_0 (with $i'_0 \geqslant 0$) denotes the value of current i' at instant $t = 0$ when K_2 is turned on, this gives

$$i' = i'_0 \exp\left(\frac{-\omega t}{Q}\right) + \frac{E_{10} N/N_{\text{nom}}}{R_a + R_f + R_S - k'_1 N}\left[1 - \exp\left(\frac{-\omega t}{Q}\right)\right] \tag{4.89}$$

with

$$Q = \frac{(L_a + L_f + L_S)\,\omega}{R_a + R_f + R_S - k'_1 N}.$$

At the end of the on period of K_2, i' reaches the following value:

$$i'_1 = i'_0 \exp\left(\frac{-2\pi\alpha}{Q}\right) + \frac{E_{10} N/N_{\text{nom}}}{R_a + R_f + R_S - k'_1 N}\left[1 - \exp\left(\frac{-2\pi\alpha}{Q}\right)\right]$$

From the moment when K_2 is turned off, in $\omega t = 2\pi\alpha$, current i' is given by

$$i' = i'_1 \exp\left(-\frac{\omega t - 2\pi\alpha}{Q}\right) + \frac{E_{10} N/N_{\text{nom}} - U}{R_a + R_f + R_S - k'_1 N} \times\left[1 - \exp\left(-\frac{\omega t - 2\pi\alpha}{Q}\right)\right] \tag{4.90}$$

- A stable regenerative operation can be obtained only if current i' increases during the period when K_2 is on and decreases during the period when K_2 is off.

- For $0 < \omega t < 2\pi\alpha$,
  the derivative of current i', deduced from (4.89) is given by

$$\frac{\mathrm{d}i'}{\mathrm{d}t} = \frac{1}{L_a + L_f + L_S}\left[E_{10}\frac{N}{N_{\text{nom}}} - (R_a + R_f + R_S - k'_1 N)i'_0\right] \times \exp\left(\frac{-\omega t}{Q}\right).$$

This derivative is positive if

$$(R_a + R_f + R_S)i'_0 < E_{10}\frac{N}{N_{nom}} + k'_1 N i'_0 = E(i'_0).$$

This condition is satisfied as soon as the speed is high enough to permit voltage build-up.

- For $2\pi\alpha < \omega t < 2\pi$,
  Equation (4.90) gives

$$\frac{di'}{dt} = \frac{1}{L_a + L_f + L_S}\left[E_{10}\frac{N}{N_{nom}} - U - (R_a + R_f + R_S - k'_1 N)i'_1\right] \times \exp\left(-\frac{\omega t - 2\pi\alpha}{Q}\right).$$

For this derivative to be negative, there must be:

$$(R_a + R_f + R_S)i'_1 + U > E_{10}\frac{N}{N_{nom}} + k'_1 N i'_1 . \qquad (4.91)$$

The two terms in the second part of this inequality are proportional to the speed. If this condition can be satisfied for the low speeds, the system is naturally unstable at high speed. Current i' tends to increase in such cases, as well when K_2 is on as when K_2 is off.

The same conclusion would be reached by considering the time constant $(L_a + L_f + L_S)/(R_a + R_f + R_S - k'_1 N)$ of the system. This becomes negative as soon as $k'_1 N$ becomes higher than $R_a + R_f + R_S$.

- For the braking with energy recovery to be stable in the case of a series machine, a control loop must be introduced. Equation (4.91) gives the condition which the peak current value in the machine must verify in order to the chopper being able to control the current.

  When the rotation speed becomes too high or the voltage supply falls to a value too low, it may be impossible to satisfy condition (4.91). In this case, the stability can be restored by series-connecting a resistance R' with the machine (Fig. 4.53). This produces mixed braking since the power delivered by the machine is partly dissipated in copper losses and partly recovered by the supply.

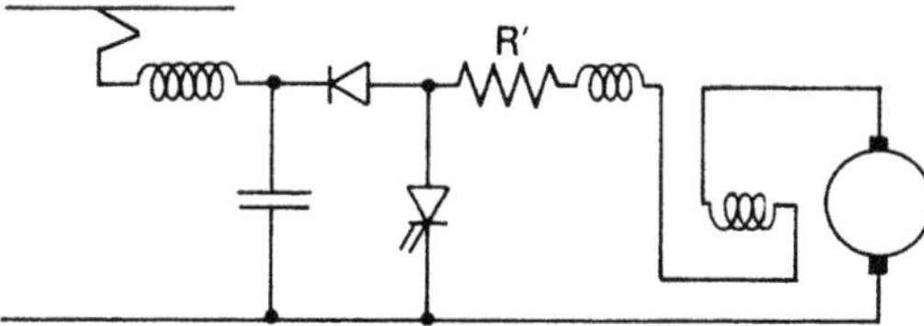

Fig. 4.53

Bibliography

The main use of choppers feeding RL E loads lies in variable-speed drives with DC motors. The majority of publications concerning choppers deal with such drives.

- In the *books* [1–5] dealing with speed variation of DC motors, equal attention is given to equipment with a rectifier powered by an AC supply and to equipment with a chopper powered by a DC supply. In books dealing with variable-speed drives with DC motor and AC motor, the part which concerns the DC motor is normally divided up in the same way.

  With the books we have denoted three articles [6–8] which give an *overall view* of variable-speed drives and which contain many useful references.

- The articles on the chopper-fed DC motor have been divided into two groups:
  - The first one [9–20] deals mainly with *operation and steady-state characteristics*. Many of the articles provide an analysis of the machine with a view to improving its characteristics determination when supplied by a chopped voltage.
  - The second [21–33] deals with *transient analysis*, stability problems as well as those of control and response time. They thus provide details of modelling and simulation techniques.

- Many of articles on variable-speed drives with chopper and DC motor concern two main fields of application.
  - Firstly, *electric traction* [34–46] with locomotives powered either directly in DC or in AC with a rectifier between the catenary and the chopper. These articles are devoted to the different operating modes and braking problems in particular.
  - Secondly, *battery-powered vehicles* [47–52] where the chopper–DC motor system is powered by a battery.

- The other types of articles have been grouped together. In the first, we have placed the studies on *input-current harmonics* [53–60], the disturbances they cause (notably in electric traction), and the input filter. The second group concerns *multiphase choppers* [61–67] and the third group brings together *other studies* [68–74] on certain specific operating modes or combinations.

- Finally, the regular appearance of articles on the use of choppers in equipment with an *AC induction motor* [75–84] should be noted. In some cases, the voltages between slip-rings are rectified and the rectified voltage is applied to a resistance; a chopper parallel-connected across the resistance allows the apparent value of the latter and the motor slip to vary. Other articles concern the chopper placed in the feedback loop of a static Scherbius system; the chopper placed between the rectifier and the inverter enables the performance to be improved.

DC Motor Drives: Books and General Studies

1. Ramshaw R (1973) *Power Electronics. Thyristor Controlled Power for Electric Motors.* Chapman and Hall, London
2. Chauprade R (1975) *Commande Electronique des Moteurs à Courant Continu.* Eyrolles, Paris
3. Sen PC (1981) *Thyristor DC Drives.* Wiley, New York
4. Leonhard W (1985) *Control of Electrical Drives.* Springer, Berlin
5. Dewan SB, Sleman GR, Straughen A (1985) *Power Semiconductor Drives.* Wiley, New York
6. Murphy GJ (1972) Consideration in the design of drive system for the on-the-road electric vehicles. *Proc. IEEE*, **20**(12): 1519–1533 (231 references)
7. Jones BL, Brown JE (1984) Electrical variable-speed drives. *Proc. Inst. Electr. Eng.*, Part A, **131**(7): 516–558 (385 references)
8. Van Wyck JD, Skudelny H Ch, Müller-Hellman A (1986) Power electronics, control of the electromechanical energy conversion process and some applications. *Proc. Inst. Electr. Eng., Part B*, **133**(6): 369–399 (354 references)

Chopper-fed DC Motor: Steady State Analysis

9. Parimelalagan R, Rajagopalan V (1971) Steady-state investigations of a chopper-fed DC motor with separate excitation. *IEEE Trans. Ind. Gen. Appl.*, **7**(1): 101–108
10. Franklin PW (1972) Theory of the DC motor controlled by power pulses. Part 1: Motor operation. *IEEE Trans. Power Appar. Syst.*, **91**(1): 249–255 Part 2: Braking methods, commutation and additional losses. *ibid.*, 256–262
11. Mellitt B, Rashid MH (1974) Analysis of DC chopper circuits by computer-based piecewise-linear technique. *Proc. Inst. Electr. Eng.*, **121**(3): 173–178
12. Doradla SR, Sen PC (1975) Solid state series motor drive. *IEEE Trans. Ind. Electron. Control Instrum.*, **22**(2): 164–171
13. Dubey GK, Shepherd W (1975) Analysis of DC series motor controlled by power pulses. *Proc. Inst. Elect. Eng.* **122**(12): 1397–1398
14. Damle PD, Dubey GK (1976) Analysis of chopper fed series motor. *IEEE Trans. Ind. Electron. Control Instrum.*, **23**(1): 92–97
15. Sornicle D (1978) Commande de moteur à courant continu quatre quadrants. *Rev. Gén. Electr.*, **87**(4): 291–297
16. Singh SN, Kohli DR (1982) Analysis and performance of a chopper controlled separately excited DC motor. *IEEE Trans. Ind. Electron.*, **29**(1): 1–6
17. Satpathi H, Dubey GK, Singh LP (1983) Performance and analysis of chopper fed DC series motor with magnetic saturation, armature reaction and eddy current effect. *IEEE Trans. Power Appar. Syst.*, **102**(4): 981–989
18. Satpathi H, Dubey GK, Singh LP (1983) A general method of analysis of chopper fed DC separately excited motor. *IEEE Trans. Power Appar. Syst.*, **102**(4): 990–997
19. Naik KB, Jain VK, Saxena NS, Dubey GK (1984) Improvement of performance of a chopper fed separately excited motor using LC output filter. *IEEE Trans. Power Appar. Syst.*, **103**(7): 1837–1846
20. Naik KB, Dubey GK, Jain VK (1985) Steady state and dynamic response analysis of a chopper controlled DC separately excited motor. *IEEE Trans. Power Appar. Syst.*, **104**(7): 1750–1782

Chopper-fed DC Motor: Transient Analysis

21. Kamaluddin AK (1974) Analog computer simulation study of a SCR controlled split-field DC series motor and the design criteria for the control circuit. *IEEE Trans. Ind. Electron. Control Instrum.*, **21**(3): 179–185
22. Williams BW (1978) Complete state-space digital computer simulation of chopper-fed DC motor. *IEEE Trans. Ind. Electron. Control Instrum.*, **25**(3): 255–260
23. Dubey GK, Shepherd W (1981) Transient analysis of chopper-fed DC series motor. *IEEE Trans. Ind. Electron. Control Instrum.*, **28**(2): 146–159
24. Rashid MH (1981) Dynamic responses of DC chopper-controlled series motor. *IEEE Trans. Ind. Electron. Control Instrum.*, **28**(4): 323–330
25. Kohli DR, Ahmad SU (1982) Performance of a chopper-controlled DC drive with elastic coupling and periodically varying load torque. *IEEE Trans. Ind. Appl.*, **18**(6): 712–727
26. Strangas EC, Hamilton HB (1983) A model for the chopper controlled DC series motor. *IEEE Trans. Power Appar. Syst.*, **102**(5): 1403–1407
27. Sri-Jayantha M, Hayhoe GF, Henry JJ (1984) SCR-controlled DC motor model for an electric vehicule propulsion system simulation. *IEEE Trans. Ind. Electron.*, **31**(1): 18–25
28. Anjaneyulu PB, Prabhu SS, Dubey GK (1984) Stability analysis, design and simulation of a closed-loop converter-controlled DC drive. *IEEE Trans. Ind. Electron.*, **31**(2): 175–180
29. Blasko V (1985) Model of chopper-controlled DC series motor. *IEEE Trans. Ind. Appl.*, **21**(1): 207–217
30. Hill RJ, Cork P (1985) Chopper control of DC disc-armature motor using power MOSFET's. *Proc. Inst. Electr. Eng., Part B*, **132**(2): 93–99
31. Caro J, Dufour J, Jakubowicz A (1986) A microprocessor-based position control of a DC drive taking into account the load's variations. *IEEE Trans. Ind. Appl.* **22**(6): 982–988
32. Nishimoto M, Dixon JW, Kulkarni AB, Ooi BT (1987) An integrated controlled-current PWM rectifier chopper link for sliding mode position control. *IEEE Trans. Ind. Appl.*, **23**(5): 894–900
33. Hong SC, Park MH (1987) Microprocessor-based high-efficiency drive of a DC motor. *IEEE Trans. Ind. Electron.*, **34**(4):433–440

Choppers in Electric Traction Application

34. Van Eck RA (1971) The separately excited DC traction motor applied to DC and single-phase AC rapid transit systems and electrified railroads. *IEEE Trans. Ind. Gen. Appl.*, **7**(5): 643–657
35. Leroy J, Guibereau S (1973) Le hacheur de courant Késar en traction. Réalisations et perspectives d'avenir pour le Métropolitain. *Rev. Gén. Electr.*, **82**(4): 243–248
36. Cossié A (1973) Compte-rendu d'essais et de mesures à la SNCF sur des engins de traction équipés de hacheurs. *Rev. Gén. Electr.*, **82**(4): 249–253
37. Tsuboi T, Izawa S, Wajima K, Ogawa T, Katta T (1973) Newly developed thyristor chopper equipment for electric railcars. *IEEE Trans. Ind. Appl.*, **9**(9): 294–301
38. Sen PC, Doradla SR (1976) Symmetrical and extinction angle control of solid-state series motor drive. *IEEE Trans. Ind. Electron. Control Instrum.*, **23**(1): 31–38
39. Farrer W (1976) DC-to-DC thyristor chopper for traction application. *Proc. Inst. Electr. Eng.*, **123**(3): 239–244
40. Kitaocha T, Ohno E, Ashiya M, Katsuki K, Katta T (1977) Automatic variable field chopper control system for electric railcars. *IEEE Trans. Ind. Appl.*, **13**(1): 18–25
41. Bailey RB, Williamson DF, Stitt TD (1978) A mode chopper propulsion system for rapid transit. Application with high regeneration capability. *IEEE Trans. Ind. Appl.*, **14**(6): 573–580
42. Brockman JJ, King JH, Kusko A (1980) Rapid transit experience with chopper-controlled DC motor propulsion. *IEEE Trans. Ind. Appl.*, **16**(3): 350–361
43. Bhadra SN, De NK, Chattopadhyay AK (1981) Regenerative braking performance analysis of a thyristor–chopper controlled DC series motor. *IEEE Trans. Ind. Electron. Control Instrum.*, **28**(4): 342–347
44. Allan J, Mellitt B, Pong MH (1984) Rheostatic stabilisation of a chopper-controlled brake for traction drives. *Proc. Inst. Electr. Eng., Part B*, **131**(5): 190–194
45. Allan J, Mellitt B, Wu WH (1985) A Gate-Turn-Off thyristor chopper for traction drives. *Proc. Inst. Electr. Eng., Part B*, **132**(5): 245–250

46. Kemp RJ (1987) Introduction of chopper controlled trains on established DC railways. *Proc. Inst. Electr. Eng., Part B*, **134**(3): 141–147

Choppers in Battery-Powered Vehicles

47. Berman B (1972) Design considerations pertaining to a battery powered regenerative system. *IEEE Trans. Ind. Appl.*, **8**(2): 184–189
48. Berman B (1972) Battery powered regeneratrice SCR drive. *IEEE Trans. Ind. Appl.*, **8**(2): 190–194
49. Berman B (1972) All solid-state method for implementing a traction drive control. *IEEE Trans. Ind. Appl.*, **8**(2): 195–202
50. Benezech J (1983) Le véhicule électrique urbain: les moteurs et le contrôle. *Rev. Gén. Electr.*, **82**(6): 413–428
51. Bose BK, Steigerwald RL (1978) A DC motor control system for electric vehicle drive. *IEEE Trans. Ind. Appl.*, **14**(6): 566–572
52. Steigerwald RL (1980) A two-quadrant transistor chopper for an electric vehicle drive. *IEEE Trans. Ind. Appl.*, **16**(4): 535–541

Input Current Harmonics

53. Feng SYM, Sander WA, Wilson TG (1970) Small-capacitance nondissipative ripple filters for DC supplies. *IEEE Trans. Magnetics*, **6**(1): 137–142
54. Tachibana K, Tsuboi T, Kariya S (1972) Harmonic currents in catenary systems from chopper control. *IEEE Trans. Ind. Appl.*, **8**(2): 203–210
55. Lowe TJ, Mellitt B (1974) Thyristor chopper control and the introduction of harmonic currents into track circuits.. *Proc. Inst. Electr. Eng.*, **121**(4): 269–275
56. Duck EW, Lawrence LS, Lowe TJ (1976) Thyristor chopper control and the introduction of harmonic currents into track circuits. *Proc. Inst. Electr. Eng.*, **123**(6): 523–530
57. Chowdhuri P, Williamson DF (1977) Electrical interference from thyristor-controlled DC propulsion system of a transit car. *IEEE Trans. Ind. Appl.*, **13**(6): 539–550
58. Michel M, Voos H (1980) Frequency and current ripple of DC to DC power converters with respect to different modulation techniques. *IEEE Trans. Ind. Appl.*, **16**(3): 452–457
59. Bonert R, Dewan SB (1980) Filter design with uncontrolled rectifier and DC chopper. *IEEE Trans. Magnetics*, **16**(5): 928–930
60. Steigerwald RL, Hopkins DC (1981) Characteristic input harmonics of DC–DC converters and their effect on input filter design. *IEEE Trans. Ind. Electron. Control Instrum.*, **28**(3): 73–82

Multiphase Choppers

61. Reimers E (1972) Design analysis of multiphase DC chopper motor drive. *IEEE Trans. Ind. Appl.*, **8**(2): 136–144
62. Reimers E (1973) An application of two-phase DC chopper motor drive. *IEEE Trans. Ind. Appl.*, **9**(3): 285–293
63. Mori H, Sawa K, Imura T (1973) Harmonic analysis of chopper controlled electric rolling stock. *IEEE Trans. Ind. Appl.*, **9**(3): 302–309
64. Arockiasamy R, Bhatia CM, Jha CS (1979) Steady-state investigations on a multi-phase chopper–motor system. *IEEE Trans. Power Appar. Syst.*, **98**(2): 379–386
65. Barton TH (1983) The three-phase chopper – Analytic solution. *IEEE Trans. Ind. Appl.*, **19**(6): 1070–1075
66. Rashid MH, Bhadra SN (1985) Filter design for multiphase DC choppers. *Proc. Inst. Electr. Eng., Part B*, **132**(2): 77–80
67. Bhadra SN, Chattopadhyay AK, De NK, Basak PC (1987) Computer-aided analysis of a multiphase chopper-driven DC series motor and test results on a microprocessor-based three-phase chopper drive. *IEEE Trans. Ind. Appl.*, **23**(2): 304–312

Other Studies, Other Combinations

68. Morman WH, Ramsey MH, Hoft RG (1972) 50-kW thyristor DC-to-DC converter. *IEEE Trans. Ind. Appl.*, **8**(5): 617–635
69. Moury P, Schoorens H, Seguier G (1975) Caractéristiques du transformateur élévateur de tension continue utilisant un hacheur en montage parallèle. *Rev. Gén. Electr.*, **84**(1): 4–12
70. Barton TH (1980) The transfer characteristics of a chopper drive. *IEEE Trans. Ind. Appl.*, **16**(4): 489–495
71. Nguyen UT (1982) Hacheur haute tension à décalage série (H.A.D.S.). *Rev. Gén. Electr.*, **91**(6): 35–42
72. Cuk S (1983) New magnetic structures for switching converter. *IEEE Trans. Magnetics*, **19**(2): 75–83
73. Chin SA, Chen DY, Lee FC (1983) Optimization of the energy-storage inductors for DC-to-DC converters. *IEEE Trans. Aerosp. Electron. Syst.*, **19**(2): 203–214
74. Ferrieux JP, Toutain E, Le-Huy H (1986) A multifunction DC–DC converter as versatile power conditioner. *IEEE Trans. Ind. Appl.*, **22**(6): 1037–1043

Choppers in Rotor Circuit of Induction Motor

75. Joshi PR, Dubey GK (1974) Optimum DC dynamic braking control of an induction motor using thyristor chopper controlled resistance. *IEEE Trans. Ind. Electron. Control Instrum.*, **21**(2): 60–65
76. Sen PC, Ma KHJ (1975) Rotor chopper control for induction motor drive: TCR strategy. *IEEE Trans. Ind. Appl.*, **11**(1): 43–49
77. Dubey GK, Pillai SK, Reddy PP (1975) Analysis and design of a doubly fed chopper for speed control of slip-ring induction motors. *IEEE Trans. Ind. Electron. Control Instrum.*, **22**(4): 531–538
78. Pillai SK, Desai KM (1977) A static Scherbius drive with chopper. *IEEE Trans. Ind. Electron. Control Instrum.*, **24**(1): 24–29
79. Wani NS, Ramamoorty M (1977) Chopper controlled slipping induction motor. *IEEE Trans. Ind. Electron. Control Instrum.*, **24**(2): 153–161
80. Ramamoorty M, Wani NS (1978) Dynamic model for a chopper-controlled slip-ring induction motor. *IEEE Trans. Ind. Control Instrum.*, **25**(3): 260–266
81. Sen PC, Ma KHJ (1978) Constant torque operation in induction motors using chopper in rotor circuit. *IEEE Trans. Ind. Appl.*, **14**(5): 408–414
82. Jean Y, Viarouge P, Le-Huy H, Dickinson EJ (1983) Autoadaptative chopper for speed regulation of a wound rotor induction machine. *IEEE Trans. Ind. Appl.*, **19**(6): 1046–1051
83. Taniguchi K, Mori H (1986) Applications of a power chopper to the thyristor Scherbius. *Proc. Inst. Electr. Eng., Part B*, **133**(4): 225–229
84. Taniguchi K, Takeda Y, Hirisa T (1987) High-performance slip-power recovery induction motor. *Proc. Inst. Electr. Eng., Part B*, **134**(4): 193–197

Chapter 5

Forced Commutation of Thyristors

This chapter deals with a comparative analysis of the main types of *turn-off circuit of thyristors* in DC choppers. In Vol. 4, of this series we shall study the types of commutation circuits found in inverters.

- The analysis of turn-off circuit operation is made much easier by the use of a certain number of simplifying *hypotheses*, which normally have little effect on this operation. These hypotheses are based on the difference between the various time-scales which must be taken into account, when studying the converters (see Sect. 1.3).

  In the analysis of turn-off circuits, the time-scale is given by the turn-off time t_q of the thyristors. In the case of fast thyristors used in forced commutation, this time is less than a hundred microseconds.

  - The voltage variations across the voltage generators (or receptors), the current variations through current receptors (or generators) will be ignored, at least during commutation.
  - The rise and fall time of the current in the semiconductor devices, as well as reverse current peaks during turn-off, will be ignored.
  - Phenomena linked to the presence of snubber circuits limiting $\mathrm{d}i/\mathrm{d}t$, $\mathrm{d}v/\mathrm{d}t$ and overvoltage will be ignored. Moreover, such snubbers will not be shown on the diagrams of the analyzed circuits, even though their presence is necessary for the semiconductor devices to operate safely.

- An examination of the main types of turn-off circuits leads to their grouping in classes of circuits whose behaviour is basically the same insofar as the turn-off process of the thyristor is concerned.

  Three basic commutation modes can be distinguished:
  - parallel commutation by capacitor,
  - parallel commutation by oscillating circuit,
  - series commutation.

  - The first mode is often known as "hard commutation". The thyristor is turned off by inserting a capacitor which temporarily diverts the current of the current generator (or receptor) and makes this current flow through a path which is parallel either to the thyristor which has to be turned off, or to the "switch" where this current must be transferred.

- The second mode is often known as "soft commutation" or resonant turn-off. The diverting circuit of the current is a slightly damped oscillating circuit.
- In the third mode, the turn-off circuit connects a voltage in series with the thyristor which is to be turned off. This causes an immediate transfer of the current from the current-source to the "switch" where it has to be transferred.

- This possible classification of turn-off circuits into classes of circuits with identical commutation processes explains the manner in which *analysis will be presented*:
  - First, the operation and basic characteristics for each of the three main commutation modes will be presented. This study will be carried out on the step-down chopper with a turn-off circuit using the mode under consideration. This will enable an overall comparison of the three modes to be made.
  - We will follow this by showing – in the case of the step-down chopper – how each of the above modes defines a class of commutation circuits. These can be deduced from each other.
    - by changing the type of some semiconductor devices,
    - and/or by modifying the topological layout of the elements, without the commutation process being affected.

    An indication will be given as to how the analysis concerning the buck converter can be applied, as a whole, to the boost converter.
  - A final note will concern current-reversible and full-bridge choppers, indicating the particular features of the turn-off circuits in these converters.

5.1 Parallel Commutation by Capacitor

Study of the principle of parallel commutation by capacitor will begin with the circuit shown in Fig. 5.1. It represents a step-down chopper, connected between a DC voltage generator U and a DC current load I'. Thyristor T_1 – the main thyristor – connects the input to the output for a time interval equal to $\alpha_1 T$ during each cycle. The free-wheeling diode D_2 enables current I' to flow when T_1 is off.

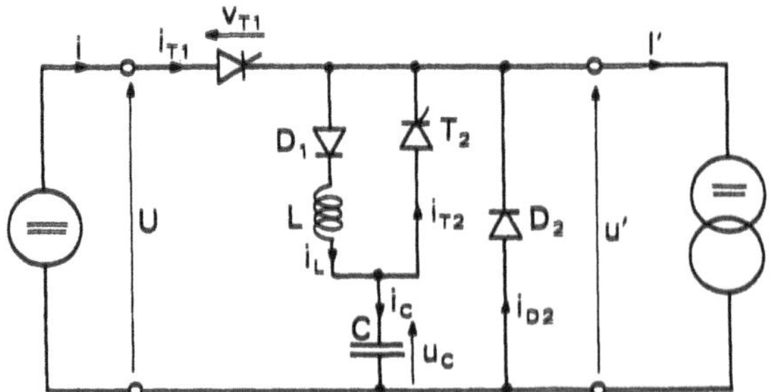

Fig. 5.1

In order to allow for the main thyristor T_1 to act as a turn-on and turn-off controlled switch, a turn-off circuit must be associated with it. It is made up of capacitor C, auxiliary turn-off thyristor T_2, inductor L and diode D_1. Turning T_2 on causes T_1 to turn off by setting the capacitor voltage u_C across the current load. Diode D_1 and inductor L enable the capacitor to charge suitably.

5.1.1 Operation

5.1.1.1 Successive Phases, Waveforms (Fig. 5.2)

The analysis begins at the end of the conducting period of diode D_2: current i_{D_2} is equal to I', thyristor T_1 is blocked and the voltage across it equal to U; currents i_{T_2}, i_C and i_L are equal to zero, as is voltage u_C across the capacitor.

- *Conduction of T_1, Charging of C*
- At the beginning of the switching cycle, for $t = 0$, thyristor T_1 is fired.[1] It diverts load current I' and applies generator voltage U to the load; the voltage across diode D_2 is thus made equal to $-U$ and D_2 is blocked.
- Simultaneously, the firing of T_1 applies, via diode D_1, a voltage step U to the resonant circuit formed by L and C. Current i_C, equal to current i_L, is given by

$$L\frac{\mathrm{d}i_C}{\mathrm{d}t} + \frac{1}{C}\int_0^t i_C\,\mathrm{d}t = U$$

with

$$i_C(0) = 0,\ u_C(0) = 0.$$

This gives

$$u_C = U(1 - \cos\omega t)$$

$$i_C = \frac{U}{L\omega}\sin\omega t \tag{5.1}$$

with ω used to denote the resonant angular frequency of LC circuit:

$$\omega = 1/\sqrt{LC}. \tag{5.2}$$

In the voltage source and in thyristor T_1, current i_C is added to current I'.

- After a half-cycle of oscillation of resonant LC circuit, for $\omega t_1 = \pi$, current i_C becomes zero, and diode D_1 is turned off. From $t = t_1$, voltage u_C is equal to $2U$, and the capacitor is ready to ensure the next turn-off of T_1.

[1] Throughout this chapter, it is assumed that the thyristors are fired on by short pulses, in any case shorter than their conduction time intervals.

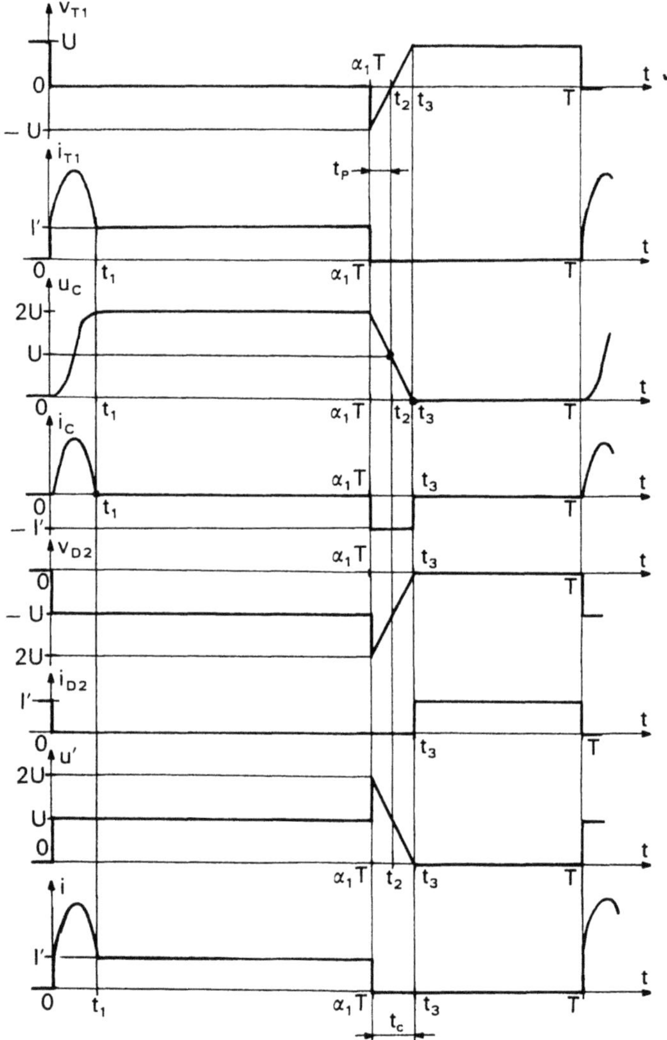

Fig. 5.2

- *Only T_1 conducting*

For t between t_1 and $\alpha_1 T$, thyristor T_1 is the only conducting device:

$$i = i_{T_1} = I'; \quad i_L = i_C = i_{T_2} = i_{D_2} = 0$$

$$u' = U; \quad u_C = 2U.$$

- *Turn-off of T_1, Discharging of C*

– For t equal to $\alpha_1 T$, a firing pulse is sent to the auxiliary thyristor T_2. The latter, which has a voltage $u_C - U$ equal to $+U$ across it, starts to conduct.

Turning T_2 on applies a voltage $U - u_C$ equal to $-U$ across T_1, causing it to turn off. Conduction of T_2 makes the voltage across D_2 equal to $-2U$ and confirms that this diode is blocked.

- Starting from $t = \alpha_1 T$, current I' is diverted by the circuit made up of thyristor T_2 and capacitor C, thus giving

$$u' = u_C = 2U - \frac{1}{C}\int_{\alpha_1 T}^{t} I'\,\mathrm{d}t;$$

voltage u_C decreases linearly:

$$u' = u_C = 2U - \frac{I'}{C}(t - \alpha_1 T). \tag{5.3}$$

- For $t = t_3$, voltage $-u'$ across the free-wheeling diode D_2 falls to zero and tends toward a positive value. D_2 becomes conducting and takes load current I', causing thyristor T_2 to turn-off: commutation is over.

- *Only D_2 conducting*

During the remainder of the switching cycle ($t_3 < t < T$), diode D_2 conducts alone

$$i_{D_2} = I'; \quad i = i_{T_1} = i_L = i_{T_2} = i_C = 0$$

$$u' = u_C = 0; \quad v_{T_1} = U.$$

Figure 5.2 gives the waveforms of v_{T_1} and i_{T_1}, of u_C and i_C, of v_{D_2} and i_{D_2}, of output voltage u' and input current i of the converter. For the circuit under study, i equals i_{T_1} and u' equals $-v_{D_2}$ throughout the cycle.

The duration of the commutations are normally very short compared to the switching cycle. In Fig. 5.2, and in *all the waveforms diagrams* in this chapter, this duration has been expanded. By this means, the various periods which make up a commutation in most cases can be distinguished more clearly.

The intersections which define the various instants have been marked by a large dot. In this case, t_1 corresponds to i_C falling to zero, t_2 to u_C reaching value U, and t_3 to u_C falling to zero.

5.1.1.2 Commutation Time. Reverse-Bias Time

During commutation, the voltage across thyristor T_1 is

$$v_{T_1} = U - u_C = -U + \frac{I'}{C}(t - \alpha_1 T). \tag{5.4}$$

It remains negative until instant $t = t_2$, which is the middle of period $(\alpha_1 T, t_3)$, such that

$$-U + \frac{I'}{C}(t_2 - \alpha_1 T) = 0.$$

This gives the *reverse-bias time* t_p of thyristor T_1:

$$t_p = t_2 - \alpha_1 T = \frac{CU}{I'}. \tag{5.5}$$

This time must be greater than the turn-off time t_q of the thyristor.

The *commutation time* t_c is defined as the time interval which separates the firing of the auxiliary thyristor T_2 from the diversion of current I' by diode D_2:

$$t_c = t_3 - \alpha_1 T = 2\frac{CU}{I'} = 2t_p. \tag{5.6}$$

- *Remarks*

- Time t_1, which is equal to $\pi\sqrt{LC}$, needed to charge the capacitor, gives the minimum time period which must be kept between the firing of T_1 and its turn-off by T_2. This time imposes the minimal value to be given to α_1.
- Similarly, commutation time t_c indicates the minimum time period between the firing of T_2 and the re-firing of T_1; it imposes the maximum value of α_1. Moreover, the quarter of LC circuit free oscillation cycle must be greater than the turn-off time of auxiliary thyristor T_2; indeed, when thyristor T_1 is fired, thyristor T_2 remains reverse-biased only during the period when voltage u_C remains less than U, since v_{T_2} equals $u_C - u'$ when T_1 is on.
- For the calculation and plotting of the waveforms, the ripple of the output current of the chopper has been ignored throughout its cycle T.

  Even if this assumption is unacceptable, the current variations during the commutation time interval $(\alpha_1 T, t_3)$ can be ignored. Time intervals t_p and t_c are then calculated, by taking I' as the value of this current when the thyristor–diode commutation occurs. It has been shown (Sect. 4.1.1.2, final remark) that this is the peak value of current i'.

5.1.1.3 Addition of an Auxiliary Discharge Circuit

Equations (5.5) and (5.6) show that the main thyristor reverse-bias time and the commutation time are inversely proportional to current I'. The dependence of t_p and t_c on I' can be reduced by adding an auxiliary path for discharging the capacitor (Fig. 5.3). This path is made up of inductor L' and diode D_3.

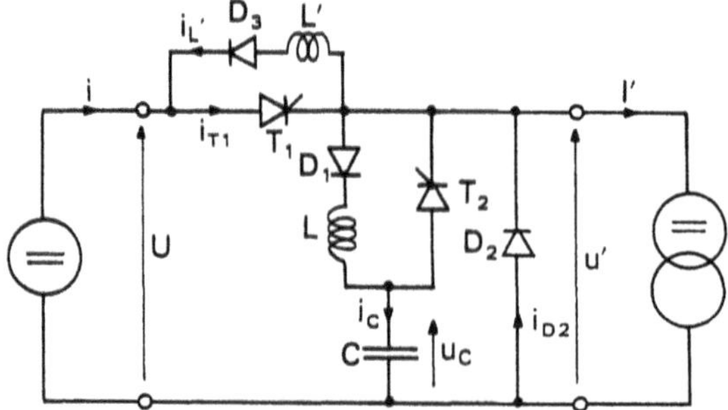

Fig. 5.3

- Adding this auxiliary circuit has no effect on the operation when the main thyristor T_1 is conducting or when the free-wheeling diode D_2 is conducting. This circuit only comes into operation during the T_1–D_2 commutation time period.

From the T_2 firing instant $t = \alpha_1 T$ onwards, the capacitor discharge is now ruled by the equations

$$u_C = u' = 2U - \frac{1}{C}\int_{\alpha_1 T}^{t} (I' + i_{L'})\,\mathrm{d}t$$

$$L'\frac{\mathrm{d}i_{L'}}{\mathrm{d}t} = u_C - U.$$

The following can be deduced:

$$u_C = u' = U(1 + \cos[\omega'(t - \alpha_1 T)]) - L'\omega' I' \sin[\omega'(t - \alpha_1 T)]$$

$$i_{L'} = \frac{U}{L'\omega'}\sin[\omega'(t - \alpha_1 T)] - I'(1 - \cos[\omega'(t - \alpha_1 T)])$$

with

$$\omega' = \frac{1}{\sqrt{L'C}} \tag{5.7}$$

The voltage across thyristor T_1 is expressed as

$$v_{T_1} = U - u_C = -U\cos[\omega'(t - \alpha_1 T)] + L'\omega' I' \sin[\omega'(t - \alpha_1 T)].$$

For $t = t_3$, commutation ends when voltage u_C and current $i_{L'}$ fall simultaneously to zero.

Figure 5.4 shows the waveforms of v_{T_1} and i_{T_1}, of u_C and i_C, of v_{D_2} and i_{D_2}, of u' and i. A comparison with Fig. 5.2 shows that the differences only appear during the commutation interval.

- The reverse-bias time t_p of the main thyristor is equal to the time interval $(\alpha_1 T, t_2)$ which separates the firing of T_2 from the moment when v_{T_1} falls to zero:

$$t_p = \frac{1}{\omega'}\arctan\frac{U}{L'\omega' I'}\cdot \tag{5.8}$$

When I' varies from infinity to zero, t_p no longer goes from zero to infinity but varies from zero to $\pi/2\omega'$.

The commutation time t_c is equal to the time interval $(\alpha_1 T, t_3)$:

$$t_c = 2t_p = \frac{2}{\omega'}\arctan\frac{U}{L'\omega' I'} \tag{5.9}$$

$$0 \leqslant t_p \leqslant \frac{\pi}{\omega'}\cdot$$

This time is always less than the oscillating half-cycle of L′C circuit.

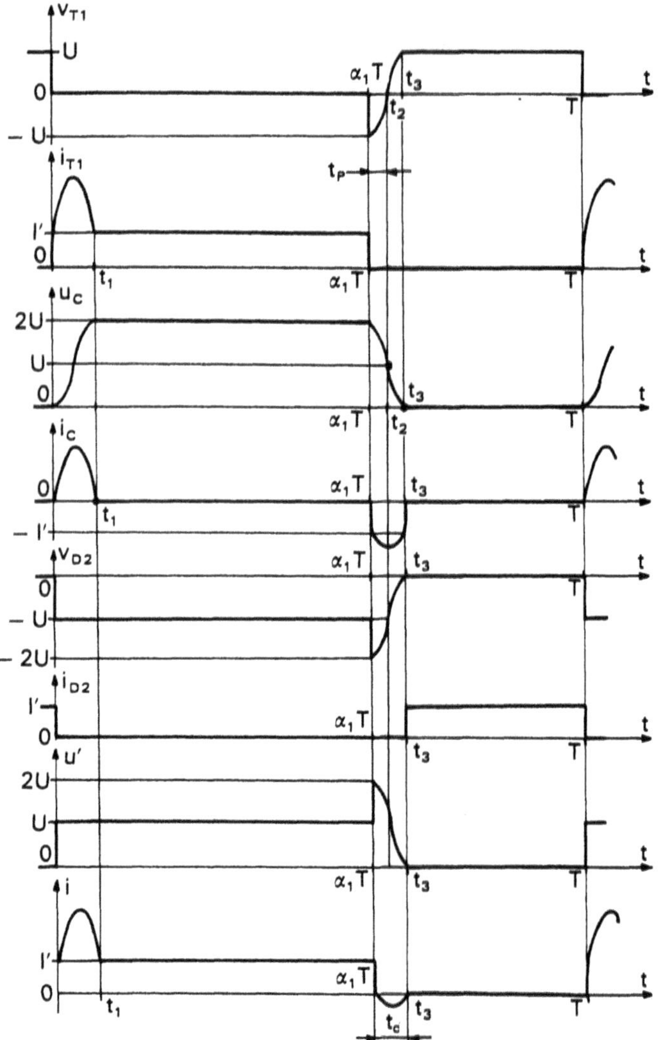

Fig. 5.4

– *Remark*: The preceding equations might be taken as indicating that the topology with an auxiliary discharge path can operate at no load. In fact, this is not the case. Although our analysis has ignored it, damping of the oscillating LC and L′C circuits inevitably occurs.

At each commutation, the current in the load forces the voltage across the capacitor to return to zero, in order for it to be transferred to diode D_2. If I' were zero, there would be a progressive decrease in the amplitude of the voltage u_C oscillations, until the voltage equalled to U. The main thyristor could no longer be turned off when the current in the load would become different from zero.

5.1.1.4 The Anti-Return Diode

When the load can become current-reversible, a diode D_{AR} – usually known as an anti-return diode – is series-connected with the main thyristor after the turn-off circuit (Fig. 5.5). This diode prevents the current of the current source from flowing through the turn-off circuit. In the case of the topology with an auxiliary discharge circuit of the capacitor, this diode prevents the current source from delivering current into the voltage source via this circuit. This current would make any chopping by T_1 impossible.

5.1.2 Effects of the Turn-off Circuit

Adding an auxiliary turn-off circuit modifies the relationships worked out in the simplified analysis of the chopper (Chap. 3) where the effects of this circuit were not taken into account.

5.1.2.1 Influence on the Voltage across the Load

- The turn-off circuit imposes a minimum conduction time for the main thyristor, equal to the oscillation half-cycle of LC circuit. The relative closing time α_1 of thyristor T_1 cannot be less than $\alpha_{1_{min}}$, such that

$$\alpha_{1_{min}} = \frac{t_1}{T} = \frac{\pi}{\omega T}. \tag{5.10}$$

The turn-off circuit similarly imposes a value of α_1 which must not be exceeded. Whenever the oscillation quarter-cycle of LC circuit is greater than the blocking time of T_2, this gives

$$\alpha_{1_{max}} = 1 - \frac{t_c}{T}.$$

As a function of the value of I', this leads to:

- for the chopper without auxiliary discharge circuit,

$$\alpha_{1_{max}} = 1 - \frac{2CU}{I'T}; \tag{5.11a}$$

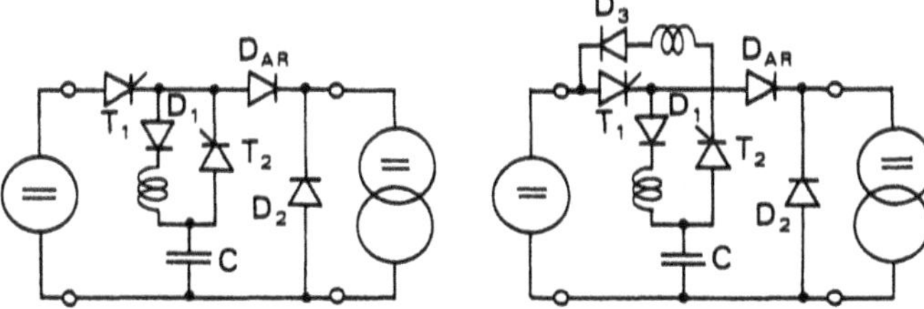

Fig. 5.5

- for the chopper with auxiliary discharge circuit,

$$\alpha_{1_{max}} = 1 - \frac{2}{\omega' T} \arctan \frac{U}{L'\omega' I'} \cdot \tag{5.11b}$$

- Commutation modifies the output voltage u'. It brings about a voltage surge equal to $2U$. Moreover, it slightly modifies the average value U' compared to its theoretical value $\alpha_1 U$. Indeed, as u' is not zero between $t = \alpha_1 T$ and $t = t_3$, U' changes from $\alpha_1 U$ to $\alpha_1 U + \Delta U'$, with

$$\Delta U' = \frac{1}{T} \int_{\alpha_1 T}^{\alpha_1 T + t_c} u' \, dt = U \frac{t_c}{T} \cdot$$

Depending on current I', this gives:

- for the chopper without auxiliary discharge circuit,

$$\Delta U' = \frac{2CU^2}{I'T}; \tag{5.12a}$$

- for the chopper with auxiliary discharge circuit,

$$\Delta U' = \frac{2U}{\omega' T} \arctan \frac{U}{L'\omega' I'} \cdot \tag{5.12b}$$

5.1.2.2 Influence on the Current Supplied by the Generator

- The generator must supply not only current I' during the time interval $(0, \alpha_1 T)$ but also the charge current of the turn-off circuit capacitor at the beginning of this period. This charge current corresponds to a half-sinewave of

  amplitude $\sqrt{\frac{C}{L}}\, U$,

  angular frequency $\frac{1}{\sqrt{LC}}$.

  The peak value of the current supplied by the voltage generator is thus

$$i_{max} = I' + \sqrt{\frac{C}{L}}\, U . \tag{5.13}$$

- In cases where the turn-off circuit has an auxiliary path for discharging capacitor C, the generator must absorb the current which goes through this path during the commutation period. The peak value of this current is

$$I' \left[\sqrt{1 + \left(\frac{U}{L'\omega' I'} \right)^2} - 1 \right] . \tag{5.14}$$

5.1.2.3 Influence on the Semiconductor Device Ratings

- In the main thyristor, the commutation circuit modifies the average value, the r.m.s. value and, especially, the peak value of the current. These values are increased as a result of adding the capacitor charge current.
- The reverse voltage across the main thyristor is no longer zero. This thyristor must sustain a reverse voltage equal to U, thus preventing the use of an asymmetrical thyristor.
- The maximum reverse voltage which diode D_2 has to withstand becomes equal to $2U$. The average and r.m.s. values of the current in this diode are slightly reduced, since I' only flows through it from $t = t_3$ onwards and not from $t = \alpha_1 T$.

5.1.3 Characteristics

- Figure 5.6 indicates the variations in the reverse-bias time t_p, as a function of current I'. The curves have been plotted for several values of the ratio

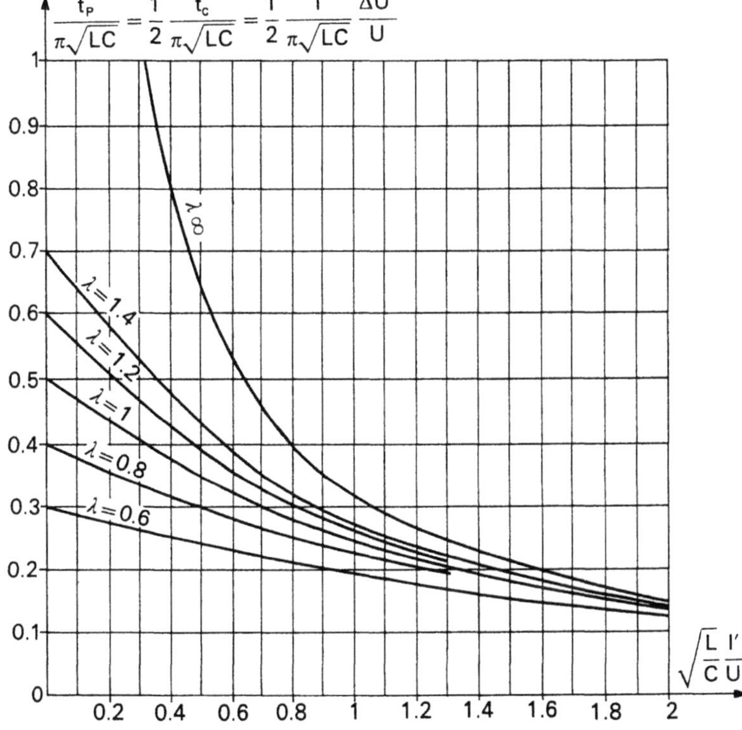

Fig. 5.6

λ between the angular frequency of resonant LC circuit and that of L'C circuit:

$$\lambda = \sqrt{L'/L}.$$

Time t_p is related to the half-cycle, $\pi\sqrt{LC}$, of LC circuit. Current I' is related to the peak value, $U\sqrt{C/L}$, of the capacitor charge current.

The curve plotted for infinite λ gives t_p when there is no auxiliary capacitor discharge circuit.

- The characteristics of Fig. 5.6 also enable the variations in commutation time t_c to be observed, since we have

$$t_c = 2t_p.$$

They also indicate the variations in the increase of the output voltage average value, since

$$\frac{\Delta U'}{U} = \frac{t_c}{T} = \frac{2t_p}{T}.$$

5.2 Parallel Commutation by Oscillating Circuit

The diagram shown in Fig. 5.7 will be used to analyze the principle of parallel commutation by oscillating circuit.

The turn-off circuit comprising inductor L, capacitor C and auxiliary thyristor T_2 has been added to the main thyristor T_1 and the free-wheeling diode D_2. Diode D_1 is used to charge C. A diode D_3 has to be connected in antiparallel across T_1. The circuit comprising resistance R_a and diode D_a is used to end the discharge of C. (We have shown the anti-return diode D_{AR}, which needs only be added if there is a risk of current I' being reversed. In the absence of D_{AR}, the operation at the end of the cycle would be slightly different.)

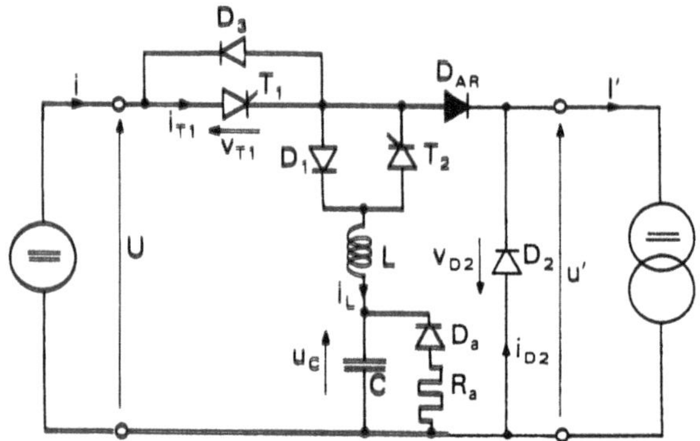

Fig. 5.7

5.2.1 Operation

5.2.1.1 Successive Phases, Waveforms

We start from the following state which, for steady-state operation, is found periodically at the end of the switching cycle:

- only diode D_2 conducts, letting current I' flow. Thyristor T_1 is off and the voltage across it is equal to the supply voltage U.
- voltage u_C and current i_L are zero.

The waveforms of Fig. 5.8 enable to see the time evolution of the main variables: voltage v_{T_1} across T_1 and current i_{T_1}through the latter, voltage u_C and current i_L of the turn-off circuit, voltage v_{D_2} across D_2 and current i_{D_2}, output voltage u' and input current i.

- *T_1 conducting, C Charging*

For $t = 0$, firing T_1 leads to D_2 being turned off and to current I' being diverted through the voltage generator. Moreover, when T_1 is conducting, this enables capacitor C to be charged, via diode D_1, during one half-cycle of oscillating LC circuit.

As in the case of the circuit in Fig. 5.1, this gives:

– for $0 < t < t_1$,

$$u_C = U(1 - \cos \omega t)$$

$$i_L = \frac{U}{L\omega} \sin \omega t \tag{5.15}$$

$$i_{T_1} = I' + \frac{U}{L\omega} \sin \omega t,$$

with $\omega = 1/\sqrt{LC}$ and $t_1 = \pi/\omega$.

– for $t_1 < t < \alpha_1 T$,

$$u_C = 2U; \quad i_L = 0; \quad i_{T_1} = I'.$$

- *Switching off T_1, C starting to discharge*

Firing T_2, for $t = \alpha_1 T$, starts the turn-off process of T_1. This firing causes LC circuit to oscillate once again. (This was interrupted by the turn-off of D_1 at the instant $t = t_1$).

From $t = \alpha_1 T$,

$$\begin{aligned} u_C &= U(1 + \cos[\omega(t - \alpha_1 T)]) \\ i_L &= -\frac{U}{L\omega} \sin[\omega(t - \alpha_1 T)]. \end{aligned} \tag{5.16}$$

A negative half-cycle of i_L begins and the capacitor starts to discharge.

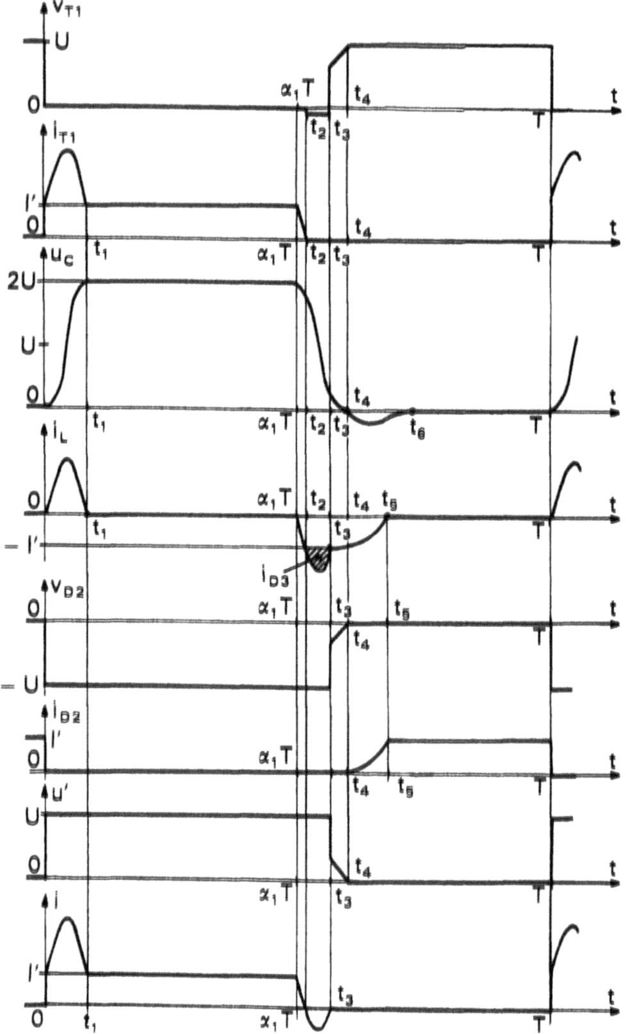

Fig. 5.8

As long as thyristor T_1 is conducting, the diode D_3 is off. The current through T_1 is such that

$$I' = i_{T_1} - i_L, \text{ with negative } i_L$$

The discharge current reduces i_{T_1} and brings it to zero for $t = t_2$. Thyristor T_1 then turns off. The value of t_2 is given by

$$\sin \omega(t_2 - \alpha_1 T) = \frac{L\omega I'}{U} \tag{5.17}$$

with $0 < \omega(t_2 - \alpha_1 T) < \pi/2$.

Switching off of T_1 is only possible if *the peak value of current* i_L *is greater than* I'.

• *D_3 conducting, C continuing to discharge*

From $t = t_2$, oscillation of LC circuit goes on through diode D_3 which diverts $I' + i_L$ (now negative) from T_1. Diode D_3 remains conducting until instant $t = t_3$. At this point, $-i_L$, having reached its peak value, once more becomes equal to I'. t_3 is thus given by

$$\sin[\omega(t_3 - \alpha_1 T)] = \frac{L\omega I'}{U} \tag{5.18}$$

with $\pi/2 < \omega(t_3 - \alpha_1 T) < \pi$.

During the time interval (t_2, t_3), thyristor T_1 is reverse-biased. The reverse voltage across it is equal to the forward voltage drop of D_3. This interval gives the *reverse-bias time* t_p of the main thyristor:

$$t_p = t_3 - t_2 = \frac{2}{\omega} \arccos \frac{L\omega I'}{U} \tag{5.19}$$

with $0 < \omega(t_3 - t_2)/2 < \pi/2$.

At the end of this interval, the voltage across the capacitor is still positive:

$$u_C(t_3) = U\left[1 - \sqrt{1 - \left(\frac{L\omega I'}{U}\right)^2}\right]$$

• *End of C discharging*

From $t = t_3$, all the current I' flows through capacitor C, inductor L and thyristor T_2. The capacitor discharges linearly:

$$u_C = U\left[1 - \sqrt{1 - \left(\frac{L\omega I'}{U}\right)^2}\right] - \frac{I'}{C}(t - t_3). \tag{5.20}$$

As i_L is equal to $-I'$ and is therefore constant, the output voltage is given by

$$u' = u_C + L\frac{di_L}{dt} = u_C$$

The voltage across the main thyristor is

$$v_{T_1} = U - u' = U - u_C$$

The voltages across diodes D_2 and D_a, given by

$$v_{D_2} = -u' = -u_C$$

$$v_{D_a} = -u_C$$

remain negative as long as u_C is positive.

This operation ends for $t = t_4$, when voltage u_C becomes zero. Time interval $t_4 - t_3$ is thus given by

$$t_4 - t_3 = \frac{CU}{I'}\left[1 - \sqrt{1 - \left(\frac{L\omega I'}{U}\right)^2}\right]. \tag{5.21}$$

- *Transfer of current I' to diode D_2*

For $t = t_4$, diodes D_2 and D_a start to conduct. Figure 5.9 shows the diagram for the only part of the circuit where the currents are other than zero.

The transfer of current I' from thyristor T_2 to diode D_2 is ruled by the following equations:

$$i_{D_2} = I' + i_L$$

$$L\frac{di_L}{dt} + u_C = 0$$

$$i_L = C\frac{du_C}{dt} + \frac{u_C}{R_a}.$$

From the second-order differential equation

$$LC\frac{d^2u_C}{dt^2} + \frac{L}{R_a}\frac{du_C}{dt} + u_C = 0$$

with

$$(u_C)_{t=t_4} = 0, \quad \left(\frac{du_C}{dt}\right)_{t=t_4} = -\frac{I'}{C}$$

it is possible to deduce, assuming that the damping is less than the critical damping, that

$$\begin{aligned} u_C &= \frac{-I'}{C\omega\sqrt{1-\zeta^2}}\exp[-\zeta\omega(t-t_4)]\sin[\omega\sqrt{1-\zeta^2}\,(t-t_4)] \\ i_L &= -I'\exp[-\zeta\omega(t-t_4)]\{\cos[\omega\sqrt{1-\zeta^2}\,(t-t_4)] \\ &\quad + \frac{\zeta}{\sqrt{1-\zeta^2}}\sin[\omega\sqrt{1-\zeta^2}\,(t-t_4)]\} \end{aligned} \tag{5.22}$$

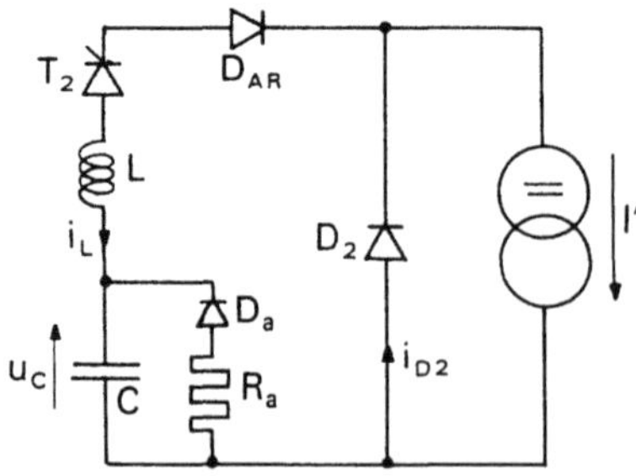

Fig. 5.9

with

$$\zeta = \frac{1}{2R_a}\sqrt{\frac{L}{C}} = \frac{1}{2R_a C\omega}. \tag{5.23}$$

The transfer of current I' to diode D_2 ends, for $t = t_5$, when current i_L becomes zero, thus causing thyristor T_2 (and diode D_{AR}) to be switched off. The duration of this transfer is given by

$$t_5 - t_4 = \frac{1}{\omega\sqrt{1-\zeta^2}} \arctan\left(\frac{-\sqrt{1-\zeta^2}}{\zeta}\right) \tag{5.24}$$

with

$$\pi/2 < \omega\sqrt{1-\zeta^2}\,(t_5 - t_4) < \pi.$$

At the end of this interval, the voltage across the capacitor is negative:

$$u_C(t_5) = -\frac{I'}{C\omega}\exp\left[\frac{-\zeta}{\sqrt{1-\zeta^2}}\arctan\left(-\frac{\sqrt{1-\zeta^2}}{\zeta}\right)\right]. \tag{5.25}$$

• *End of the cycle*

When T_2 is off, diode D_2 lets current I' flow until the beginning of the following cycle. Capacitor C discharges into resistance R_a with time constant $R_a C$:

$$u_C = u_C(t_5)\exp\left[\frac{-(t - t_5)}{R_a C}\right]. \tag{5.26}$$

On the waveforms in Fig. 5.8, it was assumed that u_C becomes negligible from $t = t_6$ onwards and that each new cycle begins with zero voltage u_C. We will discuss this point in further detail in the next section.

(This is true only if there is an anti-return diode. When there is no such a diode, the end of the discharge will contain a damped oscillation across L, D_1 and D_2).

5.2.1.2 Role of the Damping Circuit $D_a R_a$: Choice of Resistance R_a

The damping dipole, comprising resistance R_a series-connected to diode D_a, is intended to bring the voltage across capacitor C back to zero, at the end of commutation. When there is no such dipole, the amplitude of the current oscillations as well as those of the voltage of the turn-off circuit would increase continuously, theoretically, until infinity.

• *Operation without damping circuit*

To illustrate the need for a damping circuit, we shall follow the operation of the chopper when there is no such circuit. In such a case, a negative voltage remains across the capacitor at the moment when the firing of the main thyristor initiates a new switching cycle.

- The voltage u_C at the beginning of the $(n+1)^{\text{th}}$ cycle is denoted $-u_{CO,n}$. During the oscillating charge of C via D_1 and L, which begins for $t = nT$, voltage u_C can be expressed as

$$u_C = \frac{U}{L\omega}\left(1 - \cos[\omega(t - nT)]\right) - u_{CO,n}\cos[\omega(t - nT)].$$

After an oscillating half-cycle of LC circuit, when diode D_1 switches off, the value of voltage u_C is

$$2U + u_{CO,n}.$$

Firing auxiliary thyristor T_2, for $t = \alpha_1 T + nT$, causes the oscillation of resonant LC circuit during which:

$$u_C = U(1 + \cos[\omega(t - nT - \alpha_1 T)]) + u_{CO,n}\cos[\omega(t - nT - \alpha_1 T)]$$

$$i_L = -\frac{U + u_{CO,n}}{L\omega}\sin[\omega(t - nT - \alpha_1 T)].$$

The oscillation goes on until $t = t_3 + nT$, when $-i_L$, after having reached its maximum, takes once more the value I':

$$\sin[\omega(t_3 - nT - \alpha_1 T)] = \frac{L\omega I'}{U + u_{CO,n}}$$

with

$$\frac{\pi}{2} < \omega(t_3 - nT - \alpha_1 T) < \pi.$$

At instant $t = t_3 + nT$, the value of voltage u_C can be expressed as

$$u_C(t_3 + nT) = U - (U + u_{CO,n})\sqrt{1 - \left(\frac{L\omega I'}{U + u_{CO,n}}\right)^2}.$$

This voltage is positive if

$$u_{CO,n} < \sqrt{U^2 + (L\omega I')^2} - U. \tag{5.27}$$

- *If voltage $u_C(t_3 + nT)$ is positive* (type-A commutation in Fig. 5.10), a linear discharge period of C begins in $t = t_3 + nT$ and ends at the instant $t = t_4 + nT$ when u_C becomes zero and diode D_2 starts to conduct.

  The transfer of current I' from the commutation circuit to diode D_2 – from $t = t_4 + nT$ onwards – is ruled by equation

$$L\frac{di_L}{dt} + u_C = L\frac{di_L}{dt} + \frac{1}{C}\int_{t+nT}^{t} i_L\,dt = 0.$$

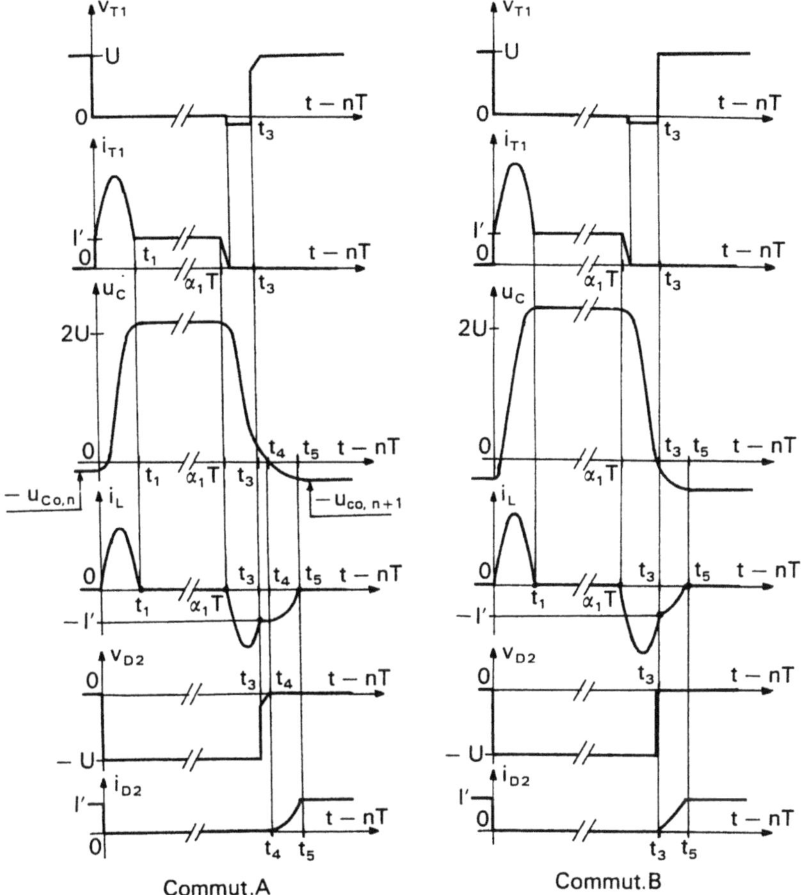

Fig. 5.10

with $i_L(t_4 + nT) = -I'$;
giving

$$i_L = -I' \cos[\omega(t - t_4 - nT)]$$

$$u_C = -L\omega I' \sin[\omega(t - t_4 - nT)].$$

At the end of the transfer for $t = t_5 + nT$, such that $\omega(t_5 - t_4 - nT) = \pi/2$, the value of voltage u_C is

$$-u_{CO,n+1} = -L\omega I' \tag{5.28}$$

which it keeps until the end of the switching cycle.

Since condition (5.27) is satisfied, $-u_{CO,n+1}$ is greater than $-u_{CO,n}$.

- *If voltage* $u_C(t_3 + nT)$ *is negative* (type-B commutation in Fig. 5.10), the transfer of current I' from the commutation circuit to diode D_2 starts from the instant $t = t_3 + nT$ when diode D_3 turns off. From this instant onwards, i_L and u_C are given by

$$i_L = -I' \cos[\omega(t - t_3 - nT)] - \frac{1}{L\omega}\left[U - (U + u_{CO,n})\sqrt{1 - \left(\frac{L\omega I'}{U + u_{CO,n}}\right)^2}\right] \times \sin[\omega(t - t_3 - nT)]$$

$$u_C = -L\omega I' \sin[\omega(t - t_3 - nT)] + \left[U - (U + u_{CO,n})\sqrt{1 - \left(\frac{L\omega I'}{U + u_{CO,n}}\right)^2}\right] \times \cos[\omega(t - t_3 - nT)].$$

Current i_L becomes zero for $t = t_5 + nT$ and the value of voltage u_C is then

$$-u_{CO,n+1} = -\left[2U^2 + u_{CO,n}^2 + 2U[u_{CO,n} - \sqrt{(U + u_{CO,n})^2 - (L\omega I')^2}]\right]^{1/2} \tag{5.29}$$

and u_C remains at this value until the end of the switching cycle.
A study of Eq. (5.29) shows that $|u_{CO,n+1}|$ is necessarily greater than $|u_{CO,n}|$.

- Equation (5.28) and, then, Eq. (5.29) enable the development of the remaining voltage across the capacitor to be followed; theoretically, this increases indefinitely.

  In fact, due to the inevitable damping of the turn-off circuit, the amplitude of the oscillations of voltage u_C – on either side of U – would eventually be stabilized. However, the amplitude of the variations of u_C obtained in this way would probably be much higher than $2U$. This would involve a considerable increase in the size of all the elements of the commutation circuit and thus it is almost essential to add a damping circuit.

- *Choice of resistance R_a*

Resistance R_a determines the damping coefficient ζ of the commutation circuit during the transfer of current I' from thyristor T_2 to diode D_2. Furthermore, this resistance provides the time constant with which the residual voltage subsisting across the capacitor then disappears.

In theory, reducing the residual voltage to zero would require an infinite period of time. In practice, however, the commutation circuit can be considered as having returned to the quiescent state when this voltage has dropped to below 5% of value $L\omega I'$ which it would have in one cycle without a damping dipole.

Figure 5.11 shows, for several values of ζ, the current i_L variations and those of voltage u_C, from instant $t = t_4$ when diode D_a becomes conducting and u_C becomes zero.

For a given value of ω, the decrease of R_a or the increase of ζ increases the duration $t_5 - t_4$ of the transfer of current I'. But the choice of R_a must depend on the total duration of the interval between the moment when u_C becomes zero and its return to a negligible value. If 0.05 $L\omega I'$ is considered to be this negligible value, the minimum of $t_6 - t_4$ can be obtained for

$$\zeta \simeq 0.7$$

and its value is approximately

$$t_6 - t_4 \simeq 3.75/\omega.$$

Under such conditions and taking account of Eq. (5.18) and (5.21) which give

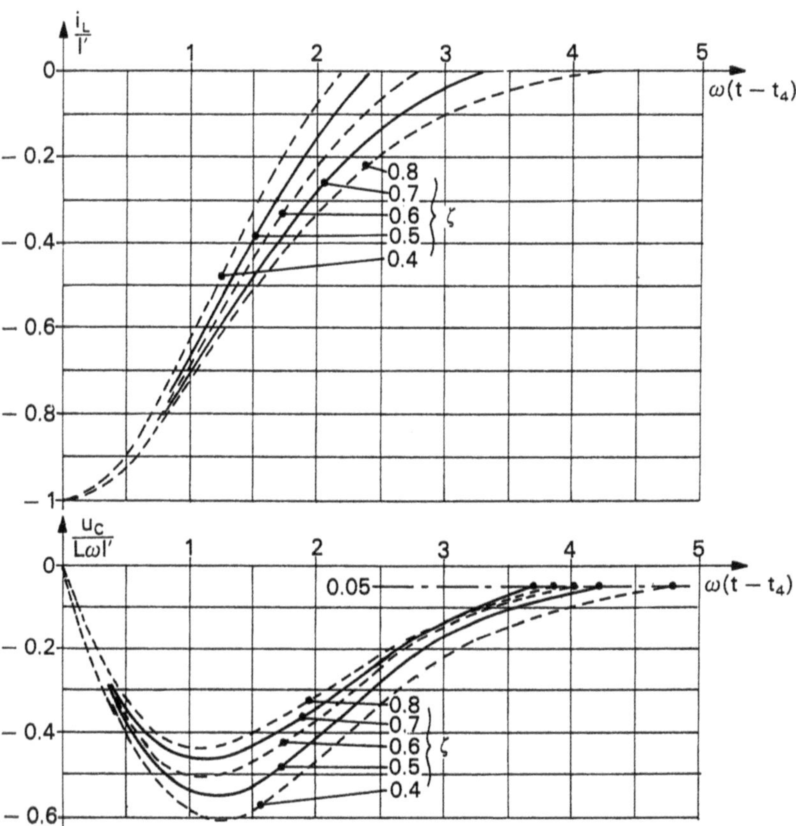

Fig. 5.11

$t_3 - \alpha_1 T$ and $t_4 - t_3$, the value of *commutation time* t_c is

$$t_c = t_6 - \alpha_1 T = \frac{1}{\omega}\left[\pi\text{-arcsin}\left(\frac{L\omega I'}{U}\right) + \frac{1-\sqrt{1-(L\omega I'/U)^2}}{L\omega I'/U} + 3.75\right]. \quad (5.30)$$

- *Remark.* For any value R_a, the energy to be dissipated for each cycle in this resistance (and in the other dissipative elements of the turn-off circuit) is equal to the energy stored in the inductor when u_C becomes zero, i.e.

$$\frac{1}{2} L\, I'^2.$$

5.2.2 Effects of the Turn-off Circuit

5.2.2.1 Influence on the Voltage across the Load

- The turn-off circuit limits the range of possible variations of α_1 and thus of the average value of output voltage $\alpha_1 U$.
  - The minimum value of α_1 is given by the charging time of the capacitor (Eq. (5.10)):

$$\alpha_{1_{\min}} = \frac{t_1}{T} = \frac{\pi}{\omega T}.$$

  - The maximum value of α_1 is given by the duration of commutations:

$$\alpha_{1_{\max}} = 1 - \frac{t_c}{T}$$
$$\simeq 1 - \frac{1}{\omega T}\left[\pi - \arcsin\left(\frac{L\omega I'}{U}\right) + \frac{1-\sqrt{1-(L\omega I'/U)^2}}{L\omega I'/U} + 3.75\right]. \quad (5.31)$$

- Voltage across the load, u', does not fall to zero as soon as $t = \alpha_1 T$ (see Fig. 5.8). The commutation does not bring about any voltage surge, but gives u' an average value which is higher than its theoretical value $\alpha_1 U$. This increase, calculated by

$$\Delta U' = \frac{1}{T}\int_{\alpha_1 T}^{t_4} u'\, \mathrm{d}t$$
$$= \frac{1}{T}\left[\int_{\alpha_1 T}^{t_3} U\,\mathrm{d}t + \int_{t_3}^{t_4}\left(U\left[1-\sqrt{1-\left(\frac{L\omega I'}{U}\right)^2}\right] - \frac{I'}{C}(t-t_3)\right)\mathrm{d}t\right],$$

has a value

$$\Delta U' = \frac{U}{\omega T}\left[\pi - \arcsin\left(\frac{L\omega I'}{U}\right) + \frac{[1 - \sqrt{1 - (L\omega I'/U)^2}]^2}{2L\omega I'/U}\right]. \tag{5.32}$$

5.2.2.2 Influence on the Current Supplied by the Generator

- When the main thyristor is fired, the voltage source U must deliver the charge current of capacitor C, in addition to current I'. The charge current corresponds to a half-sinewave of

  angular pulsation $\omega = \dfrac{1}{\sqrt{LC}}$

  amplitude $\sqrt{C/L}\,U$.

  As in the case of the first commutation mode, the peak value of the current delivered by the supply is (Eq. (5.13)):

  $$i_{max} = I' + \sqrt{C/L}\,U$$

  But, in this case, the following is necessary for T_1 to be turned off by the discharge oscillation:

  $$\sqrt{C/L}\,U > I'; \quad \text{thus,} \quad i_{max} > 2I'.$$

- When the auxiliary thyristor is fired, for $t = \alpha_1 T$, the current through the voltage source falls gradually to zero and this source must then absorb one part of a current sinewave with a peak value

  $$\sqrt{C/L}\,U - I'.$$

5.2.2.3 Influence on the Semiconductor Device Ratings

The commutation circuit causes no transient over-voltage across the load and thus has no effect on the voltage ratings imposed on the main thyristor T_1 and diode D_2.

The average, r.m.s. and peak values of the current in the main thyristor are increased by adding the charge current of the capacitor at the beginning of the ON-state period and by the current's gradually falling to zero at the end of this period.

Conversely, the commutation reduces the average and r.m.s. value of the current in diode D_2.

5.2.3 Characteristics

Depending on current I' which flows through the load, Fig. 5.12 shows the curves which give

the reverse-bias time t_p,

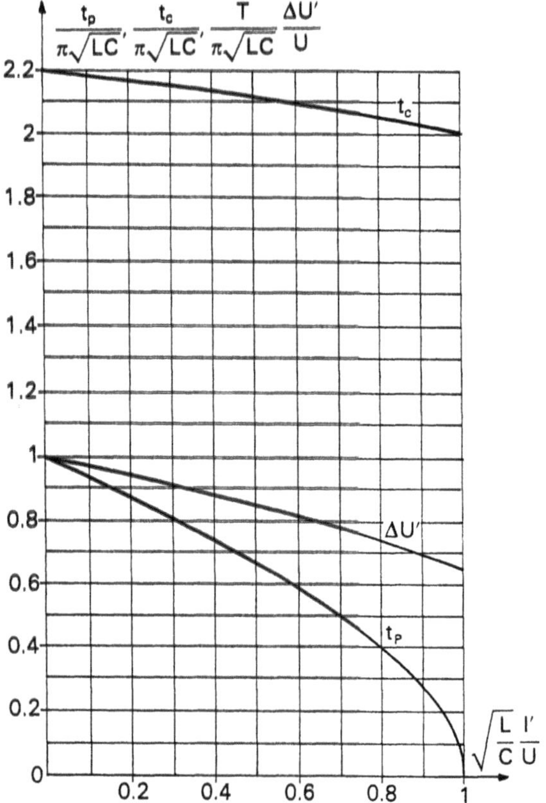

Fig. 5.12

the commutation time t_c,
the variation in the average output voltage $\Delta U'$.

Current I' is referred to the amplitude, $U\sqrt{C/L}$, of the capacitor charge (or discharge) current. The ratio $I'/(U\sqrt{C/L})$ must not be higher than 1, which corresponds to a zero reverse-bias time. If the current ripple in the load cannot be ignored, the value of I' when T_1 is switched off will be considered.

Times t_p and t_c are referred to the oscillation half-cycle, $\pi\sqrt{LC}$, of resonant LC circuit. The commutation time has been computed for ζ equal to 0.7; it shows little variation as a function of I'.

The increase $\Delta U'$ in the average output voltage has been related to U and multiplied by $(T/\pi\sqrt{LC})$.

5.3 Series Commutation: Principle

In presenting turn-off by series commutation, the diagram shown in Fig. 5.13 will be used. Capacitor C and auxiliary turn-off thyristor T_2 are added to main

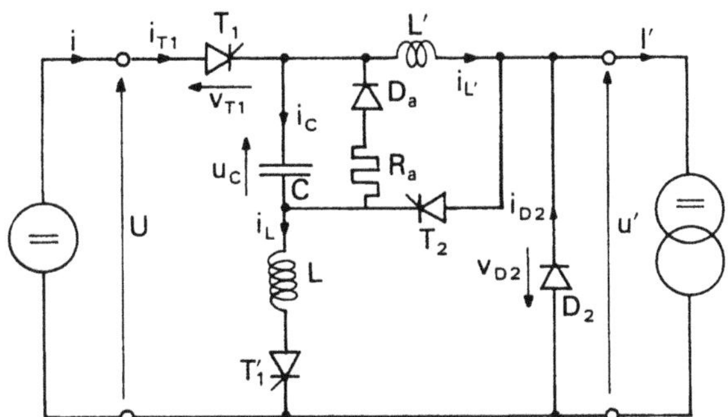

Fig. 5.13

thyristor T_1 and free-wheeling diode D_2. C can thus be directly series-connected with T_1 between the input and the output of the chopper.

An inductor L' must be added to enable T_1 to let I' through during its ON-state period. Together with inductor L, the second auxiliary thyristor T'_1 enables C to be charged. A damping circuit comprising resistance R_a and diode D_a enables voltage u_C to be brought down to zero at the end of commutation.

5.3.1 Operation

An analysis of the operation begins with the following phase which, in steady state, re-occurs at the end of each switching cycle:

- diode D_2 enables load current I' to flow; thyristor T_1 is off; the voltage across it is equal to U;
- voltage u_C across the capacitor, currents $i_{L'}$ and i_L through inductors L and L' are all zero.

Figure 5.14 indicate the time evolution of the main variables.

- *T_1 conducting*

At the beginning of the switching cycle, for $t = 0$, thyristors T_1 and T'_1 are fired simultaneously.

- Via L', the conduction of T_1 applies voltage U across the output terminals. Current $i_{L'}$ increases linearly ($L'\ \mathrm{d}i_{L'}/\mathrm{d}t = U$); current i_{D_2} decreases in the same way.

  For $t = t'_1 = L'I'/U$, the transfer of current I' from diode D_2 to the voltage source is ended: current $i_{L'}$ becomes constant and equal to I', and voltage u' becomes equal to U.

- The simultaneous conduction of T_1 and T'_1 enables the capacitor to be

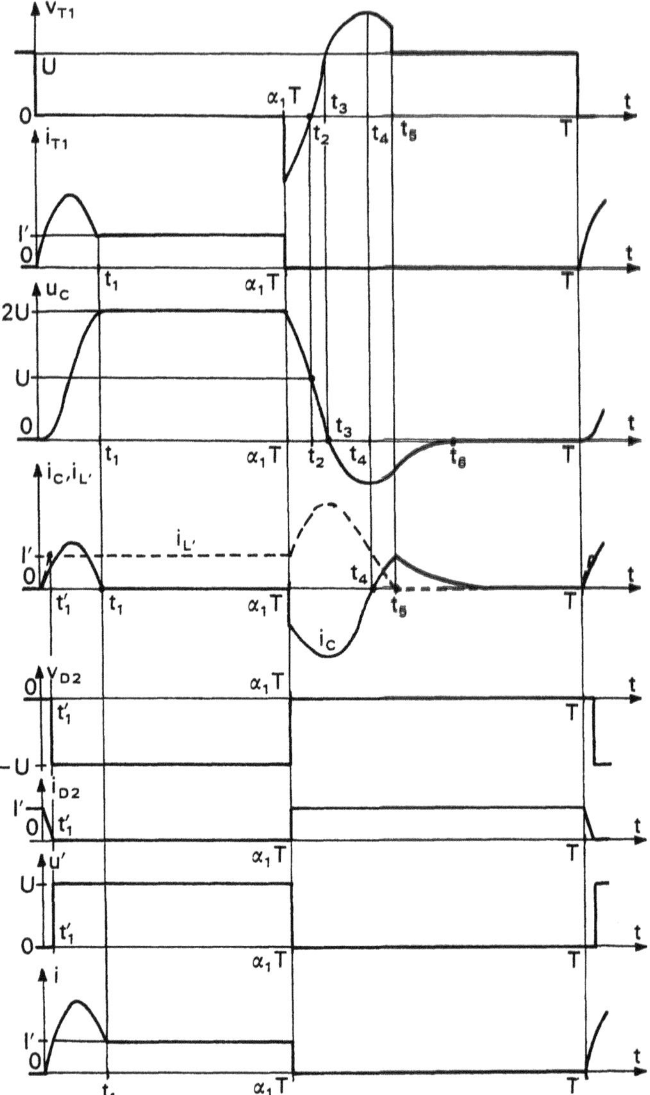

Fig. 5.14

charged via inductor L:

$$u_C = U(1 - \cos \omega t) \tag{5.33}$$

$$i_L = i_C = \frac{U}{L\omega} \sin \omega t$$

with

$$\omega = \frac{1}{\sqrt{LC}}$$

After an oscillating half-cycle, for $t = t_1 = \pi\sqrt{LC}$, thyristor T'_1 turns off: voltage u_C is then equal to $2U$.

- Current i (or i_{T_1}) is equal to $i_L + i_{L'}$. It increases from $t = 0$ onwards, reaches its maximum $I' + (U/L\omega)$, for $t = t_1/2$, and then decreases until it reaches value I'. It keeps this value until instant $t = \alpha_1 T$.

• *T_1 turn-off*

For $t = \alpha_1 T$, the turn-off of T_1 is controlled by firing T_2.

- When T_2 is fired, this applies voltage u_C, equal to $2U$, across inductance L'. Diode D_2 thus becomes conducting and T_1, across which the voltage reaches value $-U$, is turned off. The transfer of current I' from the voltage source to diode D_2 is thus carried out instantaneously.

- Starting from $t = \alpha_1 T$, the conduction of T_2 causes the resonant circuit comprising the capacitor and inductor L' to oscillate.

From $L'\,\mathrm{d}i_{L'}/\mathrm{d}t = u_C$ and $u_C = \dfrac{1}{C}\displaystyle\int_{\alpha_1 T}^{t} (-i_{L'})\,\mathrm{d}t$,

with $i_{L'}(\alpha_1 T) = I'$ and $u_C(\alpha_1 T) = 2U$,

the following can be deduced:

$$u_C = 2U\cos[\omega'(t-\alpha_1 T)] - L'\omega' I' \sin[\omega'(t-\alpha_1 T)]$$

$$i_{L'} = \frac{2U}{L'\omega'}\sin[\omega'(t-\alpha_1 T)] + I'\cos[\omega'(t-\alpha_1 T)]$$

$$\text{with } \omega' = \frac{1}{\sqrt{L'C}} \tag{5.34}$$

- Thyristor T_1 is no longer reverse-biased when voltage u_C becomes equal to U, for $t = t_2$, such that

$$U = 2U\cos[\omega'(t_2-\alpha_1 T)] - L'\omega' I' \sin[\omega'(t_2-\alpha_1 T)].$$

The *reverse-bias time* t_p of the main thyristor, equal to $t_2 - \alpha_1 T$, can be obtained by reformulating the previous equation as follows:

$$1 = 2\frac{1-\tan^2(\omega' t_p/2)}{1+\tan^2(\omega' t_p/2)} - 2\frac{L'\omega' I'}{U'}\,\frac{\tan(\omega' t_p/2)}{1+\tan^2(\omega' t_p/2)}$$

which gives

$$t_p = \frac{2}{\omega'} \arctan\left[\frac{1}{3}\left(\sqrt{\left(\frac{L'\omega' I'}{U}\right)^2 + 3} - \frac{L\omega' I'}{U}\right)\right]. \tag{5.35}$$

- For $t = t_3$, voltage u_C falls to zero. This causes diode D_a to be turned on. At that moment, current $i_{L'}$ reaches its maximum:

$$t_3 - \alpha_1 T = \frac{1}{\omega'} \arctan\left(\frac{2U}{L'\omega' I'}\right). \tag{5.36}$$

$$i_{L'}(t_3) = I_{L'\max} = I' \sqrt{1 + \left(\frac{2U}{L'\omega' I'}\right)^2}. \tag{5.37}$$

- *Inductor L' discharging, end of commutation*

From $t = t_3$ onwards, T_1 and T'_1 are off and D_2 and T_2 are on. In order to follow the evolution of u_C and $i_{L'}$ only the part of the circuit shown in Fig. 5.15 may be taken into consideration:

$$u_C - L' \frac{di_{L'}}{dt} = 0$$

$$i_{L'} = -C \frac{du_C}{dt} - \frac{u_C}{R_a}.$$

The damping is to be less than critical damping. (This is necessary to ensure that T_2 turns off). Taking these initial conditions into account ($i_{L'}(t_3) = I_{L'\max}$, $u_C(t_3) = 0$), one gets

$$u_C = -\frac{I_{L'\max}}{C\omega' \sqrt{1 - \zeta'^2}} \exp[-\zeta'\omega'(t - t_3)] \sin[\omega' \sqrt{1 - \zeta'^2}(t - t_3)]$$

$$i_{L'} = I_{L'\max} \exp[-\zeta'\omega'(t - t_3)] \left(\cos[\omega' \sqrt{1 - \zeta'^2}(t - t_3)] + \frac{\zeta'}{\sqrt{1 - \zeta'^2}} \sin[\omega' \sqrt{1 - \zeta'^2}(t - t_3)]\right)$$

$$\text{with } \zeta' = \frac{1}{2R_a C\omega'}. \tag{5.38}$$

Thyristor T_2 turns off for $t = t_5$, when $i_{L'}$ becomes zero.

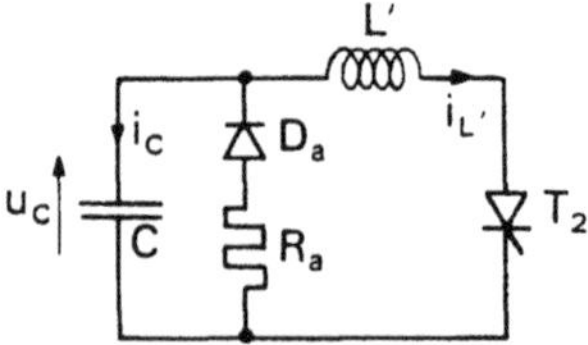

Fig. 5.15

- During time interval (t_3, t_5), for $t = t_4$, the voltage across the capacitor reaches its negative peak value. The instant $t = t_4$, for which the derivative of u_C is zero, is such that

$$\zeta' \sin[\omega' \sqrt{1-\zeta'^2}(t_4 - t_3)] - \sqrt{1-\zeta'^2}$$
$$\times \cos[\omega' \sqrt{1-\zeta'^2}(t_4 - t_3)] = 0.$$

It can thus be deduced that

$$u_{C_{\min}} = u_C(t_4) = -\frac{I_{L'\max}}{C\,\omega'} \exp[-\zeta' \arctan(\sqrt{1-\zeta'^2}/\zeta)]. \tag{5.39}$$

During the whole period when D_2 and T_2 are simultaneously conducting, the voltage across T_1 is given by

$$v_{T_1} = U - u_C.$$

Voltage v_{T_1} thus becomes more than U from $t = t_3$ onwards, and reaches its maximum for $t = t_4$:

$$v_{T_{1\max}} = U - u_{C_{\min}}.$$

- From $t = t_5$ onwards, T_2 is off and the capacitor discharges into resistance R_a. As in the case of parallel commutation by oscillating circuit (see § 2.1.2), commutation is considered to be over when, for $t = t_6$, the voltage is 5% of the value which it would have if R_a were absent. The *commutation time* t_c is minimal for

$$\zeta' \simeq 0.7$$

and then it is approximately equal to

$$t_c \simeq (t_3 - \alpha_1 T) + \frac{3.75}{\omega'}$$

$$t_c \simeq \frac{1}{\omega'} \arctan\left(\frac{2U}{L'\omega' I'}\right) + \frac{3.75}{\omega'}. \tag{5.40}$$

• *Remarks*

As in the case of parallel commutation by oscillating circuit, if there is no damping dipole D_a–R_a, an increase in the amplitude of the current and voltage oscillations of the commutation circuit would be obtained from one cycle to the next. This increase is, theoretically, unlimited but would, in practice, be limited by damping which is inevitably present in resonant LC and L′C circuits. However, it would lead to a considerable increase in component rating.

For each cycle, the dissipated energy in the damping dipole is equal to

$$\frac{1}{2} L' I^2_{L'\max}.$$

5.3.2 Effects of the Turn-off Circuit

5.3.2.1 Influence on the Voltage across the Load

- The turn-off circuit limits the range of possible variations in the relative duration of the controlled switch on-time.

$$\alpha_{1_{min}} = \frac{t_1}{T} = \frac{\pi}{\omega T}$$

$$\alpha_{1_{max}} = 1 - \frac{t_c}{T} \simeq 1 - \frac{1}{\omega' T}\left(\arctan\frac{2U}{L'\omega' I'} + 3.75\right). \tag{5.41}$$

- If voltage u' does fall to zero for $t = \alpha_1 T$, it does not reach value U for $t = 0$ but only for $t = t'_1$. This gives a variation in its average value compared to $\alpha_1 U$:

$$\Delta U' = \frac{1}{T}\int_0^{t'_1}(-U)\,dt = -\frac{L'I'}{T}. \tag{5.42}$$

Unlike as for the previous commutation modes, the average voltage is reduced.

5.3.2.2 Influence on the Current Supplied by the Generator

At the beginning of each switching cycle, the waveform of current is modified as a result of

- the gradual rise of current I',
- the supply of the capacitor charge current via T'_1 and L.

The peak value of the current supplied by the generator is given by

$$i_{max} \leqslant I' + \sqrt{\frac{C}{L}}\,U.$$

Equality occurs when the rise of current I' has a duration equal to or less than a quarter of the oscillating cycle of resonant LC circuit.

5.3.2.3 Influence on the Semiconductor Devices Ratings

In the case of the main thyristor T_1, the presence of the commutation circuit leads to

- an increase in the peak forward voltage which it must sustain;
- the application of a reverse voltage equal to U, thus precluding the use of an asymmetrical thyristor;
- an increase in the peak, average and r.m.s. values of the current, on account of the additional capacitor charge current.

The commutation circuit also leads to a slight increase in the average and r.m.s. values of the current in diode D_2, on account of the non-instantaneous transfer of current I' from D_2 to T_1, at the beginning of the switching cycle.

5.3.3 Characteristics

Figure 5.16 gives the variations in time intervals t_p and t_c and voltage $\Delta U'$, as a function of current I'. The curves have been plotted for some values of the ratio

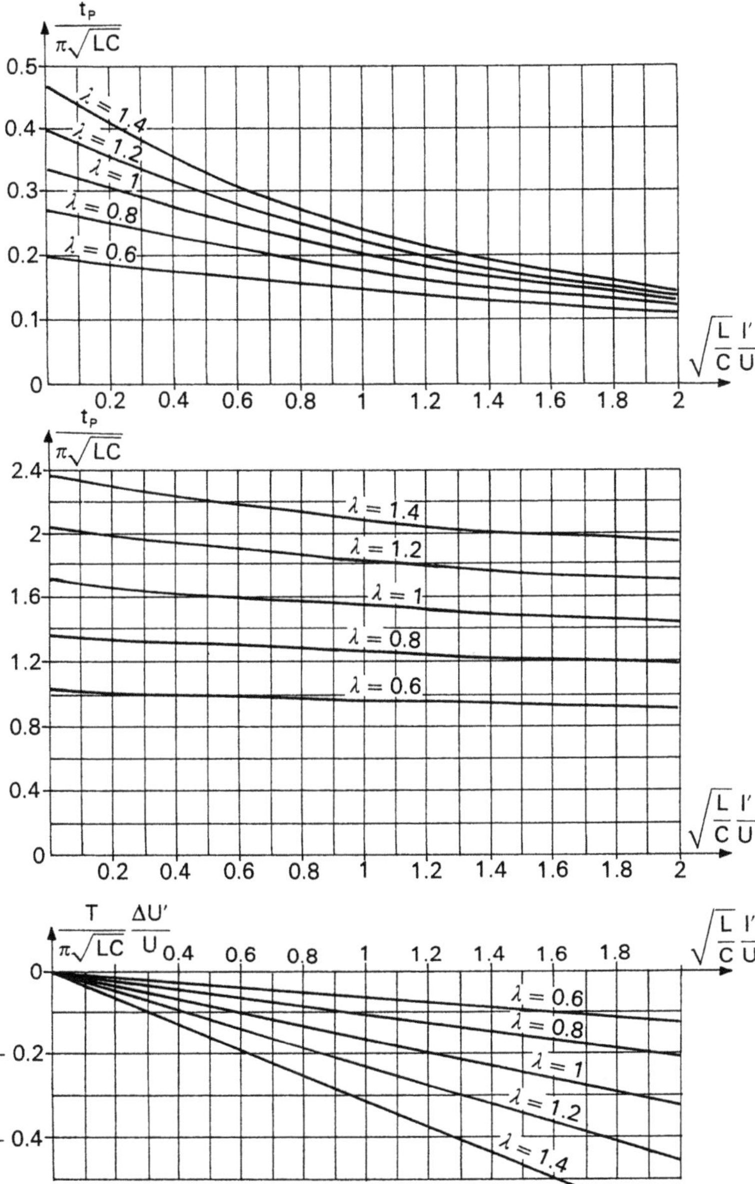

Fig. 5.16

λ between the oscillating angular frequency of resonant LC circuit and that of the L'C circuit:

$$\lambda = \sqrt{\frac{L'}{L}}.$$

Current I' is once more related to the peak value, $U\sqrt{C/L}$, of the charge current of capacitor C.

- *The reverse-bias time* t_p of the main thyristor is related to the oscillating half-cycle, $\pi\sqrt{LC}$, of resonant LC circuit.
- *The commutation time* t_c is similarly related to $\pi\sqrt{LC}$; it is calculated for a value of ζ equal to 0.7.
- *The variation in the average output voltage* $\Delta U'$ is related to voltage U of the generator and multiplied by the ratio of switching cycle T to half-cycle $\pi\sqrt{LC}$. It is proportional to I' and L' and thus to λ^2

5.4 Comparison of the Main Commutation Modes

In comparing the three main commutation modes, a distinction must be made between

- *the intended useful effect*: obtaining a sufficient reverse-bias time for the thyristor which has to act as controlled turn-off switch;
- *parasitic effects*: reduction of the control range, modification of the output voltage, current and voltage overload imposed on the semiconductor devices, and energy losses during commutations.

For all three commutation modes, the duration of the capacitor charge and the peak value of charge current remain the same. All the times have thus been related to $\pi\sqrt{LC}$ and all the currents to $U\sqrt{C/L}$.

The comparison will be based essentially on the characteristics in Figs. 5.8, 5.12 and 5.16.

5.4.1 Reverse-Bias Time

The reverse-bias time t_p of the main thyristor must be longer than its recovery time t_p.

The reduced value of t_p changes

- from 0.5 to 0.15, when normalized current I' changes from 0 to 2, in the case of parallel commutation by capacitor (with $\lambda = 1$);
- from one to zero, when normalized current I' changes from 0 to 1, in the case of parallel commutation by oscillating circuit;
- from 0.33 to 0.13, when normalized current I' changes from 0 to 1, in the case of series commutation (with $\lambda = 1$).

- A clear difference between the parallel commutation by oscillating circuit and the other two modes becomes apparent.
  - For low values of the current, at given values of L and C, it gives a much better reverse-bias time than the other two modes. In other words, it provides the same reverse-bias time with elements L and C with considerably lower values.
  - However, for the higher normalized current values, the other two modes offer further advantages. They provide a reverse-bias time which is longer and much less sensitive to variations in the normalized values of I' (i.e. to variations in the load current and the generator voltage). Furthermore they alone enable operation for normalized I' greater than or equal to 1.

Parallel commutation by oscillating circuit therefore assumes operation at low normalized I'. This corresponds to a high relative value of the current in the commutation circuit. Greater stresses are thus imposed on the semiconductor devices placed in this circuit. Moreover, the current peak supplied by the generator and flowing through the main thyristor when the capacitor is charging is relatively greater.

If parallel commutation by oscillating circuit enables better use to be made of passive components, this is carried out to the detriment of the semiconductor devices which are subjected to greater current stresses.

- As concerns comparison between parallel commutation by capacitor and series commutation, the former always gives a noticeably longer commutation time.

5.4.2 Parasitic Effects

5.4.2.1 Limitation of the Control Range

The minimum acceptable value $\alpha_{1_{min}}$ for the duty ratio is fixed by the oscillating half-cycle of resonant LC circuit. Insofar as $\alpha_{1_{min}}$ is concerned, there is thus no distinction to be made between the three commutation modes.

The maximum duty ratio $\alpha_{1_{max}}$ depends on the commutation time. When the normalized current changes from zero to 1, the normalized value of t_c changes
- from 1 to 0.5, for parallel commutation by capacitor (with $\lambda = 1$),
- from 2.2 to 2, for parallel commutation by oscillating circuit (with $\zeta = 0.7$),
- from 1.7 to 1.53 for series commutation (with $\lambda = 1$ and $\zeta' = 0.7$).

For given values of I', L and C, parallel commutation by capacitor gives by far the best commutation time.

However, if the commutation time is referred to the reverse-bias time, it can be seen that, in the zone of low normalized values of I', parallel commutation by oscillating circuit gives a t_c/t_p ratio which is virtually the same as that in parallel commutation by capacitor.

Series commutation once more gives the least satisfactory performances.

5.4.2.2 Modification of the Average Value of the Output Voltage

- In the case of the variation $\Delta U'$ of the output voltage compared with its theoretical value $\alpha_1 U$, the two parallel commutation modes – by capacitor or by oscillating circuit – are totally different from series commutation.
  - For the first two modes, current I' is not transferred instantaneously from the main thyristor to diode D_2: there is a resulting increase in the average voltage applied across the load.
  - For series commutation, there is no instantaneous rise of current I' through the main thyristor; there is a resulting decrease in the average value of the output voltage.

Series commutation would thus seem well adapted to cases where the average voltage across the load must have a very low value, without the switching frequency having to be excessively reduced.

- The waveform of the output voltage is only affected in a noticeable way in case of parallel commutation by capacitor. The voltage peak $2U$ at the end of the conduction period normally causes few problems.

5.4.2.3 Overloads Imposed on Semiconductor Devices

For given values of L, C and I', the current peak in the main thyristor T_1, caused by the capacitor charge, is the same in all the three commutation modes.

- When thyristor T_1 is switched off, parallel commutation by capacitor applies across it a voltage equal to $-U$. The maximum reverse voltage across diode D_2 is equal to $2U$ and not U. The two current I' transfers – from T_1 to D_2 and from D_2 to T_1 – are theoretically instantaneous.
- Parallel commutation by oscillating circuit is "softer". The reverse voltage applied across T_1 in order to turn it off is low. The reverse voltage across D_2 does not exceed U. The current turn-off in T_1 takes place gradually.
- At turn-off, the same stresses are imposed on thyristor T_1 by series commutation as by parallel commutation by capacitor. But, for series commutation, the forward voltage across T_1 is greater than U. This requires the voltage rating of this device to be increased. The maximum reverse voltage across D_2 remains equal to U.

Overall, this comparison of the different stresses indicates the disadvantages of series commutation.

5.4.2.4 Commutation Losses

The main disadvantage of parallel commutation by oscillating circuit and series commutation is that both require the energy stored in an inductor to be dissipated for each cycle.

The value of this energy is

$$\frac{1}{2} L I'^2, \text{ for parallel commutation by oscillating circuit,}$$

$$\frac{1}{2}L'I'^2\left[1+\left(\frac{2U}{L'\omega'I'}\right)^2\right], \text{ for series commutation.}$$

This drawback is thus greater in the latter commutation mode.

5.4.3 Concluding Remarks

It is difficult to draw any general conclusions from this comparison of the main commutation modes. As will be seen, this difficulty is increased since each of the basic configurations studied above may lead to modified circuits with specific properties.

It can however be noted that:

1. *Series commutation* requires the addition of an inductor through which flows a current with a high r.m.s. value, since this inductor is connected in the link periodically established between the generator and the load.

   Moreover, all except one of the criteria used in the comparison are unfavourable to this commutation mode. Its use is limited to a few specific applications.
2. The advantages of *parallel commutation by oscillating* circuit are increased and the drawbacks lessened when there is an increase in the relative value of the current oscillation supplied by the oscillating circuit.

   This commutation mode enables the passive components – capacitor and inductor – to be reduced, providing that the stresses on the semiconductor devices are increased.
3. The *parallel commutation by capacitor* has the same types of application as the previous commutation mode. It requires larger commutation components; but it reduces the increase in the semiconductor device ratings, especially in current.

5.5 Modification of the Forced Commutation Circuits in the Case of the Step-Down Chopper

As indicated at the beginning of this chapter, each of the commutation modes studied defines a class of circuits which can be deduced from each other by modification.

- of the type of semiconductor switches,
- of the topological layout of the commutation circuit elements, without there being any effect on the turn-off process of the main thyristor.

Starting from circuits used in the analysis of the general properties of the three main commutation modes, it is thus possible to find the majority of the turn-off circuits which are used. These can be compared, within each class, concerning the advantages and drawbacks of the different variants.

Firstly the modification will be presented and discussed in the case of the step-down chopper or buck converter. We will show then that this analysis can be applied to the case of the step-up chopper or boost converter.

The modifications are presented in the following order:

- type of semiconductor switches,
- position of the elements:

this order is normal inasmuch as certain topological modifications are linked to changes in the type of semiconductor switches.

We have chosen to leave out the minor modifications (the permutations of two elements directly series-connected, for example) in order to concentrate on modifications which affect certain aspects of the operation.

5.5.1 Modification of the Type of Semiconductor Switch

5.5.1.1 Replacing a Diode by a Thyristor

A diode can always be replaced by a thyristor as long as a suitable control is used for the thyristor.

Starting with the circuit in Fig. 5.3, it is possible to obtain the three circuits in Fig. 5.17 (thyristors replacing diodes are drawn in black).

Starting with the circuit in Fig. 5.7, it is possible to obtain that in Fig. 5.18.

(The diagram in Fig. 5.13 has no diode in the turn-off circuit and thus the thyristor-diode replacement is irrelevant.)

- In the case of the circuit using parallel commutation by capacitor, in which thyristor T'_3 replaces diode D_3 (Fig. 5.17b and 5.17c), the *capacitor discharge auxiliary circuit* can be either introduced or not, depending on whether T'_3 is controlled at the same time as T_2 or not.

  This enables this circuit to be removed as soon as current I' has reached a value such that it is no longer necessary to shorten the commutation time (infinite value of λ in Fig. 5.6). The advantage of this possibility lies in the fact

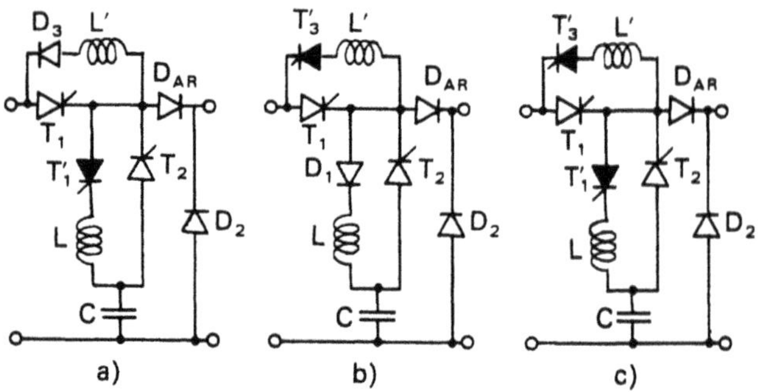

Fig. 5.17

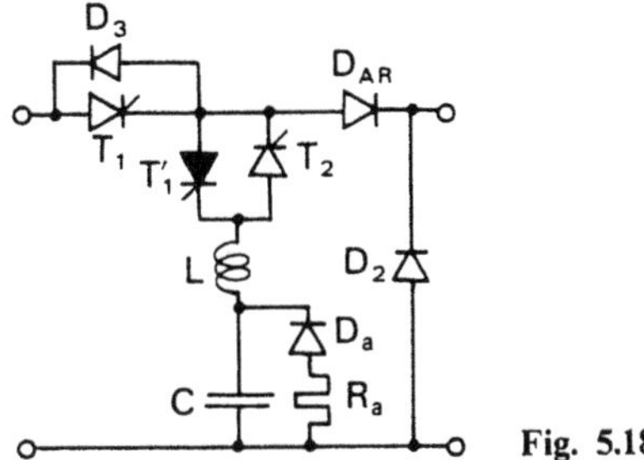

Fig. 5.18

that the removal of the auxiliary discharge circuit leads to an increase in the reverse-bias time of the main thyristor.

- In the circuits in Fig. 5.17a, 5.17c and 5.18, capacitor charging diode D_1 is replaced by thyristor T'_1. This enables the *charging half-cycle of the turn-off capacitor to be moved* within the conduction period of the main thyristor, providing that T_1 and T'_1 are not fired simultaneously.

  Figure 5.19 represents this possibility in the case of
  - parallel commutation by capacitor,
  - parallel commutation by oscillating circuit,
  - series commutation.

  If T'_1 is fired in $t = \alpha_1 T - \pi\sqrt{LC}$, the turn-off of T_1 begins as soon as capacitor charging is over.

The interest of this control appears if T_1 is required to be kept conducting for a relatively long period of time: it avoids the capacitor discharging during this period via the inevitable leakage conductances.

5.5.1.2 Replacing a Thyristor by a Diode

Replacing a thyristor by a diode requires that the thyristor be not subjected to any forward bias during the switching cycle.

- Such is the case of the thyristor T_2 in the circuit in Fig. 5.18, when the turn-on of thyristor T'_1 is controlled for $t = \alpha_1 T - \pi\sqrt{LC}$.

  If thyristor T_2 is replaced by diode D'_2 (Fig. 5.20), turning on T'_1 starts the turn-off process of T_1. In comparison with the operation of the circuit in Fig. 5.7, this brings about an extra delay, equal to the oscillating half-cycle of resonant LC circuit, between the starting of the turn-off process of T_1 and the beginning of the transfer of current I' to diode D_2. This delay can cause problems when working with a theoretical conduction time of T_1, which is close to the oscillation half-cycle of the turn-off circuit.

- With the circuit in Fig. 5.7, if thyristor T_2 is controlled in $t = \pi\sqrt{LC}$ (i.e. at the end of the charge period of capacitor C) it is not subjected to any reverse

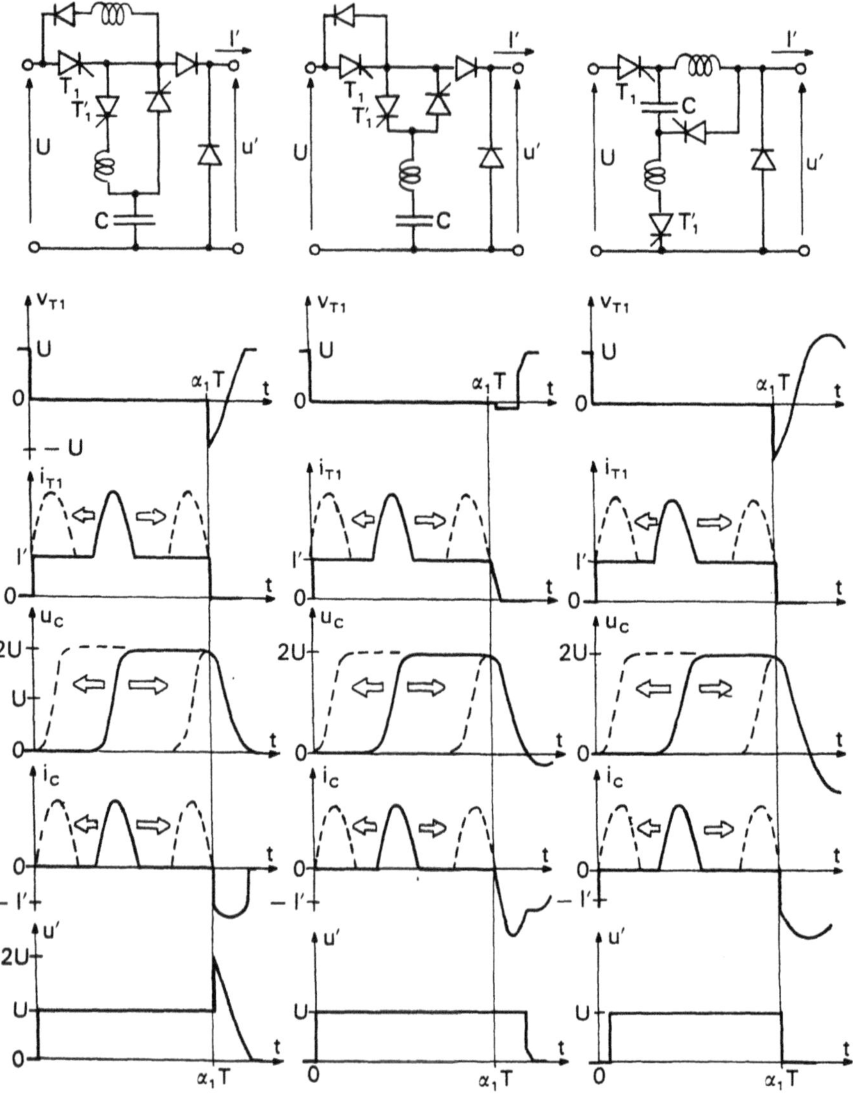

Fig. 5.19

bias. It can thus be replaced by a diode. The latter, together with diode D_1, are connected in antiparallel and act as a short-circuit. In replacing them by a direct link between inductor L and the cathode of T_1, the result is the "*single-thyristor*" chopper (Fig. 5.21) used in certain low power applications.[1]

[1] Circuit in Fig. 5.21 is the basic diagram of switch-mode power supplies using "resonant switch". The single controlled switch is often a transistor operating as a thyristor, i.e. its controlled turn-off ability is not used.

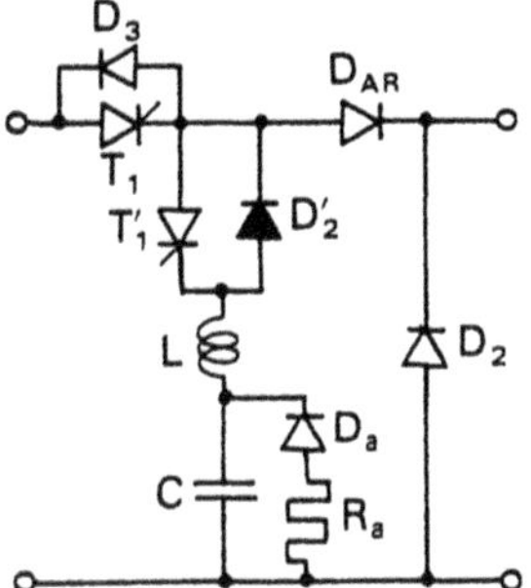

Fig. 5.20

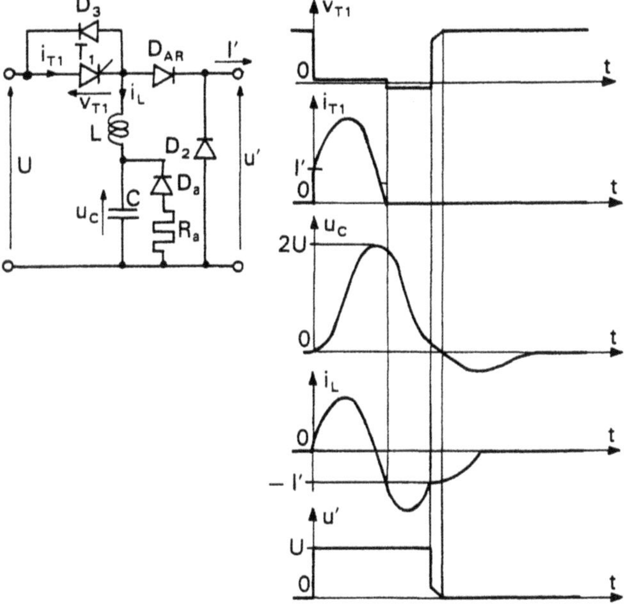

Fig. 5.21

In this configuration, the conduction time of the main thyristor cannot be controlled: a variation in the duty cycle can only be obtained by varying the operating frequency.

- *Remark:* The study of the modifications in the type of semiconductor device clearly indicates that the possibilities of control are directly linked to the number of thyristors in the turn-off circuit. This will be confirmed by the study of the modification in the layout of the elements.

5.5.2 Modifications in the Topological Disposition of the Elements

5.5.2.1 First Modification

The first modification in the disposition of the elements concerns the circuits where a thyristor T'_1 is used for the charge of the turn-off capacitor. It consists in *no longer making the charge current flow through the main thyristor* T_1.

For the circuit using parallel commutation by capacitor it is possible to change from the configuration in Fig. 5.17a to that in Fig. 5.22a, by moving the anode connection of thyristor T'_1 from the cathode of T_1 to its anode.

In the case of the circuit using commutation by oscillating circuit, the same change in connection makes the circuits in Fig. 5.18 or 5.20 change to those in Fig. 5.22b or 5.22c.

For the series commutation circuit, moving the whole of commutation circuit from after T_1 to before it brings about the change from Fig. 5.7 to Fig. 5.22d.

This modification causes a decrease in the peak value of the current in the main thyristor. The average and r.m.s. values of this current decrease to an extent which varies with the frequency and the duty cycle.

For the topology shown in Fig. 5.22c the firing of T'_1 once more starts the turn-off process of T_1. However, for the other three circuits in Fig. 5.22, the modification enables the turn-off *capacitor to be charged during the period when the main thyristor is off.* This enables the minimum duty ratio $\alpha_{1_{min}}$ to be reduced

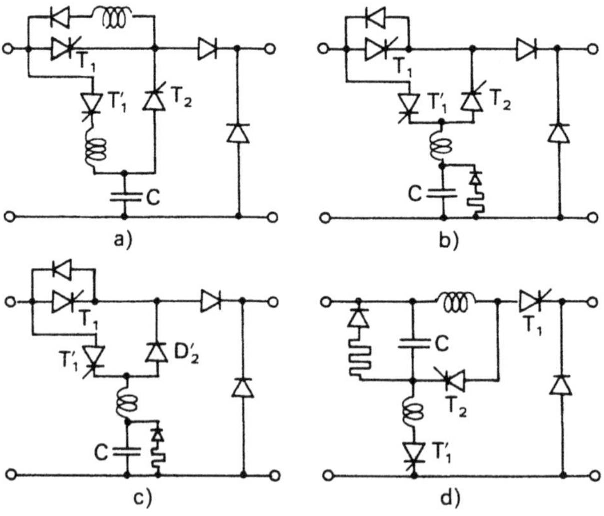

Fig. 5.22

to zero. This is of particular interest in the case of the series commutation circuit, since the transfer of current I' from thyristor T_1 to diode D_2 is instantaneous. The output voltage can then be reduced to zero.

5.5.2.2 Second Modification

The second modification in the position of the elements can *only be applied to parallel commutation circuits.* It consists in *moving the terminal of the turn-off capacitor* which is connected to the voltage generator, from the negative terminal to the positive terminal of this generator.

- For example, when applied to the parallel commutation by capacitor circuit shown in Fig. 5.3, this modification gives the diagram shown at the top of Fig. 5.23. When applied to the circuit using parallel commutation by oscillating circuit shown in Fig. 5.7, it provides the configuration in the upper part of Fig. 5.24. In order to show the effects of this modification, Figs. 5.23 and 5.24 provide the waveforms of the main variables (to be compared with Figs. 5.4 and 5.8).

- With this modification, when the main thyristor is switched off, the current flowing through the latter is diverted to a path parallel to the main thyristor instead to a path parallel to diode D_2.

  The voltage u_C across the turn-off capacitor is shifted by a quantity equal to $-U$. Instead of oscillating between 0 and $+2U$, it oscillates between $-U$ and $+U$; the maximum energy stored in the capacitor is thus divided by 4.

- The voltage generator no longer has to supply the charge current of the capacitor, since this charge can be obtained by reversing voltage u_C in a half-oscillation of resonant LC circuit. The generator has only to supply current I' during time interval $(0, \alpha_1 T)$ and, during the commutation interval of T_1, the difference between I' and the current flowing through diode D_2. For circuits using parallel commutation by oscillating circuit, the generator must also take the current going through the damping circuit at the end of commutation.

  The result is a less disturbed waveform of the current supplied by the generator. This has a positive effect on the behaviour of the input filter which is normally placed at the chopper input.

- The two clear advantages presented by this modification explain its frequent use. One precaution, however, must be taken: as the capacitor charge is obtained by reversing the voltage across it, *the capacitor must be precharged* before the circuit is started. This can be carried out, through the load, by previously firing the turn-off thyristor T_2 or by adding a specific auxiliary circuit.

5.5.2.3 Other Possible Modifications

- In the circuit using parallel commutation by oscillating circuit, the damping dipole can be eliminated by connecting diode D_2 in parallel across the turn-off capacitor C (Fig. 5.25).

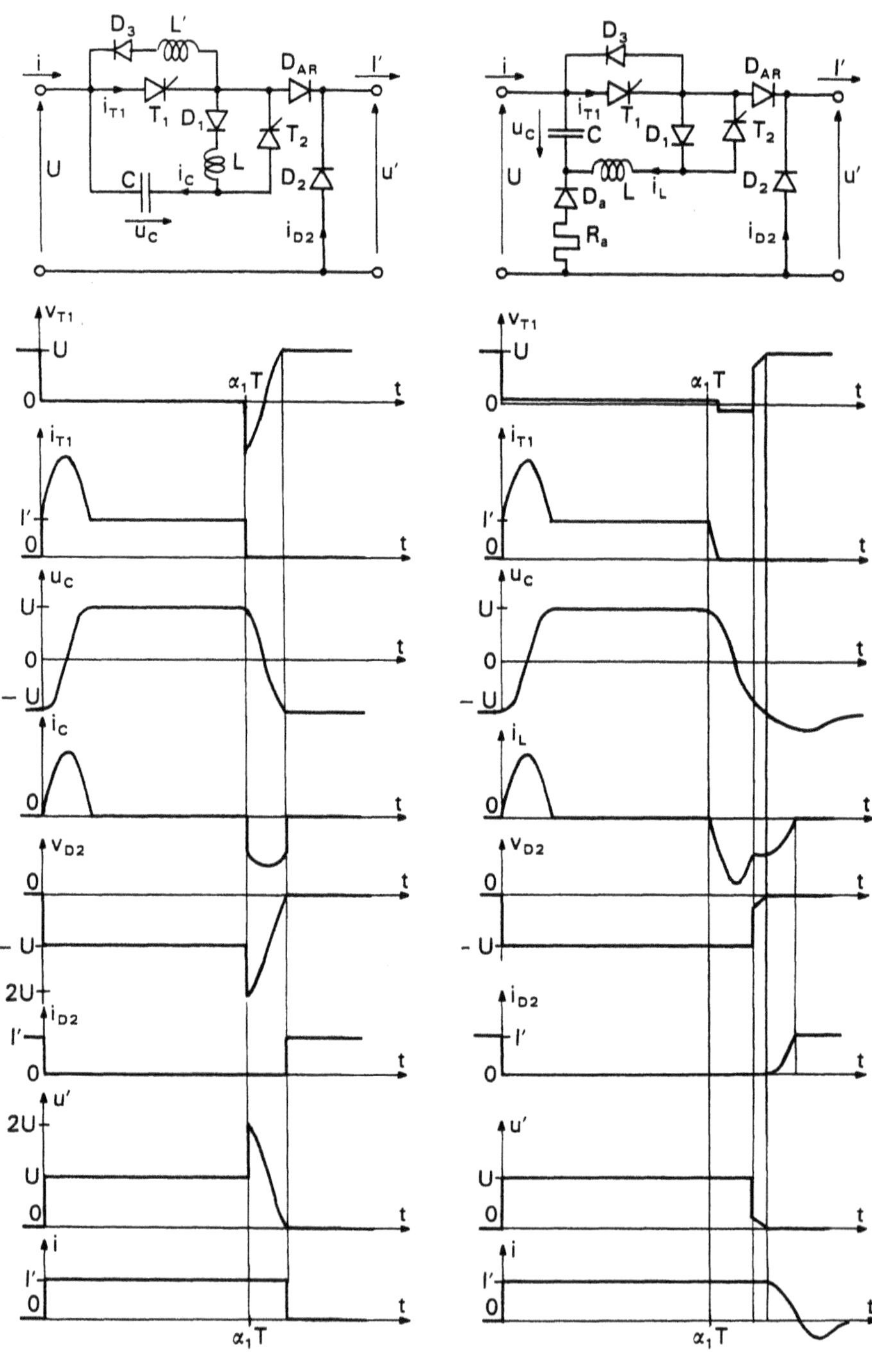

Fig. 5.23

Fig. 5.24

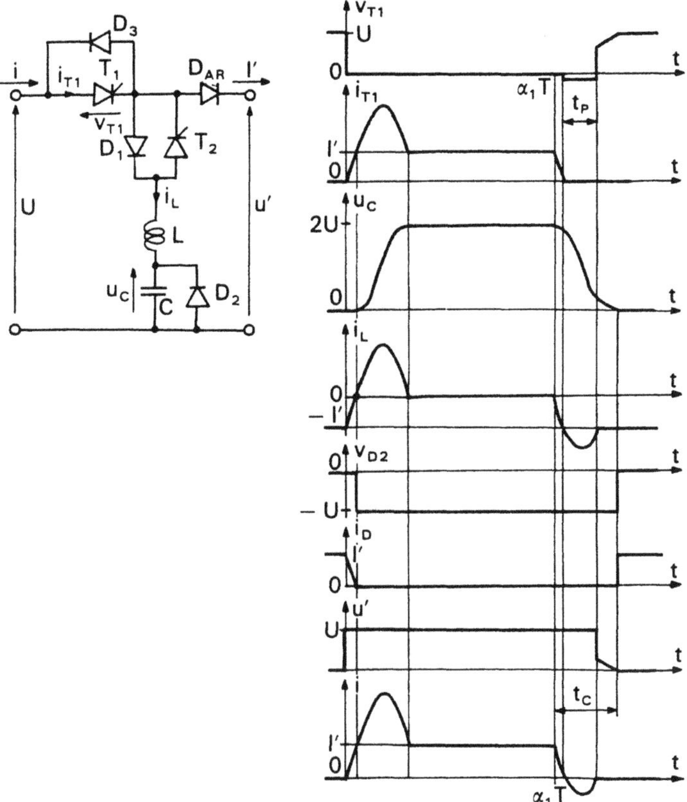

Fig. 5.25

In this case, commutation ends as soon as voltage u_C falls to zero. But current I' flows through the commutation inductor and turn-off thyristor T_2 during the whole period when diode D_2 is conducting.

Moreover, when T_1 is fired, current I' is not instantaneously taken by it. This must be written as

$$i_{T_1} = i = \frac{U}{L} t, \quad i_{D_2} = I' - i_{T_1};$$

it is only when i_{T_1} equals I' that voltage U is applied to the turn-off circuit.

This modification greatly increases the r.m.s. value of the current in both inductor L and the thyristor T_2.

- In the case of circuits with parallel commutation where the voltage across the commutation capacitor oscillates between $+U$ and $-U$, the path of this capacitor charging current can be eliminated. In order to achieve this,

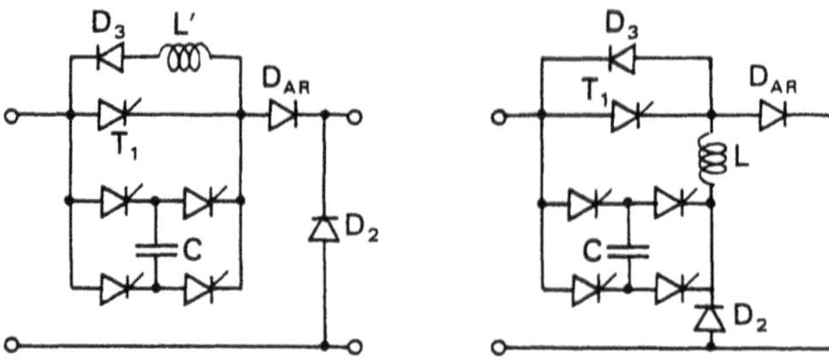

Fig. 5.26

a thyristor bridge is used (Fig. 5.26). This enables the connections of the capacitor to be reversed, i.e. to reverse its charge.

It should, however, be noted that, in the case of parallel commutation by oscillating circuit, it is no longer possible to use a damping dipole. The damping of the oscillating circuit itself must be sufficient to limit the increase in oscillations. Otherwise, the position of the free-wheeling diode must be modified, as shown in the second diagram of Fig. 5.26.

In this case, the increase in the r.m.s. value of the current in inductor L is added to drawbacks linked to the increase in the number of thyristors.

5.6 Extension of the Study to the Step-Up Chopper

5.6.1 Presentation of the Method

In Chap. 3, we initially represented the step-up chopper (see Fig. 3.7) with the position of elements and the conventional signs indicated in Fig. 5.27a. However, it was already noted (see Sect. 5.2.2.3) that no basic difference exists between the buck and the boost choppers. These remarks will enable *all the results of the analysis of commutation circuits carried out for the step-down chopper to be used in the case of the step-up chopper.*

Diode D can obviously be moved from the positive terminal to the negative terminal of voltage source U, and the arrows of voltage u' and current i can be reversed (these will simply have negative average values). This is the difference between Figs. 5.27a and 5.27b.

If the right-hand side of this figure is then turned around, this gives the configuration in Fig. 5.27c. It can be seen that the step-up chopper is derived from the step-down chopper by simply moving the current source terminal which is directly connected to the voltage source negative terminal from this terminal to the positive terminal of the latter. If A was connected to B', this would give a buck chopper; A connected to B gives a boost chopper.

Compared with the results obtained for the buck chopper, this change in connections causes

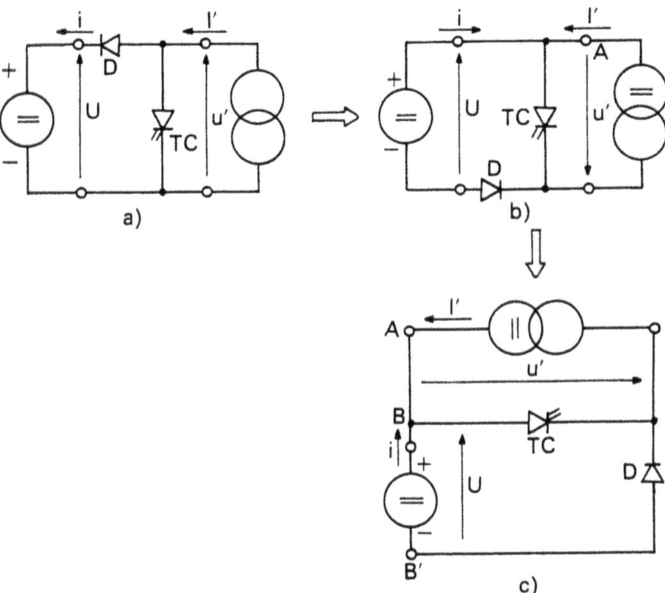

Fig. 5.27

- a shift of voltage u' by a constant value equal to $-U$,
- a shift of current i by a constant value equal to $-I'$.

5.6.2 Application to the Three Basic Commutation Circuits

- For *parallel commutation by capacitor*, changing the connection of the current source makes the circuit change from the one in Fig. 5.3 to that shown in Fig. 5.28. In the latter, the waveforms of the main variables have been plotted. Comparing these with the waveforms in Fig. 5.4 shows that v_{T_1}, i_{T_1}, u_C, v_{D_2}, and i_{D_2} remain unchanged, that u' is reduced by U and that i is reduced by I'.
- In the case of *parallel commutation by oscillating circuit*, the change of step-down chopper to step-up chopper is represented by the changes between, on the one hand, Fig. 5.7 and the waveforms of Fig. 5.8 and, on the other, Fig. 5.29.
- In the case of *series commutation*, Fig. 5.30 gives the circuit diagram and the waveforms of the main variables for the boost chopper.

All the equations presented in Sects. 5.1–5.3 can be applied, especially those giving

the reverse-bias time t_p,
the commutation time t_c,

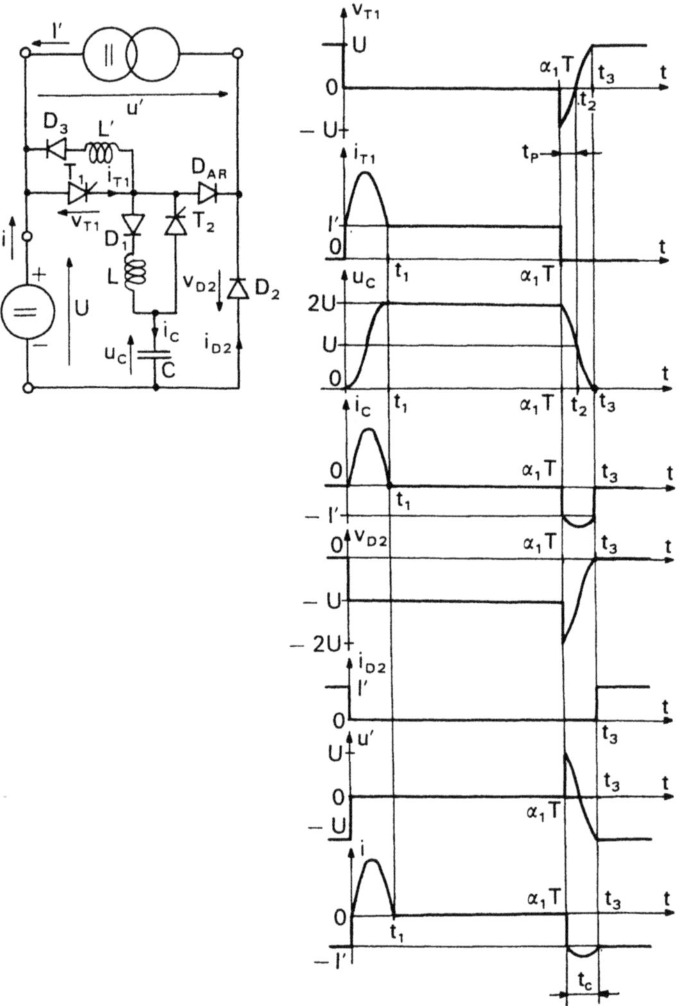

Fig. 5.28

the variation $\Delta U'$ of the average voltage value u'; the latter is now added to the average theoretical value $-(1-\alpha_1)\,U$, no longer to $\alpha_1 U$.

All the stresses applied to semiconductor devices remain the same.

The comparison proposed in Sect. 5.4 between the three basic commutation modes remains valid.

The same applies to the modifications concerning the types of semiconductor switches in the commutation circuit or the topological layout of its elements.

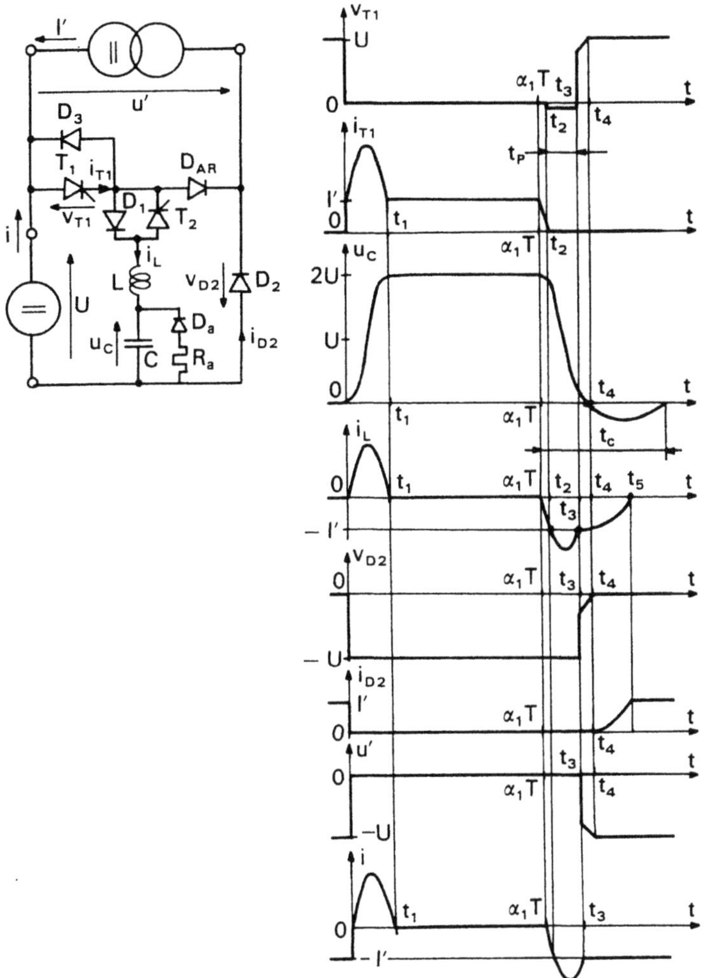

Fig. 5.29

5.7 Notes on the Turn-off Circuits in the Current-Reversible Chopper and the Full-Bridge Chopper

In the current-reversible chopper in Fig. 3.10, the two thyristors which have to operate as fully controlled on/off switches are connected in series under voltage U. In the full-bridge chopper shown in Fig. 3.12, two such pairs of thyristors can be found.

Both thyristors in a same pair can share a common turn-off circuit.

These thyristors can also turn each other off, in the case of complementary control.

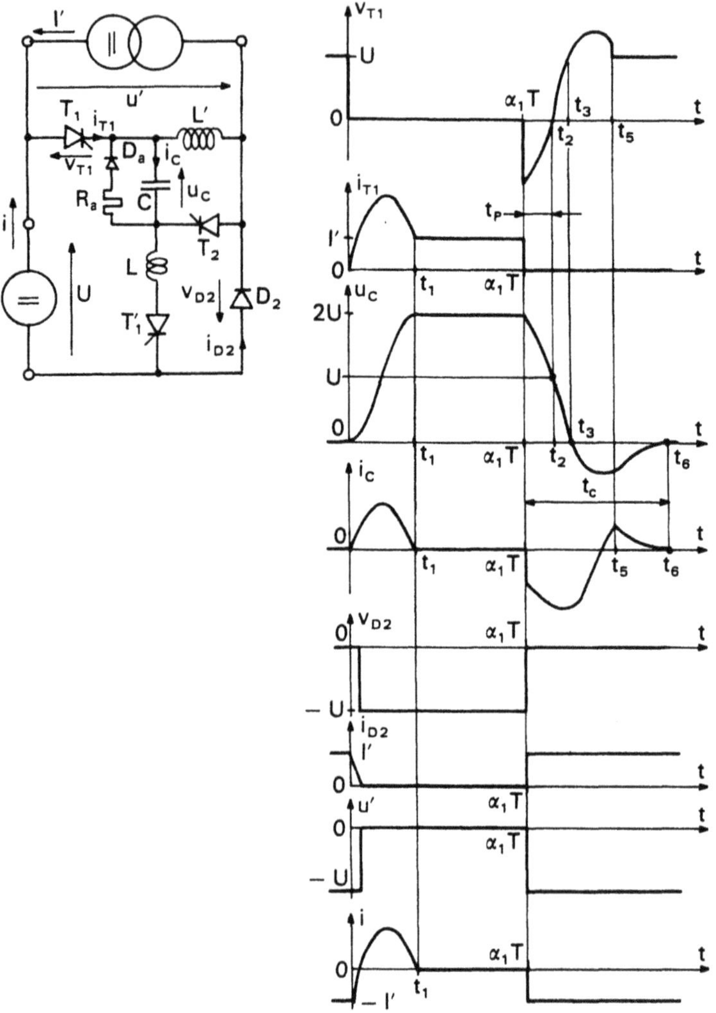

Fig. 5.30

We will provide an example of how each of these possibilities can be implemented. However, this analysis will not go into further detail as complementary control mainly concerns inverters. A more detailed analysis of turn-off circuits for a pair of complementary-controlled thyristors will be given in the fourth volume of the present series.

5.7.1 Two Thyristors Sharing the Same Forced-Commutation Circuit

Figure 5.31 shows a current-reversible chopper with two "switches" and a forced-commutation circuit using an oscillating circuit.

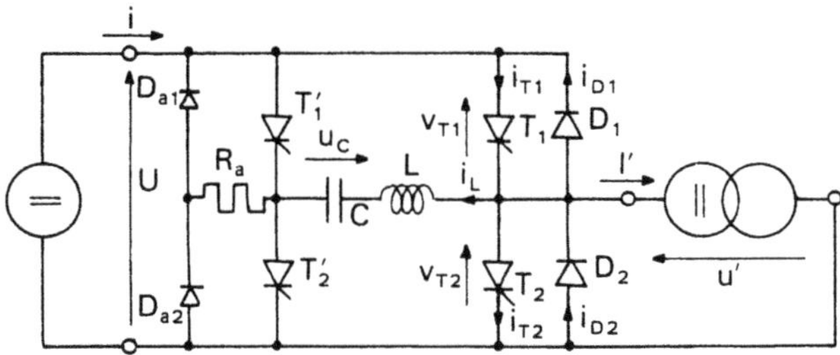

Fig. 5.31

When operating as a step-down chopper ($I' > 0$), the semiconductor devices which carry the current are T_1 and D_2; when operating as a step-up chopper ($I' < 0$), they are T_2 and D_1. To these are added the common oscillating LC circuit, and auxiliary thyristor T'_1 and T'_2 which act as a charge thyristor or a turn-off thyristor, depending on the operating condition. Only one damping resistance R_a may also be used operating either with diode D_{a1} or diode D_{a2}.

The behaviour of this circuit is very similar to that of the circuit shown in Fig. 5.23b.

5.7.1.1 Sequential Control

Firing signals are applied only to one of the main thyristors. As an example, we shall describe the operating mode in which, with I' positive, T_1 is periodically fired. The waveforms of the main variables are those shown in Fig. 5.32

- *T_1 turn-on*

We start from the state reached at the end of the switching cycle: the only conducting semiconductor device is D_2, through which I' flows. Voltages u_C and v_{T_1} are equal to $+U$. For $t = 0$, T_1 is fired: current I' is instantaneously transferred from D_2 to T_1.

- *T_1 turn-off*

For $t = \alpha_1 T$, turn-off thyristor T'_1 is fired. The LC circuit begins to oscillate, current i_L is negative.

Current i_{T_1}, equal to $I' + i_L$, falls to zero for $t = t_1$. At this instant T_1 turns off.

Diode D_1 then lets current $-(I' + i_L)$ flow until instant $t = t_2$, when it falls to zero.

From $t = t_2$, thyristor T'_1 is the only semiconductor device conducting; current i_L equals $-I'$; voltage u_C decreases linearly. This voltage reaches value $-U$ for $t = t_3$ and diode D_2 then begins conducting.

From $t = t_3$ to $t = t_4$, thyristor T'_1 lets $-i_L$ flow and diode D_2 lets $I' + i_L$ flow. When i_L falls to zero, T'_1 turns off.

The damping circuit then begins to function: voltage u_C initially lower than $-U$ rises up and reverts to $-U$, owing to a current which is sent back to the

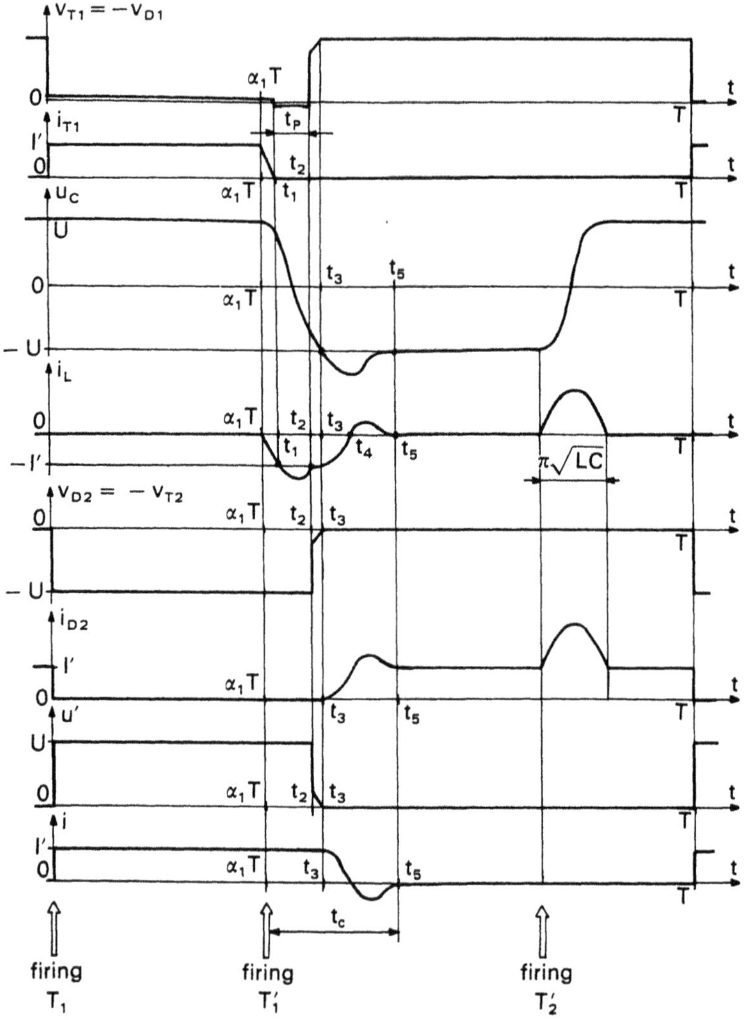

Fig. 5.32

voltage source and which goes via D_2, L, C, R_a and D_{a1}. Commutation is ended for $t = t_5$.

- *C recharging*

During the off-state period of T_1, T'_2 is fired. In the circuit comprising D_2, L, C and T'_2 an oscillation changes voltage u_C from $-U$ to $+U$. This oscillation is interrupted after half a period when i_L becomes to zero.

Remark. For negative I', the waveforms are as in Fig. 5.33: the main thyristor is T_2; the current is transferred from diode D_1 to this thyristor, the turn-off thrysitor is T'_2 and the u_C-reversing thyristor is T'_1.

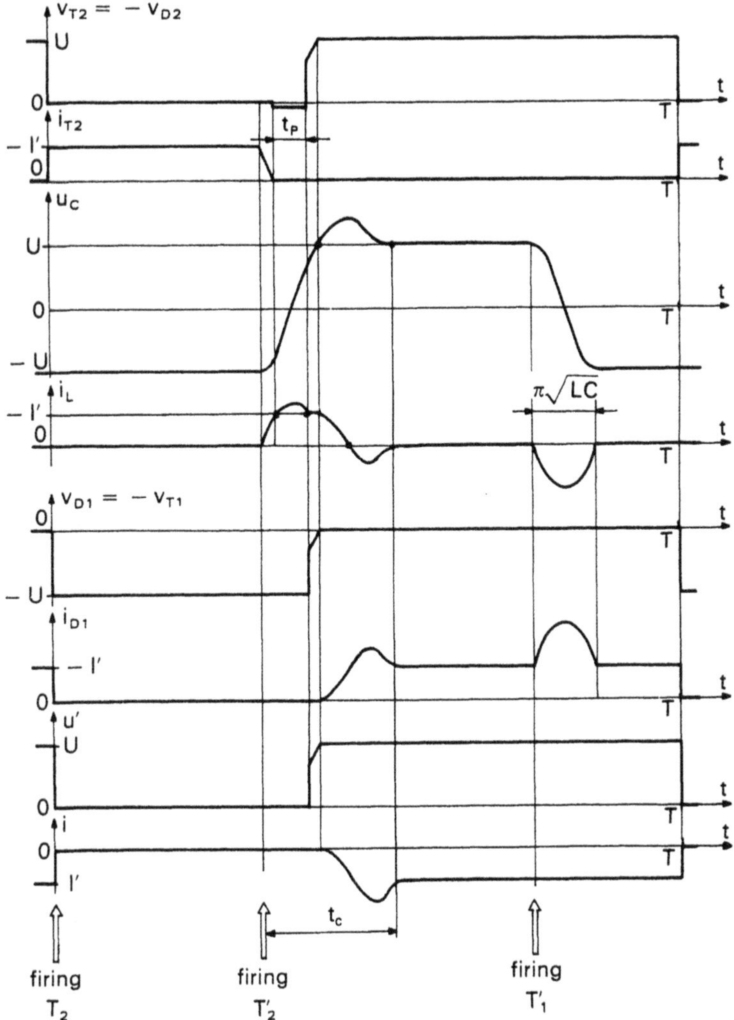

Fig. 5.33

5.7.1.2 Complementary Control

If the complementary control is used for thyristors T_1 and T_2, the commutation of T_1 by firing T'_1 functions as for sequential control. The only difference is that the time interval between the firing of T'_1 and the firing of T_2 must be greater than that between the firing of T'_1 and the moment when D_2 starts to conduct.

- If the current in the load remains positive, thyristor T_2 will never become conducting. When T'_2 is fired, with the aim of bringing this current to zero, the only effect is a reversal of the voltage across the capacitor, after T_1 has been turned off. The operation is exactly the same as with sequential control.

- If the current in the load has changed its polarity during the switching period, voltage u_C has reversed its polarity during the turn-off of T_1; it has therefore the required polarity to turn off T_2 by firing T_2'. Turning off T_2 reverses u_C and gives it back a suitable polarity for a subsequent turning off of T_1 by firing T_1'.

The explanation given for the commutation of T_1 by firing T_1' can apply to the commutation of T_2 by T_2'. T_1 has merely to be replaced by T_2, T_1' by T_2', T_2' by T_1', D_2 by D_1. The time interval between the firing of T_2' and that of T_1 must be longer than that between the firing of T_2' and the moment when D_1 becomes conducting.

5.7.2 Mutual Turn-off Procedure of the Main Thyristors: Example

In the case of complementary control, one main thyristor can be turned off by firing the other, and vice versa. To achieve this, a centre-tapped inductor is series-connected between the two thyristors.

Figure 5.34 provides the most simple of these circuits. Two closely-coupled inductances – L' and L'' – and two capacitors – C_1 and C_2 – are added to the two thyristors – T_1 and T_2 – and the two diodes – D_1 and D_2 – in order to carry out the commutations.

The two closely-coupled inductances – L' and L'' – have the same value L and are assumed to be perfectly coupled. Both capacitors – C_1 and C_2 – have the same capacitance $C/2$.

The commutation process is fairly similar to that of the series commutation choppers studied in the third part of this chapter.

5.7.2.1 Operation, Waveforms

As an example, in the case where I' is positive, the operation during a cycle, will be described. The waveforms are those shown in Fig. 5.35.

The analysis begins at the end of the switching cycle when only diode D_2 lets I' flow:

$$i_{D_2} = I'; \quad i_{T_1} = i_{T_2} = i_{D_1} = 0; \quad i_{C_1} = i_{C_2} = 0$$

$$v_{T_1} = u_{C_1} = -v_{D_1} \simeq U; \quad v_{T_2} = u_{C_2} = -v_{D_2} = u' \simeq 0.$$

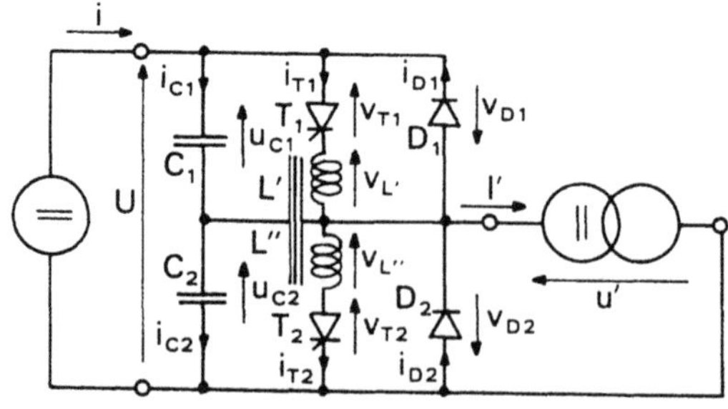

Fig. 5.34

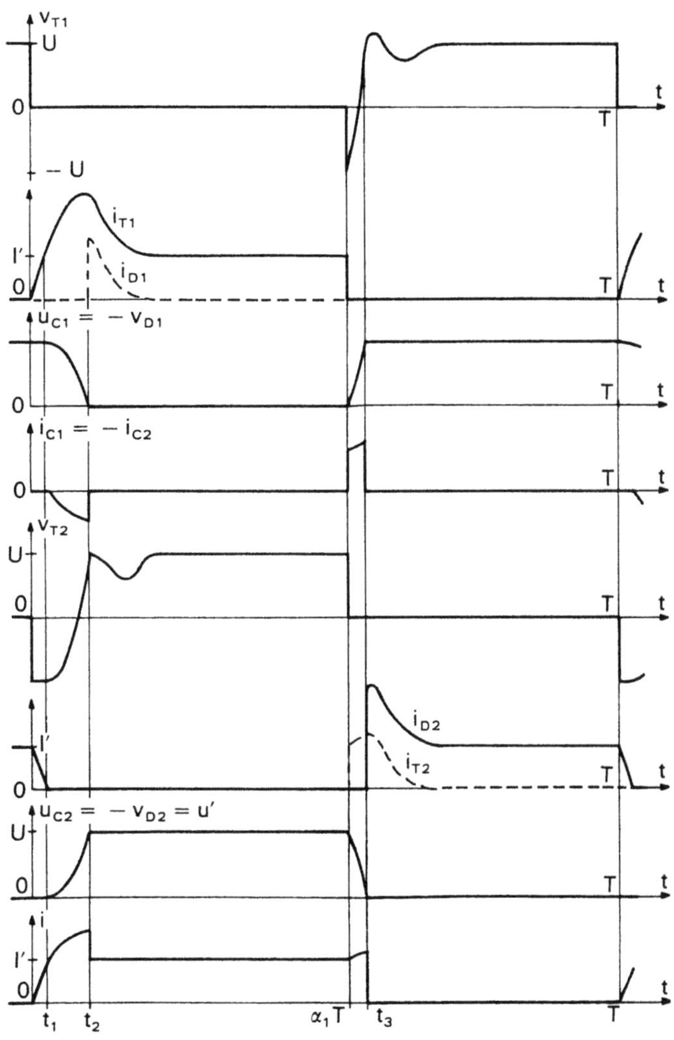

Fig. 5.35

- *Commutation from D_2 to T_1*

- For $t = 0$, T_1 is fired. As voltage u_{C_2} cannot suffer any discontinuity, turning on T_1 applies a voltage $v_{L'}$ equal to $+U$ across L': switching on T_1 applies thus a voltage $-U$ across T_2 and confirms that the latter is off.

- From $t = 0$ onwards, current i_{T_1} through T_1 increases ($L \, di_{T_1}/dt = U$), current i_{D_2} through D_2, equal to $I' - i_{T_1}$, decreases. For $t = t_1$, i_{T_1} reaches value I' and D_2 turns off.

- During time period (t_1, t_2), only T_1 is conducting. Capacitor C_2 is charging and capacitor C_1 discharging:

$$U = L\frac{di_{T_1}}{dt} + u_{C_2} \tag{5.43}$$

with

$$I_{T_1} = I' + i_{C_2} - i_{C_1}.$$

Since the sum $u_{C_1} + u_{C_2}$, equal to U, is constant,

$$\frac{C}{2}\frac{du_{C_1}}{dt} = -\frac{C}{2}\frac{du_{C_2}}{dt}$$

$$i_{C_1} = -i_{C_2}.$$

Equation (5.43) thus gives

$$L\frac{d}{dt}(2\,i_{C_2}) + \frac{2}{C}\int_0^t i_{C_2}\,dt = U \tag{5.44}$$

$$i_{T_1} = I' + 2\,i_{C_2}.$$

An oscillation of angular frequency ω, equal to $1/\sqrt{LC}$, begins; it ceases, for $t_2 = t_1 + (\pi\sqrt{LC})/2$, when u_{C_2} reaches value $+U$. Voltage u_{C_1} is then equal to zero and diode D_1 begins conducting.

- For $t = t_2$, the conduction of D_1 interrupts the variations of u_{C_1} and u_{C_2}, which are respectively equal to 0 and $+U$. Current i_{T_1} cannot be discontinuous because of L' and current i_{D_1} instantaneously takes on value $2i_{C_2}(t_2)$.
- From $t = t_2$ onwards, thyristor T_1 and diode D_1 are simultaneously conducting, with i_{T_1} being equal to $I' + i_{D_1}$. Current i_{D_1}, with an initial value of $2i_{C_2}(t_2)$, flows through the circuit formed by D_1, T_1 and L'. This current falls to zero on account of the resistance of the coil L', and of the forward voltage drops of T_1 and D_1.

  However, it is usually necessary to increase this damping. A small resistance can be series-connected with each diode. The excess energy stored in inductor L' is equal to:

$$\frac{1}{2}L[I' + 2i_{C_2}(t_2)]^2 - \frac{1}{2}LI'^2.$$

  This energy must be dissipated before firing T_2 occurs.

- *Commutation from T_1 to D_2*

- For $t = \alpha_1 T$, in order to turn off thyristor T_1, thyristor T_2, with a voltage v_{T_2} equal to $+U$ across it, is fired. As voltage u_{C_2} is equal to $+U$, turning on T_2 applies a voltage $v_{L''}$ equal to $+U$ across L''; the same voltage appears across L'. Firing T_2 turns off T_1 by applying across it a voltage $-U$.

The instantaneous transfer of I' from T_1 to T_2 is made possible by the coupling, assumed to be perfect, between the two coils L' and L''. The total flux across the core is not modified.

- At instant $t = \alpha_1 T$, C_2 begins to discharge and C_1 to charge:

$$U = L\frac{\mathrm{d}i_{T_2}}{\mathrm{d}t} + u_{C_1}.$$

Since i_{C_1} equals i_{C_2}, that the initial value of i_{T_2} is I' and that $i_{C_1} - i_{C_2} - i_{T_2}$ must be equal to current I' in the load, current i_{C_1} goes from zero to I' and current i_{C_2} from zero to $-I'$, at instant $t = \alpha_1 T$.

- During the time interval $(\alpha_1 T, t_3)$ current i_{C_1} makes u_{C_1} change from zero to $+U$, and current i_{C_2} makes u_{C_2} change from $+U$ to zero. When u_{C_2} falls to zero, for $t = t_3$, diode D_2 begins conducting.

  The turn-on of D_2 interrupts the variations of u_{C_1} and u_{C_2}. Current i_{D_2} takes on value $2i_{C_1}(t_3)$, if $i_{C_1}(t_3)$ denotes the value of i_{C_1} immediately before its interruption.

- From $t = t_3$ onwards, diode D_2 and thyristor T_2 conduct simultaneously:

$$i_{D_2} = I' + i_{T_2}.$$

In order to damp i_{T_2} rapidly, the same procedures are used as to accelerate the decay of i_{D_1} at the end of the previous commutation.

The type of commutation used: short current pulse through T_2 to ensure the transfer of I' from T_1 to D_2, is called *indirect* or *pulse commutation.*

5.7.2.2 Other Operational Modes

- If current I' is negative, firing T_2 leads to the transfer of the current from D_1 to T_2. Firing T_1 ensures the transfer of $-I'$ from T_2 to D_1 by indirect commutation.

- If current i' of the current source changes its polarity during the switching period, there is natural commutation between two semiconductor devices connected in anti-parallel during this change. This merely requires that the control signals applied to the thyristors gates last a sufficiently long time.

  The only possible forced commutations are those to be found with positive I' and negative I'. Whatever the polarity of i', firing T_1 makes u' equal to U (commutation D_2–T_1 or T_2–D_1), firing T_2 makes u' fall to zero (commutation T_1–D_2 or D_1–T_2).

5.7.2.3 Remark

The complementary control of two "switches" connected in series under voltage U requires that, for each thyristor, a firing signal remains applied during the whole interval of closure of the corresponding switch.

If the thyristor is not conducting, this gate signal poses no problem since across the device there is low negative voltage which corresponds to the forward voltage drop of the diode connected across it in anti-parallel.

As soon as the thyristor has to be fired or re-fired, the gate is certain to be supplied.

As all the commutations are of the thyristor–diode or diode–thyristor type, a dead time can be left, between the control signals of the two thyristors, which is sufficiently long to avoid the risk of simultaneous conduction.

Bibliography

1. Zabar Z, Alexandrovitz A (1970) Guidelines on adaptation of thyristorized switch for DC motor speed control. *IEEE Trans. Ind. Electron. Control Instrum.*, **17**(1): 10–13
2. Mc Murray W (1970) Analysis of thyristor DC chopper power converters including nonlinear commutating reactors. *IEEE Trans. Magnetics*, **6**(1): 16–21
3. Dewan SB, Duff DL (1970) Analysis of energy recovery transformer in DC choppers and inverters. *IEEE Trans. Magnetics*, **6**(1): 21–26
4. Alexandrovitz A, Zabar Z (1971) Analog computer simulation of thyristorized static switch as applied to DC motor speed control. *IEEE Trans. Ind. Electron. Control Instrum.* **18**(1): 1–5
5. Revankar GN, Palsetia PK (1972) Design criteria of commutation circuit in a DC chopper. *IEEE Trans. Ind. Electron. Control Instrum.*, **19**(3): 86–89
6. Hoft RG, Patel HS, Dote Y (1972) Thyristor series resonant DC–DC chopper. *IEEE Trans. Magnetics*, **8**(3): 286–288
7. Mc Lellan PR (1975) Thyristor choppers using a bridge-connected capacitor for commutation. *Proc. Inst. Electr. Eng.*, **122**(5): 514–516
8. Ray M, Datta AK (1976) Optimum design of commutation circuit in a thyristor chopper for DC motor control. *IEEE Trans. Ind. Electron. Control Instrum.*, **23**(2): 129–132
9. Davis RM, Melling JR (1977) Quantitative comparison of commutation circuits for bridge inverters. *Proc. Inst. Electr. Eng.*, **124**(3): 237–246
10. Williams BW (1977) Impulse-commutated thyristor chopper. *Proc. Inst. Electr. Eng.*, **124**(9): 793–795
11. Mc Murray W (1978) Thyristor commutation in DC choppers. A comparative study. *IEEE Trans. Ind. Appl.*, **14**(6): 547–558
12. Jinzenji T, Kanzaki T, Koga T (1980) Characteristics of a two-phase double-surimposed auxiliary impulse commutated chopper. *IEEE Trans. Ind. Appl.*, **16**(1): 87–94
13. Dubey GK (1983) Classification of thyristor commutation methods. *IEEE Trans. Ind. Appl.*, **19**(4): 600–606
14. Williams BW (1984) Current-impulse-displacement thyristor commutation with controlled trapped energy. *Proc. Inst. Electr. Eng. Part B*, **131**(2): 21–37
15. Williamson S, Cann RG, Williams BW (1984) Calculation of power losses in thyristor converters. *IEEE Trans. Ind. Electron.*, **31**(2): 192–200

CHAPTER 6

Switch-Mode Power Supplies

6.1 Introduction

6.1.1 Development of DC Power Supplies

DC power supplies – either regulated or not – are needed in industrial equipment, in order to supply systems or part of systems with a suitable DC voltage. In the majority of cases, these supplies are fed from the single-phase industrial network.

- Figure 6.1a shows the conventional diagram of an unregulated DC power supply. It comprises
  - a transformer, with a primary supplied by the network and a secondary with a voltage adapted to the required output voltage;
  - a diode rectifier, using diodes in a bridge configuration (circuit P.D.2) or two diodes with a mid-point secondary (circuit P.2);
  - a filter, comprising an inductor and two capacitors, for instance.

- When the output voltage has to be regulated, either to keep it constant (despite the variations in the AC input voltage or output loading) or to vary it, the conventional solution of replacing diodes by thyristors is not commonly adopted as in high-power equipments.

  In the case of low-power-regulated DC supplies, the diagram in Fig. 6.1b has first been introduced. The filter inductance is replaced by a controlled device – usually a bipolar transistor operating in its active region. This "*ballast*" transistor is crossed by a mean current equal to current I' delivered to the load; the voltage across it is equal to the difference between output voltage U of the rectifier and voltage U' across the load. The power losses in the transitor, $(U - U')I'$, are considerable. This requires the use of a big heat sink and considerably reduces the efficiency of the whole circuit.

- *Switch-mode power supplies* have been developed with the aim of bringing the efficiency back to an acceptable level, as soon as the power to be supplied to the load exceeds about ten watts. The transistor operates as a switch in a configuration such as the one in Fig. 6.1c, in which a buck converter is connected to the rectifier output.

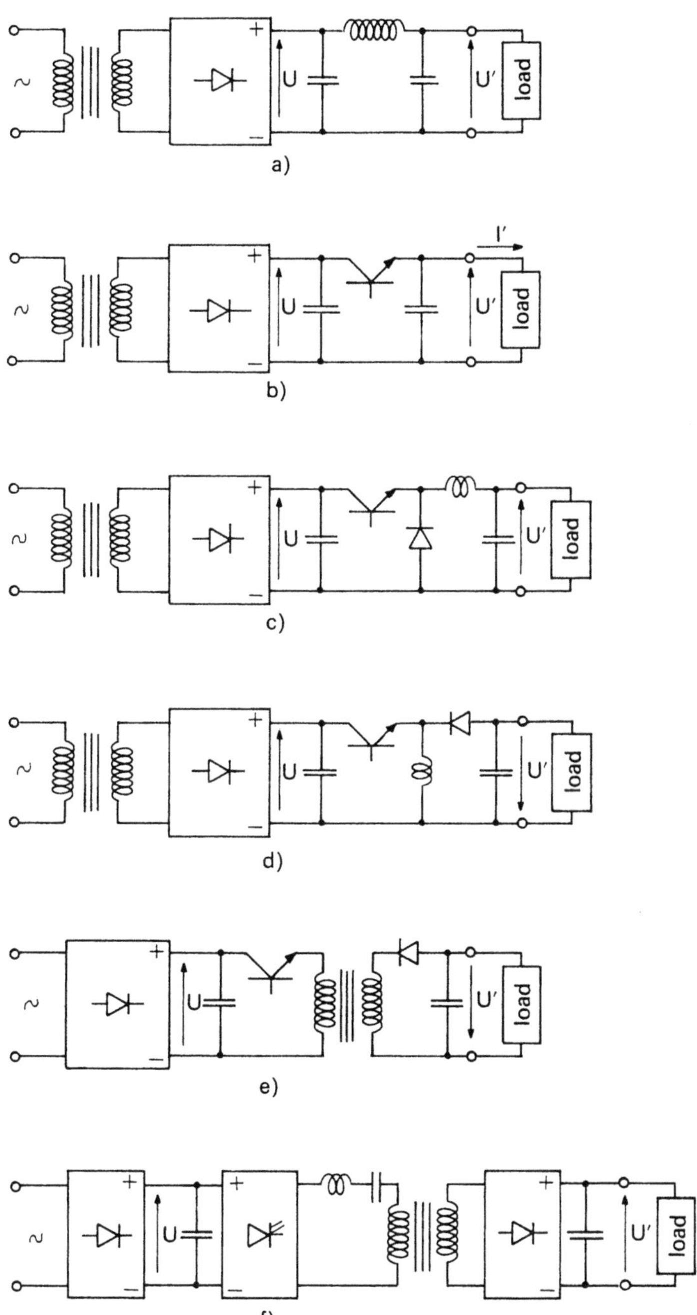

Fig. 6.1

- A large number of variations on this basic circuit have been proposed. The buck converter can be replaced by a boost converter, or by a chopper with inductive energy storage, as indicated in Fig. 6.1d. It is then possible to use the *transformer* both as a voltage adaptator and as a storage element: this gives the diagram of Fig. 6.1e.

  In this case, the operating frequency of the transformer is no longer imposed by the industrial network. In order to avoid acoustic noise, switching frequencies higher than 20 kHz are used: such frequencies may reach 50 kHz in the case of bipolar transistors, or 150 kHz in the case of field effect transistors. This increase in transformer operating frequency enables the transformer size to be considerably reduced.

  By use of multiple secondary windings, several output voltages may be obtained.

- Finally, among the various switching DC power supplies, mention must be made of supplies using a *resonant converter* which have undergone a recent and rapid development in low and medium power. Such a supply (Fig. 6.1f) may be obtained by connecting in series a rectifier, a resonant inverter built around an LC oscillating circuit, a transformer and a second rectifier. Use of such a configuration enables the controlled switches of the inverter to be operated in natural commutation and allows the interferences which are inherent in high-frequency operation to be reduced.

- In each circuit of Fig. 6.1, DC voltage U can be the DC voltage supplied, for example, by a battery instead of being the voltage supplied by a diode rectifier fed from the AC network. In fact, in many cases, the switch-mode power supplies receive their energy from a battery.

6.1.2 Organisation of the Analysis

- This chapter will confine itself to an analysis of switch-mode DC power supplies in which a chopper carries out the conversion between the DC or rectified voltage U and the output voltage U'. In the case of resonant DC–DC converters, the inverter carries out the essential part of this conversion: the analysis of such supplies will thus be considered in Volume 4 of this series which concerns the DC–AC conversion.

  Two main groups of configuration can be identified:

- For the first group, *the transformer* – if there is one – is not part of the chopper, as in Fig. 6.1c and d, for example. This group comprises:
  - *the buck converter* (series chopper),
  - *the boost converter* (parallel chopper),
  - *the buck-boost converter* (chopper with inductive energy storage),
  - *the Cùk converter* (chopper with capacitive energy storage).
- For the second group, *the transformer is an integral part of the chopper*, as in

Fig. 6.1e. This group comprises
- *the isolated inductive storage circuit* (flyback converter),
- *the isolated buck circuit* (forward converter),
- *variations or associations of the above.*

- Given the applications of the switched-mode power supplies, the *load is assumed to be purely resistive.* To reduce the ripple in the output voltage, *a capacitor* is always connected between the output terminals.

  The different nature of the load explains the need – in the case of the buck or boost converter – to reconsider the analysis of the effects of the load, studied in Chap. 4, § 1, under different conditions.

The *voltage U supply* feeding the chopper is assumed to be perfect. We will not discuss the calculation of the input filter once again but refer the reader to the relevant calculations in Sect. 2 of chapter 4.

6.2 Circuits Without Transformer

In analyzing such circuits, the semiconductor devices are assumed to be perfect. Some approximations will be made in order to greatly simplify the analysis without affecting the accuracy of the results in any noticeable way.

6.2.1 Buck Converter or Series Chopper

This circuit has already been presented. Fed by a voltage source, it must be connected to a load which acts as a current source. An inductance with parameters r, l, must therefore be inserted between the set of semiconductor devices and the output. The load is made up of resistance R shunted by capacitor C. This gives the diagram in Fig. 6.2.

Under normal operation, current i_l in the inductance l is different from zero during the whole cycle. Transistor Q is saturated from $t = 0$ to $t = \alpha T$; during the remainder of the switching cycle T, diode D lets i_l flow.

Figure 6.2 shows the waveforms of the different variables; the ripple in the output voltage u' has been deliberately exaggerated.

6.2.1.1 Average Values of Voltage u' and Current i_l

The evolutions of voltage u' and current i_l are ruled by the following equations:

$$v_D + ri_l + l\frac{di_l}{dt} + u' = 0$$

$$i_l = C\frac{du'}{dt} + \frac{u'}{R}.$$

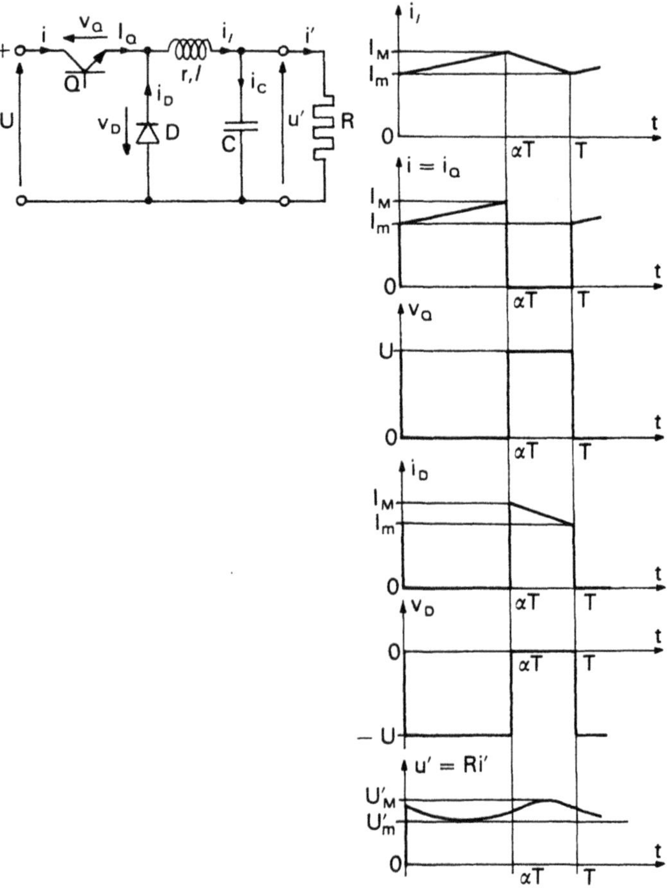

Fig. 6.2

In terms of mean values, this gives

$$v_{D\,\text{mean}} + ri_{l\,\text{mean}} + l\left(\frac{di_l}{dt}\right)_{\text{mean}} + u'_{\text{mean}} = 0$$

$$i_{l\,\text{mean}} = C\left(\frac{du'}{dt}\right)_{\text{mean}} + \frac{1}{R}u'_{\text{mean}}.$$

u' and i_l cannot suffer any discontinuity and, in steady-state operation, their values in $t = 0$ and $t = T$ are the same: the mean values of du'/dt and di_l/dt are thus zero.

Since the mean value of v_D is $-\alpha U$, by denoting U' and I_l the mean values of u' and i_l one gets

$$-\alpha U + rI_l + U' = 0 \quad \text{and} \quad I_l = \frac{U'}{R}$$

from which the following can be deduced:

$$U' = \frac{\alpha U}{1 + r/R}; \quad I_l = \frac{U'}{R} = \frac{\alpha U}{R + r}. \tag{6.1}$$

U' increases linearly as a function of α: when α changes from 0 to 1, U' changes from 0 to $UR/(R + r)$. Current i_l has the same average value as the output current, i.e. U'/R.

If r is ignored, the Eqs. (6.1) become

$$U' = \alpha U; \quad I_l = \alpha U/R. \tag{6.1'}$$

6.2.1.2 Current i_l and Voltage u' Ripples

- *Current i_l ripple*

In order to determine the ripple Δi_l of the current in the inductor, it is possible – as a first approximation – to ignore the variations in voltage u' and make no distinction between voltage u' and its average value U', throughout the whole switching cycle T.

As a result, this gives

- for $0 < t < \alpha T$, $ri_l + l\dfrac{di_l}{dt} = U - U'$, since Q is conducting;
- for $\alpha T < t < T$, $ri_l + l\dfrac{di_l}{dt} = -U'$, since D is conducting.

The whole of the analysis carried out in Sects. 4.1.1 and 4.1.2 can be directly applied: the R, L, E load, which was considered in that case now becomes r, l, U', with U' constant.

For a rapid estimate of Δi_l, the effect of r can also be neglected. U' then is equal to αU and this gives:

– for $0 < t < \alpha T$,

current i_l increases from its minimum I_m to its maximum I_M:

$$\frac{di_l}{dt} = \frac{U - \alpha U}{l} \text{ gives } i_l = I_m + \frac{U}{l}(1 - \alpha)t;$$

the maximum I_M is thus such that

$$I_M = I_m + \frac{U}{l}(1 - \alpha)\alpha T;$$

– for $\alpha T < t < T$,

the decrease in i_l is given by

$$\frac{di_l}{dt} = -\frac{\alpha U}{l}; \quad \text{thus, } i_l = I_M - \frac{U}{l}\alpha(t - \alpha T);$$

current i_l reaches again the value I_m for $t = T$.

This gives the approximate expression of Δi_l:

$$\Delta i_l = I_M - I_m = \frac{U}{l}(1 - \alpha)\alpha T \tag{6.2}$$

Δi_l is maximum for $\alpha = 0.5$ and then has the value $UT/4l$

The upper part of Fig. 6.3 shows the linear variations of i_l on either side of its average value.

If resistance r had been taken into account, the ripple would have been given by

$$\Delta i_l = \frac{U}{r}\frac{\exp(rT/l) - \exp(r\alpha T/l)}{\exp(rT/l) - 1}[1 - \exp(-r\alpha T/l)].$$

- *Voltage u' ripple*

The ripple $\Delta u'$ of voltage u' – between its minimum U'_m and its maximum U'_M – can be deduced from the approximate expressions of current i_l which have just been proposed.

If u' shows little variation, the same applies to current U'/R in resistance R. The variations in current i_l on either side of its mean value correspond to the charge and discharge current i_C of the capacitor. This current can be expressed as:

$$\text{for } 0 < t < \alpha T, \quad i_C = -\frac{\Delta i_l}{2} + \Delta i_l \frac{t}{\alpha T};$$

$$\text{for } \alpha T < t < T, \quad i_C = +\frac{\Delta i_l}{2} - \Delta i_l \frac{t - \alpha T}{T - \alpha T};$$

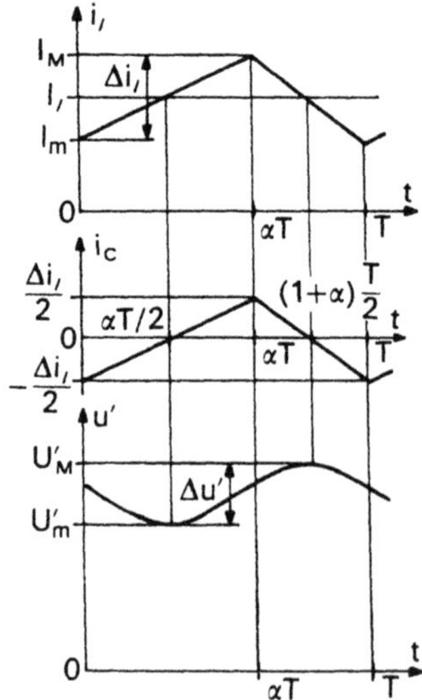

Fig. 6.3

it is shown in the middle of Fig. 6.3 and gives the waveform of voltage u' plotted underneath.

Since i_C, equal to $C\,du'/dt$, is positive during period $[\alpha T/2, (1+\alpha)T/2]$, the amplitude of the variations of u' is given by

$$\Delta u' = U'_M - U'_m = \frac{1}{C}\int_{\alpha T/2}^{(1+\alpha)T/2} i_C\,dt = \frac{1}{C}\Delta i_l \frac{T}{8}$$

or, taking Eq. (6.2) into account, by

$$\Delta u' = \frac{U}{8lC}(1-\alpha)\alpha T^2. \tag{6.3}$$

It can be seen that, for a given α, $\Delta u'/U$ is inversely proportional to l, to C and to the square of the switching frequency $1/T$.

For given values of l, C and T, the ratio $\Delta u'/U$ is at its maximum for $\alpha = 0.5$ and then has the value

$$\frac{\Delta u'}{U} = \frac{T^2}{32lC}$$

It can be noted that Δi_l and $\Delta u'$ vary depending on α according to the same law. Figure 6.4 indicates how these ripples vary when α varies from 0 to 0.5; for α going from 1 to 0.5, the same curve would be obtained.

6.2.1.3 Semiconductor Device Ratings

The waveforms in Fig. 6.2 and the equations which have been proposed show the stresses imposed on the transistor and the diode.

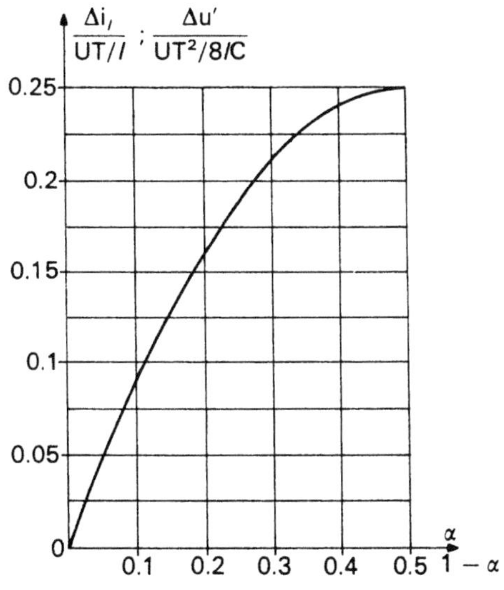

Fig. 6.4

- The maximum forward voltage across Q and the maximum reverse voltage across D are equal to U.

$$v_{Q\,max} = -v_{D\,max} = U.$$

- The average current in the transistor has a value

$$I_Q = \alpha I_l = \alpha^2 \frac{U}{R+r};$$

it reaches its maximum value for $\alpha = 1$ and is then

$$I_{Q\,max} = \frac{U}{R+r}$$

- The average current in the diode has a value

$$I_D = (1-\alpha) I_l = (1-\alpha)\alpha \frac{U}{R+r};$$

it reaches its maximum value for $\alpha = 0.5$ and is then

$$I_{D\,max} = \frac{1}{4}\frac{U}{R+r} = \frac{1}{4} I_{Q\,max}.$$

- Currents i_Q and i_D have the same peak value I_M.

$$I_M = I_l + \frac{\Delta i_l}{2}.$$

If r is ignored in the expression of I_l, as it was ignored in the calculation of Δi_l, this gives

$$I_M = \alpha\frac{U}{R} + \frac{1}{2}\frac{U}{l}(1-\alpha)\alpha T = \alpha U\left[\frac{1}{R} + (1-\alpha)\frac{T}{2l}\right].$$

The derivative of I_M with respect to α falls to zero for

$$\alpha = \frac{1}{2} + \frac{l}{RT}$$

- if T is less than $2l/R$, I_M constantly increases when α changes from 0 to 1. For $\alpha = 1$, this gives

$$I_{M\,max} = U/R;$$

- if T is greater than $2l/R$, I_M reaches its maximum for the value of α which cancels out $dI_M/d\alpha$. It then has the value

$$I_{M\,max} = \frac{U}{4R}\frac{(2l/R+T)^2}{\frac{2l}{R}T}.$$

When T changes from $2l/R$ to five times $2l/R$, $I_{M\,max}$ rises from U/R to 1.8 U/R.

6.2.1.4 Notes on Discontinuous Conduction

The operational mode which has just been analyzed assumes a current i_l which is constantly positive. If resistance r is ignored, this condition is fulfilled, according to Eqs. (6.1′) and (6.2), only if

$$I' = \frac{\alpha U}{R} \geqslant \frac{\Delta i_l}{2} = \frac{U}{2l}(1-\alpha)\alpha T,$$

i.e. as long as the load resistance is smaller than a boundary value:

$$R \leqslant \frac{2l}{(1-\alpha)T} \tag{6.4}$$

- When resistance R is above this value, the current flow in the inductor becomes discontinuous: i_l falls to zero, for $t = \beta T$, before the end of the switching cycle. The waveforms of the main variables become those shown in Fig. 6.5.
  In particular, the average value U' of the output voltage becomes greater than the value given by Eq. (6.1′), and is no longer independent of the load current.
- If the variations in voltage u' are ignored, this gives:
  - for $0 < t < \alpha T$,

$$\frac{di_l}{dt} = \frac{U - U'}{l}, \quad \text{i.e. } i_l = \frac{U - U'}{l} t;$$

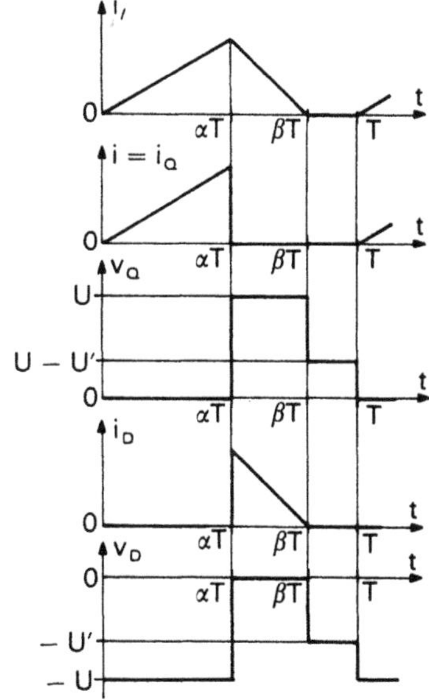

Fig. 6.5

- for $\alpha T < t < \beta T$,

$$\frac{\mathrm{d}i_l}{\mathrm{d}t} = -\frac{U'}{l}, \quad \text{i.e. } i_l = \frac{U - U'}{l}\alpha T - \frac{U'}{l}(t - \alpha T);$$

- for $\beta T < t < T$,

$$i_l = 0.$$

Since the instant $t = \beta T$ corresponds to the moment when i_l reaches zero, this gives the equation

$$\frac{U - U'}{l}\alpha T - \frac{U'}{l}(\beta T - \alpha T) = 0.$$

From this, the following can be deduced:

$$U' = \frac{\alpha}{\beta}U. \tag{6.5}$$

- The average value of current i_l is equal to that I' of the current supplied to the resistance:

$$I' = \frac{1}{T}\left[\int_0^{\alpha T}\frac{U - U'}{l}t\,\mathrm{d}t + \int_{\alpha T}^{\beta T}\left(\frac{U - U'}{l}\alpha T - \frac{U'}{l}(t - \alpha T)\right)\mathrm{d}t\right]$$

$$I' = \frac{U}{2l}(\alpha\beta - \alpha^2)T. \tag{6.6}$$

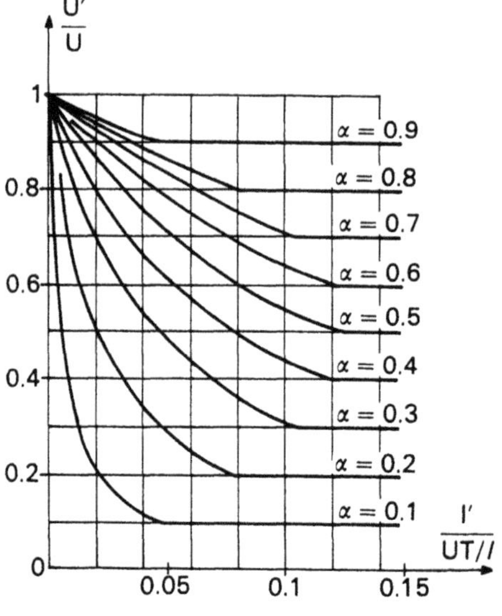

Fig. 6.6

If β is eliminated between Eqs. (6.5) and (6.6), this gives the equation which enables, at given α, the characteristics $U'(I')$ to be plotted in the zone where conduction is discontinuous (Fig. 6.6.):

$$U' = \frac{\alpha^2 U^2}{\alpha^2 U + 2l \dfrac{I'}{T}} \tag{6.7}$$

6.2.2 Boost Converter or Parallel Chopper

As the boost converter is intended to connect a current generator to a voltage receptor; an inductor must thus be added between the voltage source U and the semiconductor devices arrangement (Fig. 6.7).

In normal operating mode, current i is constantly positive. Transistor Q is saturated from $t = 0$ to $t = \alpha T$; diode D conducts during the remainder of the switching cycle T. Figure 6.7 gives the waveforms for the different variables (the ripple in u' is once more exaggerated).

6.2.2.1 Average Values of Voltage *u'* and Current *i*

Output voltage u' and input current i are related by the following equations:

$$\text{for } 0 < t < \alpha T, \quad U = ri + l\frac{di}{dt} \text{ and } C\frac{du'}{dt} + \frac{u'}{R} = 0\,;$$

$$\text{for } \alpha T < t < T, \quad U = ri + l\frac{di}{dt} + u' \text{ and } C\frac{du'}{dt} + \frac{u'}{R} = i.$$

If the average values of u' and i are denoted by U' and I, the following is obtained:

$$U = rI + l\left(\frac{di}{dt}\right)_{\text{mean}} + \frac{1}{T}\int_{\alpha T}^{T} u'\,dt$$

$$C\left(\frac{du'}{dt}\right)_{\text{mean}} + \frac{U'}{R} = \frac{1}{T}\int_{\alpha T}^{T} i\,dt.$$

The mean values of du'/dt and di/dt are zero in steady state. As will be seen, the variations of u' and i are quasi-linear: it can be assumed that the average values of u' and i on the period $(\alpha T, T)$ are equal to the average values of these variables on the whole switching cycle. This gives

$$U = rI + (1 - \alpha)U'$$

$$U/R = (1 - \alpha)I.$$

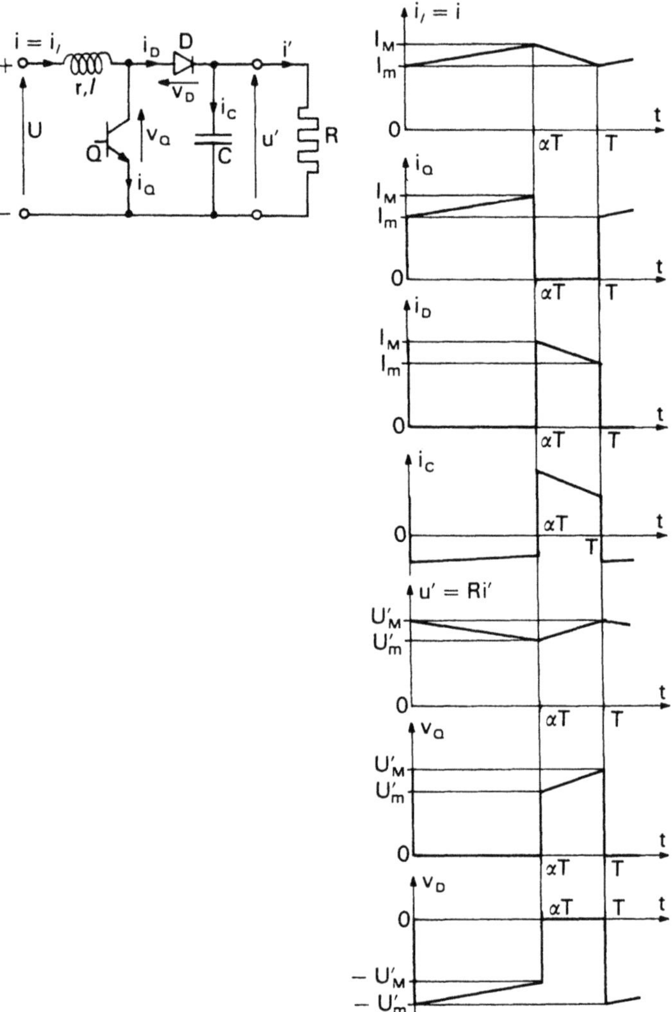

Fig. 6.7

From this the following are deduced:

$$U' = \frac{U(1-\alpha)}{r/R + (1-\alpha)^2} \tag{6.8}$$

$$I = \frac{U'}{R(1-\alpha)} = \frac{I'}{1-\alpha} = \frac{U}{r + R(1-\alpha)^2} \tag{6.9}$$

If r is ignored, these equations become

$$U' = \frac{U}{1-\alpha} \tag{6.8'}$$

$$I = \frac{U'}{R(1-\alpha)} = \frac{I'}{1-\alpha} = \frac{U}{R(1-\alpha)^2}. \tag{6.9'}$$

- The *average voltage* U' is equal to $UR/(R+r)$ for zero α. When α increases, U' increases at first and reaches its maximum U'_{max}, for $\alpha = \alpha_{max}$:

$$\alpha_{max} = 1 - \sqrt{\frac{r}{R}}; \quad U'_{max} = \frac{U}{2}\sqrt{\frac{R}{r}};$$

U' then decreases and falls to zero for α equal to 1.

The curves in full lines in Fig. 6.8 show the variations of U'/U depending on α, for r/R equal to 0.01 and 0.05. It can be observed that r must have a very low value with respect to R if a high U'/U ratio must be obtained.

- The *average current* I is equal to $U/(R+r)$ for zero α. It constantly increases depending on α and reaches value U/r when α equals 1.

Figure 6.9 shows the variations of I as related to U/R, depending on α, for r/R equal to 0.01 and 0.05. Note the sharp increase in the current delivered by the voltage source when α approaches 1.

(In Figs. 6.8 and 6.9, the curves in dashed lines correspond to zero r. They show the considerable influence of resistance r on U' and I.)

6.2.2.2 Current i and Voltage u' Ripples

- *Current i ripple*

In order to calculate the ripple Δi in current i which is delivered by the voltage source U and flows through the inductor, the output voltage ripple and resist-

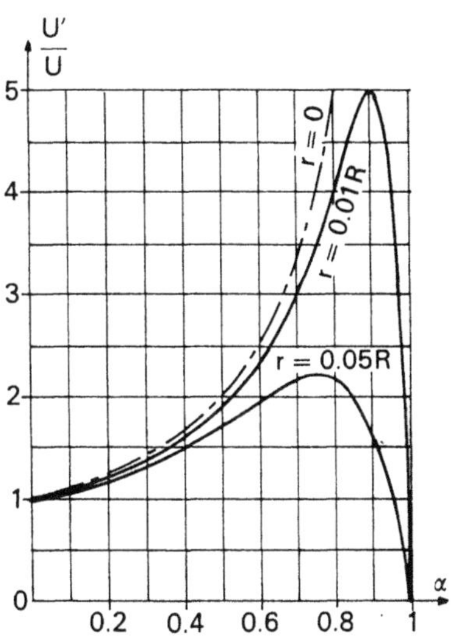

Fig. 6.8

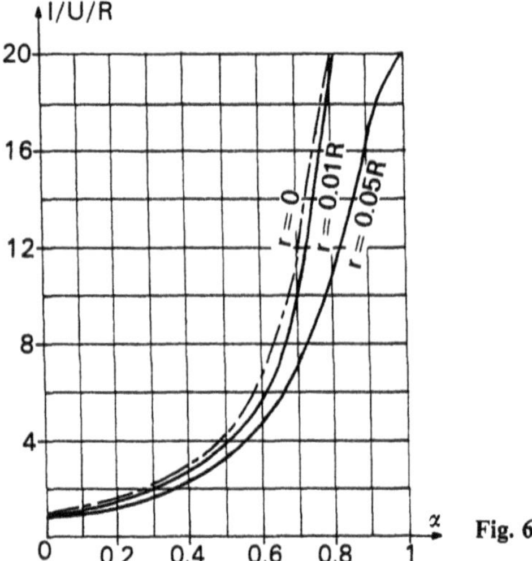

Fig. 6.9

ance r of the coil can be ignored. In such conditions:

$$\text{for } 0 < t < \alpha T, \quad l\frac{\mathrm{d}i}{\mathrm{d}t} = U \text{ gives}$$

$$i = I_{\mathrm{m}} + \frac{U}{l}t\text{; thus}$$

$$I_{\mathrm{M}} = I_{\mathrm{m}} + \frac{U}{l}\alpha T;$$

$$\text{for } \alpha T < t < T, \quad l\frac{\mathrm{d}i}{\mathrm{d}t} = U - U' = -\frac{\alpha U}{1-\alpha} \text{ gives}$$

$$i = I_{\mathrm{M}} - \frac{U}{l}\frac{\alpha}{1-\alpha}(t - \alpha T).$$

Hence, the approximate expression of the ripple:

$$\Delta i = I_{\mathrm{M}} - I_{\mathrm{m}} = \frac{U}{l}\alpha T. \tag{6.10}$$

Δi increases proportionately to α; for α equal to 1, its value is UT/l.

If r had been taken into account, this would have given

$$\Delta i = \frac{U'}{r}\frac{1 - \exp[-(1-\alpha)rT/l]}{1 - \exp(-rT/l)}[1 - \exp(-\alpha rT/l)];$$

there is little difference if the value of the time constant l/r of the coil referred to the switching cycle T is high.

• *Voltage u′ ripple*

In analyzing the output-voltage ripple, the voltage source U in series with inductor is considered as a perfect current generator delivering its current I either in the transistor or, via the diode, in R and C connected in parallel.

- For $0 < T < \alpha T$, Q is conducting:

$$0 = C\frac{du'}{dt} + \frac{u'}{R} \quad \text{or} \quad RC\frac{du'}{dt} + u' = 0\,;$$

u' is thus expressed as

$$u' = U'_M \exp(-t/RC).$$

At the end of this period, u' has moved from its maximum to its minimum. The value of the latter is

$$U'_m = U'_M \exp(-\alpha T/RC).$$

- For $\alpha T < t < T$, D is conducting:

$$I = C\frac{du'}{dt} + \frac{u'}{R} \quad \text{or} \quad RC\frac{du'}{dt} + u' = RI$$

giving the expression of u':

$$\begin{aligned} u' &= (U'_m - RI)\exp[-(t - \alpha T)/RC] + RI \\ &= U'_M \exp(-t/RC) + RI(1 - \exp[-(t - \alpha T)/RC]). \end{aligned}$$

As u' takes again value U'_M at instant $t = T$, this gives

$$U'_M = RI\frac{1 - \exp[-(1-\alpha)T/RC]}{1 - \exp(-T/RC)}.$$

- The ripple of u' is thus given by

$$\Delta u' = U'_M - U'_m = U'_M[1 - \exp(-\alpha T/RC)]$$

$$\Delta u' = RI\frac{1 - \exp[-(1-\alpha)T/RC]}{1 - \exp(-T/RC)}[1 - \exp(-\alpha T/RC)].$$

If the time constant RC is much greater than the switching cycle,

$$\Delta u' \simeq RI(1-\alpha)\alpha\frac{T}{RC}$$

or, since I equals $U'/R\,(1-\alpha)$,

$$\Delta u' \simeq U' \frac{\alpha T}{RC} \tag{6.11}$$

The ratio $\Delta u'/U'$ is proportional to α and inversely proportional to the ratio RC/T.

Equations (6.8) or (6.8′) enable $\Delta u'$ to be related to supply voltage U. If the expression of U' used is that obtained by neglecting r, this gives

$$\Delta u' \simeq U \frac{\alpha}{1-\alpha} \frac{T}{RC} \tag{6.11'}$$

6.2.2.3 Semiconductor Device Ratings

The stresses – in current as well as in voltage – which semiconductor devices have to withstand, increase with α. In practice, however, the value α_{max} of α which gives U' its maximum, is rarely exceeded.

$$\alpha_{max} = 1 - \sqrt{\frac{r}{R}}$$

- The forward voltage across the transistor and the reverse voltage across the diode have the same peak value:

$$v_{Q\,max} = -v_{D\,max} = U'_M = U' + \frac{\Delta u'}{2} = U'\left(1 + \alpha \frac{T}{2RC}\right).$$

For α_{max}, this gives

$$U'_{M\,max} = U'_{max}\left(1 + \alpha_{max} \frac{T}{2RC}\right).$$

- The current in the transistor has the following average and maximum values:

$$I_Q = \alpha I, \quad i_{Q\,max} = I + \frac{\Delta i}{2}.$$

For α_{max}, these values become, respectively,

$$\alpha_{max} \frac{U'_{max}}{R(1-\alpha_{max})} \quad \text{and} \quad \frac{U'_{max}}{R(1-\alpha_{max})} + \frac{U}{2l} \alpha_{max} T.$$

- The current in the diode has the same maximum value as the current in the transistor. Its average value is

$$I_D = (1-\alpha) I.$$

For α_{max}, it is equal to U'_{max}/R.

6.2.2.4 Notes on Discontinuous Conduction

If resistance r is ignored and taking Eq. (6.9′) and (6.10) into account, current i can only be constantly positive if

$$I = \frac{U}{R(1-\alpha)^2} \geqslant \frac{\Delta i}{2} = \frac{U}{2l}\alpha T,$$

i.e. if resistance R is such that

$$R \leqslant \frac{2l}{(1-u)^2 \alpha T}. \tag{6.12}$$

- When the load resistance exceeds this value, the current flow in inductor l becomes discontinuous: current i falls to zero fo $t = \beta T$, before the end of the switching cycle, as shown by the waveforms in Fig. 6.10.

  The average value U' of the output voltage exceeds the value given by Eq. (6.8) and becomes closely dependent on current I'.
- If the voltage u' ripple is neglected, one gets:
  - for $0 < t < \alpha T$,

$$l\frac{di_l}{dt} = U, \quad \text{giving} \quad i_l = \frac{U}{l}t;$$

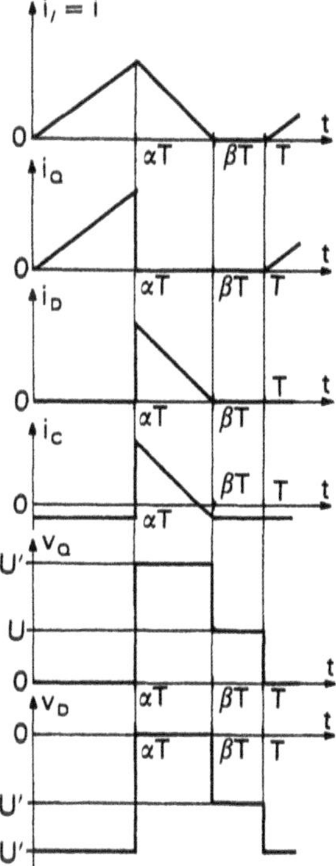

Fig. 6.10

- for $\alpha T < t < \beta T$,

$$l\frac{\mathrm{d}i_l}{\mathrm{d}t} = U - U', \quad \text{giving} \quad i_l = \frac{U}{l}\alpha T + \frac{U - U'}{l}(t - \alpha T);$$

- for $\beta T < t < T$,
  i_l is equal to zero.
  Since i_l falls to zero for $t = \beta T$, this gives

$$\frac{U}{l}\alpha T + \frac{U - U'}{l}(\beta - \alpha)T = 0$$

or

$$U' = \frac{\beta}{\beta - \alpha}U. \tag{6.13}$$

– As the mean value of the current in the capacitor is zero, this provides the equation linking β to the mean value I' of the load current:

$$C\left(\frac{\mathrm{d}u'}{\mathrm{d}t}\right)_{\text{mean}} = \frac{1}{T}\int_{\alpha T}^{\beta T} i_l \,\mathrm{d}t - I' = 0$$

yielding, if expression of i_l is taken into account,

$$I' = \frac{U}{2l}\alpha(\beta - \alpha)T. \tag{6.14}$$

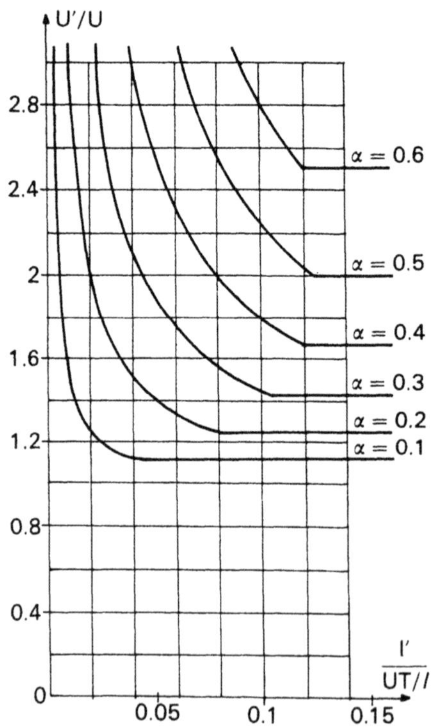

Fig. 6.11

By eliminating β between Eqs. (6.13) and (6.14), one gets

$$U' = U\,\frac{2lI' + \alpha^2 UT}{2lI'}. \tag{6.15}$$

The characteristics $U'(I')$, which correspond to different values of α in the zone where the current flow through the inductor is discontinuous (Fig. 6.11), can thus be plotted. U' is referred to U and I' to UT/l.

6.2.3 Buck–Boost Converter

The principle of the non-reversible chopper with inductive energy storage was presented in (Sect. 3.4.1). Inductor l acts as an intermediary current source between the voltage generator U at the input and the voltage receptor consisting of C connected in parallel with R at the output (Fig. 6.12).

This chopper can be regarded as the cascade connection of a buck chopper, connected between the voltage generator and the inductor, and a boost chopper, connected between the inductor and the voltage receptor. It is thus often referred as a buck–boost converter.

Figure 6.12 shows the waveforms of different variables in normal operating condition, i.e. when current i_l is always positive. The transistor is saturated from $t = 0$ to $t = \alpha T$, and in the OFF state from $t = \alpha T$ to $t = T$.

6.2.3.1 Average Values of Voltage u′ and Current i_l

Voltage u' and current i_l are given by:

$$\text{for } 0 < t < \alpha T, \quad ri_l + l\frac{di_l}{dt} = U \quad \text{and} \quad C\frac{du'}{dt} + \frac{u'}{R} = 0$$

$$\text{for } \alpha T < t < T, \quad ri_l + l\frac{di_l}{dt} + u' = 0 \quad \text{and} \quad C\frac{du'}{dt} + \frac{u'}{R} = i_l.$$

If the average values of u' and i_l are denoted by U' and I_l, this gives

$$rI_l + l\left(\frac{di_l}{dt}\right)_{\text{mean}} + \frac{1}{T}\int_{\alpha T}^{T} u'\,dt = \alpha U$$

$$C\left(\frac{du'}{dt}\right)_{\text{mean}} + \frac{U'}{R} = \frac{1}{T}\int_{\alpha T}^{T} i_l\,dt.$$

In steady state, the average values of di_l/dt and du'/dt are zero. Moreover, as the variations of i_l and u' are quasi-linear during each of the periods $(0, \alpha T)$ and $(\alpha T, T)$ the mean values of i_l and u' over the whole of the cycle can be taken as mean values during period $(\alpha T, T)$. This can be written as

$$rI_l + (1 - \alpha)U' = \alpha U$$

$$U'/R = I_l(1 - \alpha).$$

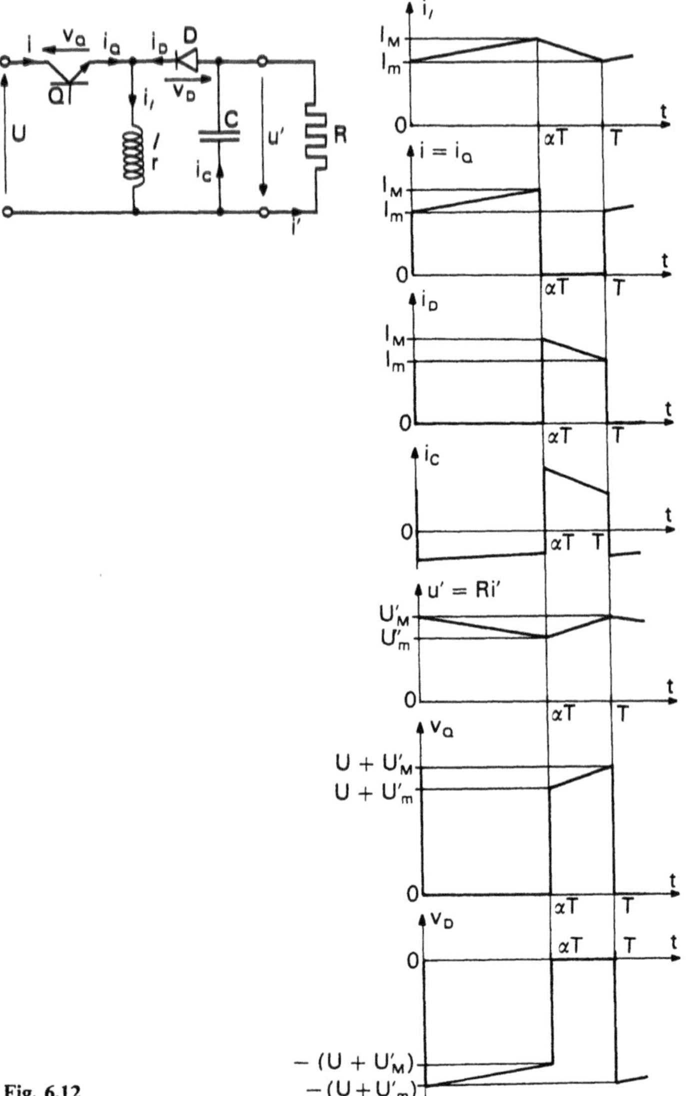

Fig. 6.12

From the above, the following can be deduced:

$$U' = U \frac{\alpha(1-\alpha)}{r/R + (1-\alpha)^2} \tag{6.16}$$

$$I_l = \frac{U'}{R(1-\alpha)} = \frac{I'}{(1-\alpha)} = \frac{U\alpha}{r + R(1-\alpha)^2} \tag{6.17}$$

If resistance r had been ignored, this would have given

$$U' = U\frac{\alpha}{(1-\alpha)} \tag{6.16'}$$

$$I_l = \frac{I'}{(1-\alpha)} = \frac{U\alpha}{R(1-\alpha)^2} \tag{6.17'}$$

- *The average voltage* U', which is equal to zero for zero α, increases firstly with α, reaches its maximum value and then decreases and falls back to zero for α equal to 1.

  The derivative of U' with respect to α is equal to zero for

$$\alpha_{\text{max}} = 1 + \frac{r}{R} - \sqrt{\frac{r}{R}\left(1 + \frac{r}{R}\right)}.$$

  U' is then at its maximum and has a value

$$U'_{\text{max}} = \frac{U}{2}\left(\sqrt{1 + \frac{R}{r}} - 1\right).$$

Figure 6.13 shows the variations of U'/U as a function of α, for r/R equal to 0.01 and 0.05 (The curve corresponding to zero r shows the great influence of r on the maximum value of U'.)

- *The average current* I_l increases from 0 to U/r when α changes from 0 to 1 (Fig. 6.14).

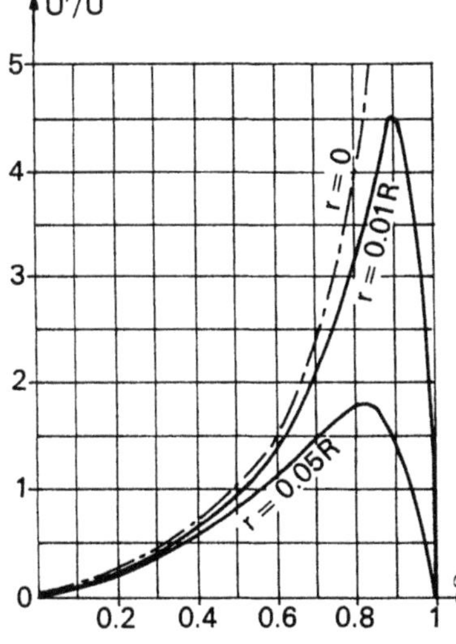

Fig. 6.13

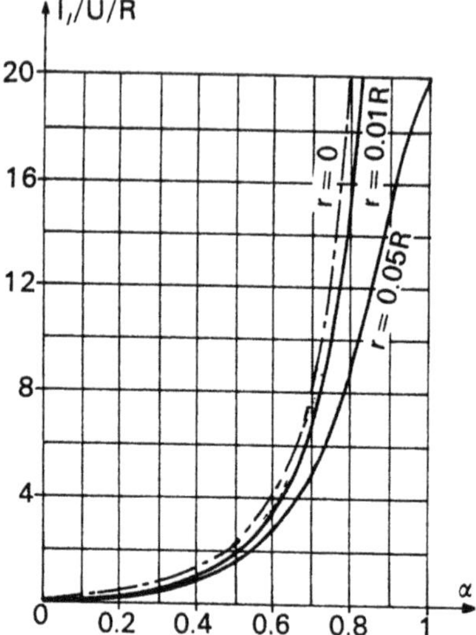

Fig. 6.14

6.2.3.2 Current i_l and Voltage u' Ripples

- *Current i_l ripple*

In order to calculate the current ripple $I_M - I_m$ in the inductor, the output voltage ripple is neglected. Furthermore, if resistance r is neglected, this gives:

- for $0 < t < \alpha T$, $l\dfrac{di_l}{t} = U$;

  thus, $i_l = I_m + \dfrac{U}{l}t$

  and $I_M = I_m + \dfrac{U}{l}\alpha T$;

- for $\alpha T < t < T$, $l\dfrac{di_l}{dt} = -U'$;

  thus, $i_l = I_M - \dfrac{U'}{l}(t - \alpha T)$.

The approximate value of the ripple is given by

$$\Delta i_l = I_M - I_m = \frac{U}{l}\alpha T. \tag{6.18}$$

When r is taken into account, one gets

$$\Delta i_l = \frac{U + U'}{r}\frac{1 - \exp[-(1-\alpha)rT/l]}{1 - \exp(-rT/l)}[1 - \exp(-\alpha rT/l)].$$

The difference is negligible if the value of l/r is high compared to T.

- *Voltage u′ ripple*

In evaluating $\Delta u'$, the inductor is considered as a perfect current generator:

$$\text{for } 0 < t < \alpha T, \quad 0 = C\frac{du'}{dt} + \frac{u'}{R};$$

$$\text{for } \alpha T < t < T, \quad I_l = C\frac{du'}{dt} + \frac{u'}{R}.$$

These equations are exactly the same as those found in the analysis of the boost chopper (Sect. 6.2.2.2). They thus give

$$\Delta u' = RI_l \frac{1 - \exp[-(1-\alpha)T/RC]}{1 - \exp(-T/RC)}[1 - \exp(-\alpha T/RC)].$$

If the value of RC is much higher than T, the following can be used:

$$\Delta u' = RI_l(1-\alpha)\frac{\alpha T}{RC}$$

or, since I_l equals $U'/R(1-\alpha)$,

$$\Delta u' = U'\frac{\alpha T}{RC} \tag{6.19}$$

6.2.3.3 Semiconductor Device Ratings

The comparison of the waveforms of i_Q and i_D in Fig. 6.12 with those in Fig. 6.7 shows how alike they are: for voltages v_Q and v_D, U must be added. Furthermore, Eqs. (6.19) and (6.18) which give $\Delta u'$ and Δi_l for the chopper with inductive energy storage are the same as Eqs. (6.10) and (6.11), which give these ripples for the boost chopper.

In determining the stresses applied to the semiconductor devices as well as to their variations depending on α, the same procedure can be adopted as for the boost chopper (Sect 6.2.2.3); however, for voltage stresses, U must be added.

6.2.3.4 Notes on Discontinuous Operation

If resistance r is neglected and Eqs. (6.17′) and (6.18) are taken into account, current i_l remains constantly positive only if

$$I_l = \frac{U\alpha}{R(1-\alpha)^2} \geqslant \frac{\Delta i_l}{2} = \frac{U}{2l}\alpha T$$

i.e. if resistance R is such that

$$R \leqslant \frac{2l}{(1-\alpha)^2 T} \tag{6.20}$$

– When value of resistance R is above this limit, current i_l becomes discontinuous. It increases from $t = 0$ to $t = \alpha T$, decreases from $t = \alpha T$ to $t = \beta T$ and remains zero from $t = \beta T$ to T (Fig. 6.15). The average output voltage U' is greater than that given by Eq. (6.16′) and becomes closely dependent on the average current I' delivered.

If voltage u' ripple is ignored,

• for $0 < t < \alpha T$,

$$l \frac{di_l}{dt} = U \quad \text{gives} \quad i_l = \frac{U}{l} t \, ;$$

• for $\alpha T < t < \beta T$,

$$l \frac{di_l}{dt} = -U' \quad \text{gives} \quad i_l = \frac{U}{l} \alpha T - \frac{U'}{l} (t - \alpha T) \, ;$$

• for $\beta T < t < T$,
i_l is zero.

As i_l falls to zero for $t = \beta T$, this gives

$$\frac{U}{l} \alpha T = \frac{U'}{l} (\beta - \alpha) T \, ;$$

$$\text{i.e. } U' = U \frac{\alpha}{\beta - \alpha}. \tag{6.21}$$

– The value of β is linked to that of current I' by

$$C \left(\frac{du'}{dt} \right)_{\text{mean}} = \frac{1}{T} \int_{\alpha T}^{\beta T} i_l \, dt - I' = 0$$

which gives

$$I' = \frac{U}{2l} \alpha (\beta - \alpha) T. \tag{6.22}$$

If β is eliminated between Eqs. (6.21) and (6.22), this leaves

$$U' = \frac{T(\alpha U)^2}{2lI'}. \tag{6.23}$$

Using these equations, it becomes possible to plot (Fig. 6.16), in the plane $I'/(UT/l)$, U'/U, the curves which show how, for a given α, U' increases when the delivered current decreases.

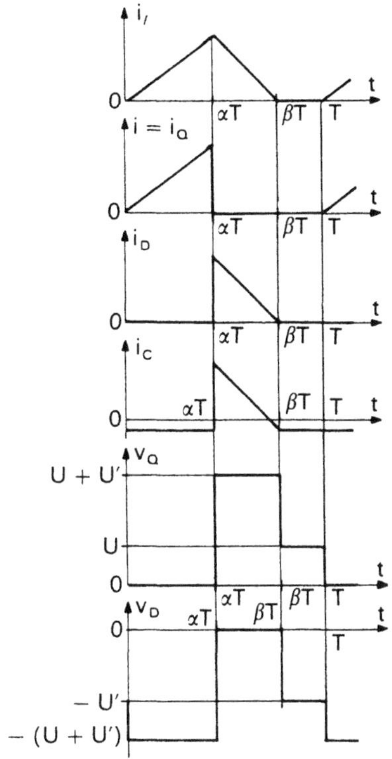

Fig. 6.15

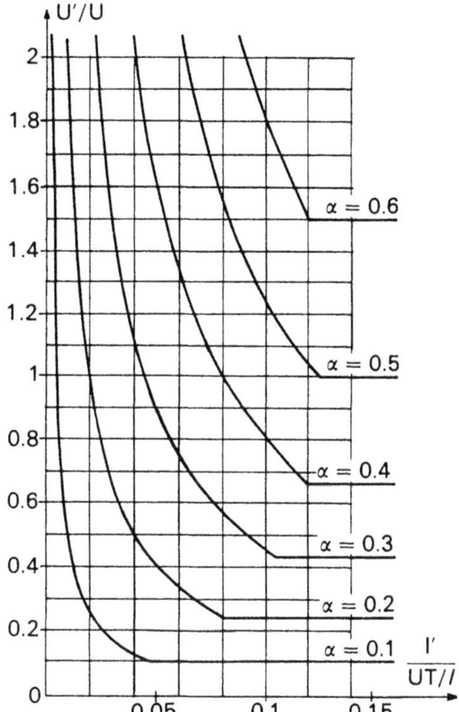

Fig. 6.16

6.2.4 Cuk Converter

The chopper with capacitive energy storage – generally referred as the Cuk converter – enables a current generator and a current receptor to be linked (see Chap. 3, § 4.3).

In the case of a switch-mode power supply, the RC type load must be changed into a current load by adding an inductor with parameters l_2, r_2. Similarly, the voltage source U must be changed into a current generator by adding a coil with parameters l_1, r_1. This provides the diagram shown at the top of Fig. 6.17, on which we have indicated the notations used.

Capacitor C_1 acts as an auxiliary voltage source. While C_1 is charging, the current source connected at the input of the convertor delivers its current to a load acting as a voltage source, as in a boost chopper. When C_1 is discharging, it acts as a voltage source supplying a current receptor, as in a buck chopper. The circuit can thus be seen as the cascade connection of a boost chopper and of a buck chopper. Therefore it is sometimes referred as the *boost–buck converter*.

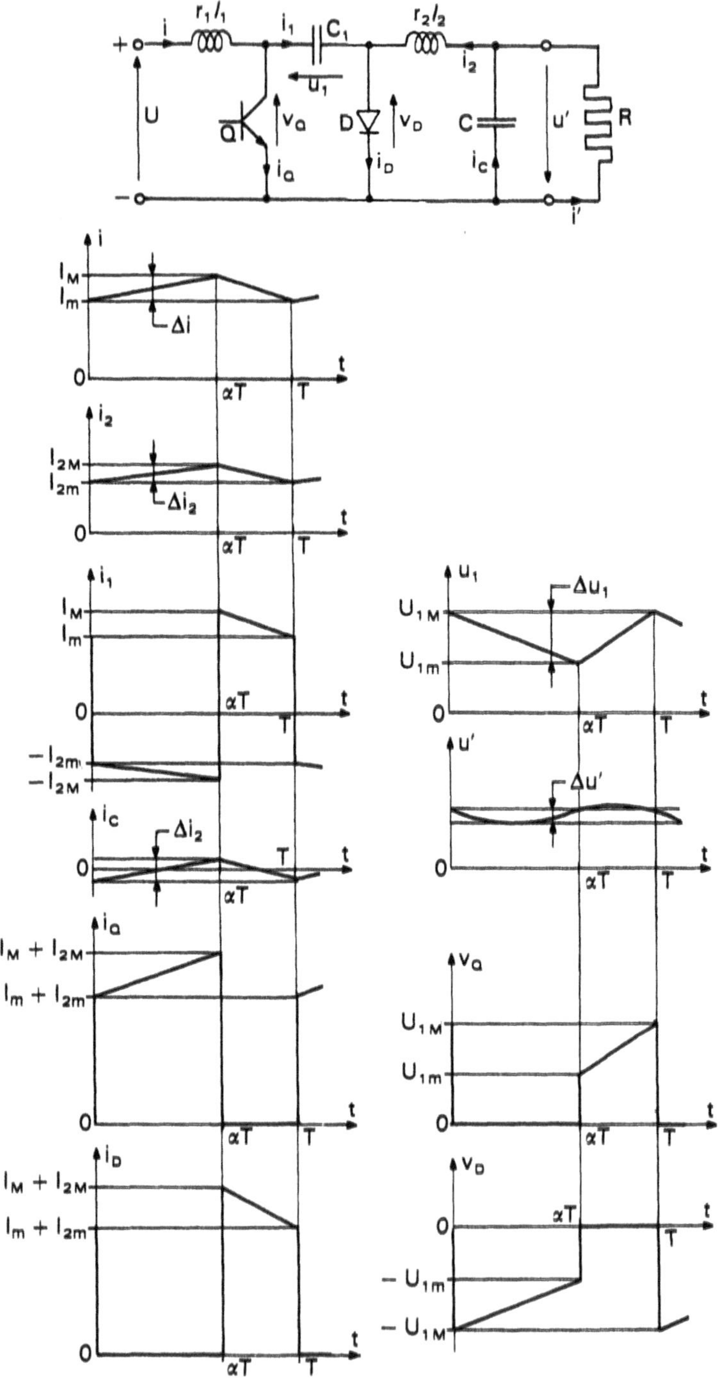

Fig. 6.17

In normal operating conditions, voltage u_1 across C_1 remains constantly positive, thus preventing any simultaneous conduction by transistor Q and diode D. The transistor is saturated between $t = 0$ and $t = \alpha T$, and, in the OFF state, from $t = \alpha T$ to $t = T$; the diode conducts during this second period. Figure 6.17 shows the waveforms of the different currents and voltages.

6.2.4.1 Average Values of Voltage u' and u_1, of Current i and i_2

Voltage u' and u_1 and currents i and i_2 are linked by

$$U = \begin{cases} r_1 i + l_1 \dfrac{di}{dt}, & \text{for } 0 < t < \alpha T, \\ r_1 i + l_1 \dfrac{di}{dt} + u_1, & \text{for } \alpha T < t < T; \end{cases}$$

$$i_2 = C\frac{du'}{dt} + \frac{u'}{R}, \qquad \text{for } \alpha 0 < t < T;$$

$$r_2 i_2 + l_2 \frac{di_2}{dt} + u' = \begin{cases} u_1, & \text{for } 0 < t < \alpha T, \\ 0, & \text{for } \alpha T < t < T; \end{cases}$$

$$C_1 \frac{du_1}{dt} = \begin{cases} -i_2, & \text{for } 0 < t < \alpha T, \\ i, & \text{for } \alpha T < t < T. \end{cases}$$

If U', U_1, I and I_2 are used to denote the average values of u', u_1, i and i_2, by integrating these over one cycle and taking the mean values, this gives

$$U = r_1 I + l_1 \left(\frac{di}{dt}\right)_{\text{mean}} + \frac{1}{T}\int_{\alpha T}^{T} u_1 \, dt$$

$$I_2 = C\left(\frac{du'}{dt}\right)_{\text{mean}} + \frac{U'}{R}$$

$$r_2 I_2 + l_2\left(\frac{di_2}{dt}\right)_{\text{mean}} + U' = \frac{1}{T}\int_0^{\alpha T} u_1 \, dt$$

$$C_1\left(\frac{du_1}{dt}\right)_{\text{mean}} = \frac{1}{T}\int_0^{\alpha T} (-i_2)\,dt + \frac{1}{T}\int_{\alpha T}^{T} i\,dt.$$

The mean values of di/dt, du'/dt, di_2/dt and du_1/dt are equal to zero. Assuming that the mean values of u_1, i_2 and i during the conduction interval of the transistor or during the conduction interval of the diode are the same as for the whole switching cycle, one gets

$$U = r_1 I + (1 - \alpha) U_1$$

$$I_2 = U'/R = I'$$

$$r_2 I_2 + U' = \alpha U_1$$

$$0 = -\alpha I_2 + (1 - \alpha) I.$$

From this it can be deduced that

$$U' = \frac{U}{(r_1/R)\alpha/(1-\alpha) + ((1-\alpha)/\alpha)(1 + r_2/R)} = RI_2 \tag{6.24}$$

$$U_1 = \frac{U'}{\alpha}\left(1 + \frac{r_2}{R}\right) = \frac{U}{(r_1/(R + r_2))\alpha^2/(1-\alpha) + 1 - \alpha} \tag{6.25}$$

$$I = \frac{\alpha}{1-\alpha} I_2 = \frac{\alpha}{1-\alpha} I' = \frac{U}{r_1 + (R + r_2)(1-\alpha)^2/\alpha^2}. \tag{6.26}$$

If resistances r_1 and r_2 had been ignored, this would have given

$$U' = U\frac{\alpha}{1-\alpha} = RI_2 \tag{6.24'}$$

$$U_1 = \frac{U'}{\alpha} = \frac{U}{1-\alpha} \tag{6.25'}$$

$$I = \frac{\alpha}{1-\alpha} I_2 = \frac{\alpha}{1-\alpha} I' = \frac{\alpha^2}{(1-\alpha)^2}\frac{U}{R}. \tag{6.26'}$$

- The average *output voltage* U' is equal to zero for zero α; it initially increases with α, reaches its maximum and then decreases and falls to zero when α equals 1. U' is at its maximum for α_{max} such as $dU'/d\alpha = 0$:

$$\alpha_{max} = \frac{1}{1 + \sqrt{r_1/(R + r_2)}}$$

and the maximum of U' equals

$$U'_{max} = U\frac{R}{2\sqrt{r_1(R + r_2)}}.$$

Resistance r_2 is usually very small compared with R; if $r_2 + R$ is replaced by R in the previous equations, this gives

$$U' = \frac{U}{(r_1/R)\alpha/(1-\alpha) + (1-\alpha)/\alpha}; \quad \alpha_{max} = \frac{1}{1 + \sqrt{r_1/R}};$$

$$U'_{max} = \frac{U}{2}\sqrt{\frac{R}{r_1}}$$

Starting from this simplified expression of U', Fig. 6.18 shows the variations in U' a function of α, for r_1 equal to zero, $0.01R$ and $0.05R$, respectively. The values found for U'_{max} are the same as in the case of the boost chopper, but U' starts from zero, as in the case of the chopper with inductive energy storage.

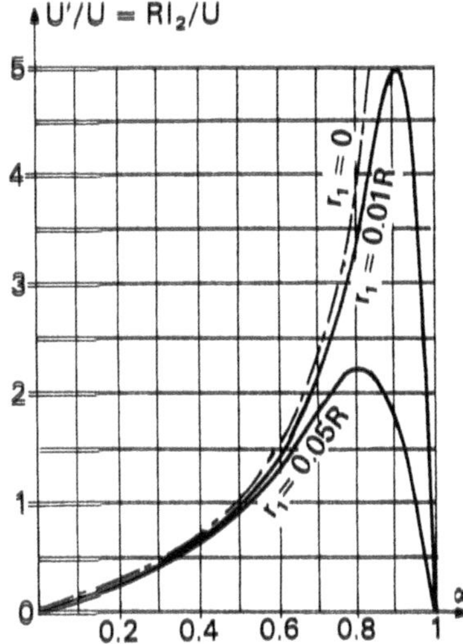

Fig. 6.18

- The average *voltage* U_1 *across the capacitor* is equal to U for α equal to zero. When α increases, this voltage also increases, reaches its maximum and then decreases. It is zero for α equal to 1.

  Its derivative with respect to α shows that its maximum is reached for

$$\alpha = 1 - \sqrt{\frac{r_1}{R + r_2 + r_1}}$$

and has a value

$$U_{1_{\max}} = \frac{U}{2} \frac{R + r_2}{\sqrt{r_1(R + r_2 + r_1)} - r_1}$$

If $R + r_2$ is replaced by R, as a first approximation, one gets

$$U_1 = \frac{U}{(r_1/R)\alpha^2/(1 - \alpha) + 1 - \alpha}; \qquad U_{\max} = \frac{U}{2} \frac{R}{\sqrt{r_1(R + r_1)} - r_1}$$

for $\alpha = 1 - \sqrt{\dfrac{r_1}{R + r_1}}$.

Using the latter equations, it is possible to plot (Fig. 6.19) the variations of U_1 as a function of α, for r_1 equal to zero, $0.01R$ and $0.05R$, respectively.

When the output voltage increases, the voltage across capacitor C_1 also increases.

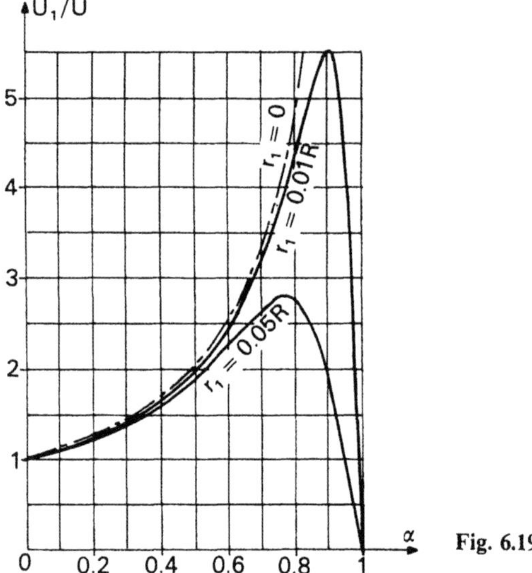

Fig. 6.19

- The average current I delivered by the voltage source U and flowing through inductor l_1, increases continuously when α changes from 0 to 1. It is zero for zero α and equals U/r_1 when α is unity.

  The curves in Fig. 6.20 have been plotted for $r_1 = 0$, $r_1 = 0.01R$ and $r_1 = 0.05R$, assuming that r_2 is negligible with respect to R.

  Average current I_2 in l_2 can be found in the curves which give U', since RI_2 equals U'.

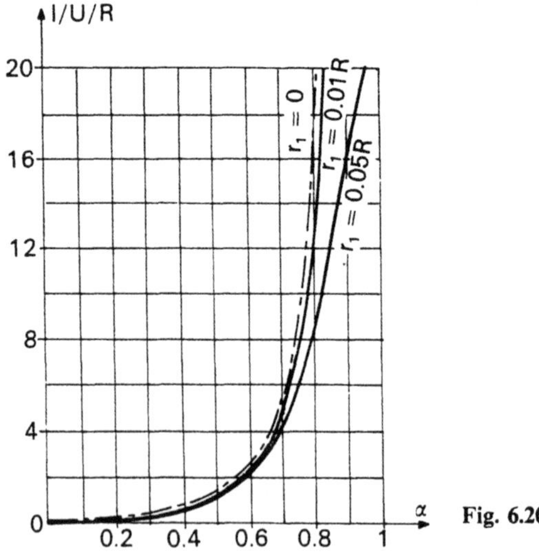

Fig. 6.20

6.2.4.2 Current i and i_2, Voltage u' and u_1 Ripples

- *Current i and i_2 ripples*

In this part of the analysis, the RC load and the capacitor C_1 are taken as perfect voltage sources of voltages U' and U_1, respectively. The equivalent diagrams corresponding to the two configurations found during each switching cycle are thus those shown in Fig. 6.21.

From $t = 0$ to $t = \alpha T$,

$$r_1 i + l_1 \frac{di}{dt} = U, \quad i \text{ increases from } I_m \text{ to } I_M$$

$$r_2 i_2 + l_2 \frac{di_2}{dt} = U_1 - U', \quad i_2 \text{ increases from } I_{2m} \text{ to } I_{2M}$$

From $t = \alpha T$ to $t = T$,

$$r_1 i + l_1 \frac{di}{dt} = U - U_1, \quad i \text{ decreases from } I_M \text{ to } I_m$$

$$r_2 i_2 + l_2 \frac{di_2}{dt} = -U', \quad i_2 \text{ decreases from } I_{2M} \text{ to } I_{2m}.$$

– if resistances r_1 and r_2 are neglected, the currents are linear functions of the time. Their ripples can be expressed as

$$\Delta i = \frac{U}{l_1}\alpha T; \quad \Delta i_2 = \frac{U_1 - U'}{l_2}\alpha T = \frac{U}{l_2}\alpha T. \tag{6.27}$$

– If resistances r_1 and r_2 are taken into account, this gives

$$\Delta i = \frac{U_1}{r_1}\frac{1 - \exp[-(1-\alpha)Tr_1/l_1]}{1 - \exp(-Tr_1/l_1)}[1 - \exp(-\alpha Tr_1/l_1)]$$

$$\Delta i_2 = \frac{U_1}{r_2}\left(1 - \frac{\exp(\alpha Tr_2/l_2) - 1}{\exp(Tr_2/l_2) - 1}\right)[1 - \exp(-\alpha Tr_2/l_2)].$$

If l_1/r_1 and l_2/r_2 are high compared with the cycle T, the difference between

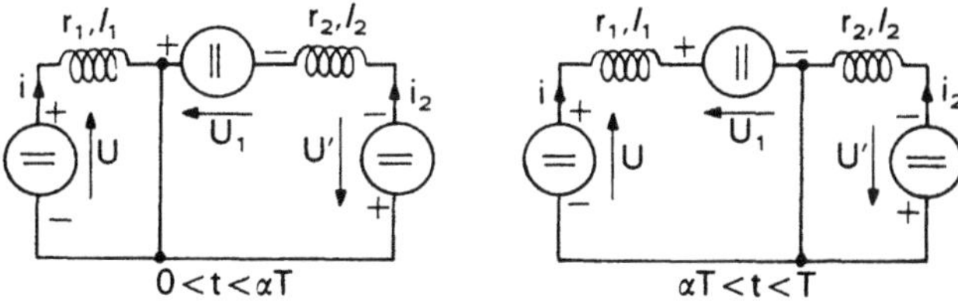

Fig. 6.21

the values of Δi and Δi_2 given by these equations and those provided by Eqs. (6.27) is negligible. The latter show that the ripples are proportional to α.

- *Voltage u' and u_1 ripples*

- The converter output circuit is the same as for the buck chopper. The ripple $\Delta u'$ of voltage u' across the RC circuit can be deduced from the ripple Δi_2 of the current which flows through it (see Sect. 6.2.1.2):

$$\Delta u' = \frac{1}{C} \Delta i_2 \frac{T}{8};$$

thus, taking Eq. (6.27) into account,

$$\Delta u' = U \frac{\alpha T^2}{8 l_2 C} \tag{6.28}$$

$\Delta u'$ increases proportionately to α.

- The ripple Δu_1 of voltage u_1 across capacitor C_1 can be calculated from its charge or discharge current i_1.

Between $t = 0$ and $t = \alpha T$, i_1 equal to $-i_2$ is negative. The variation in u_1 is:

$$-C_1 \Delta u_1 = -\int_0^{\alpha T} i_2 \, dt.$$

If the variation in i_2 is assumed to be quasi-linear:

$$C_1 \Delta u_1 = I_2 \alpha T = \frac{U'}{R} \alpha T$$

$$\Delta u_1 = U' \frac{\alpha T}{R C_1}. \tag{6.29}$$

Since U' decreases when the value of α is close to 1, ripple Δu_1 also reaches a maximum before falling, like U', to zero for $\alpha = 1$. But this maximum occurs for a value of α which is greater than that, α_{max}, which gives U' its maximum value.

6.2.4.3 Semiconductor Device Ratings

- As shown in Fig. 6.7, the forward voltage across Q and the reverse voltage across D have the same peak value:

$$v_{Q\,max} = -v_{D\,max} = U_{1M} = U_1 + \frac{\Delta u_1}{2}.$$

The transistor and diode currents also have the same peak value:

$$i_{Q\,max} = i_{D\,max} = I_M + I_{2M} = I + \frac{\Delta i}{2} + I_2 + \frac{\Delta i_2}{2}.$$

The transistor current has an average value

$$I_Q = \alpha(I + I_2) = \alpha\left(I + \frac{1-\alpha}{\alpha} I\right) = I.$$

The diode current has an average value

$$I_D = (1-\alpha)(I + I_2) = (1-\alpha)\left(\frac{\alpha I_2}{1-\alpha} + I_2\right) = I_2.$$

- Since $U_1, \Delta u_1, I, \Delta i, I_2, \Delta i_2$ increase when α changes from zero to α_{max}, the preceding equations are used to determine – in the case of α_{max} – the maximum voltage and current stresses which the transistor and the diode have to withstand.

6.2.4.4 Remark

The operation described assumes a voltage u_1 which remains constantly positive so that the diode may remain in the OFF state when the transistor is conducting.

For this to be the case, $\Delta u_1/2$ must be less than U_1. Using the simplified Eqs. (6.25′) and (6.29), this condition can be written as

$$\frac{1}{2} U' \frac{\alpha T}{RC_1} = \frac{1}{2} RI' \frac{\alpha T}{RC_1} \leqslant \frac{U}{1-\alpha}$$

giving

$$I' \leqslant \frac{2C_1 U}{\alpha(1-\alpha)T} \tag{6.30}$$

Equation (6.30) indicates the minimal value to be given to the storage capacitor C_1, in view of the maximum current the converter must be able to deliver. Since $\alpha(1-\alpha)$ is at its maximum for $\alpha = 0.5$ and has then a value of 0.25, C_1 must be such that

$$C_1 \geqslant \frac{1}{8} \frac{I'T}{U}.$$

6.3 Asymmetrical Circuits with Transformer

6.3.1 General Remarks

- The previous four circuits do not provide a galvanic insulation between input and output. Moreover, they do not allow for an average output voltage being greatly different from the input one. For this to be the case, they would have to operate with values of α very close to 0 or to 1:
  - For these values of α the duration of the commutations can no longer be ignored. Nor can the minimal values which they impose on the on-time and the off-time of the controlled switch.

- The closer α is to 1, the greater the ripples of the different variables: this reduces the quality of the output voltage, increases the interferences in the source supplying the converter, leads the switch current and voltage ratings to rise.
- In the case of the boost converter or of the step-up operation of the converters with intermediate energy storage, it has been seen that the ratio of the output voltage to the input voltage could not reach very high values, on account of the resistance r or r_1 of the coil of inductance l or l_1, respectively.

If galvanic insulation is necessary and/or if the output voltage has to be noticeably different from the input voltage U, a transformer must be used. The latter enables the desired value of U' (or the desired range of values of U') to be situated near the middle of the range of possible output voltages.

- Two types of circuit with transformer can be distinguished:
  - in a "symmetrical" circuit, an AC current supplies the transformer;
  - in an "asymmetrical" circuit, this current is unidirectional.

The first type requires two controlled switches. It must be analyzed as an inverter. We will limit our description to a single topology – the push–pull circuit.

We will thus concentrate essentially on the more usual types of asymmetrical circuit:
- the flyback circuit derived from the chopper with inductive energy storage;
- the forward circuit derived from the buck chopper;
- the different combinations and variants of the above ones.

6.3.2 Flyback Converter

The flyback converter (Fig. 6.22) can be deduced from Fig. 6.12. The inductive storage element is replaced by a transformer; the transistor is connected to the primary winding, the diode and the load to the secondary winding. This simple circuit is used for output voltages above 10 V and for power up to 200 W.

The operation of the transformer can be analyzed by assigning inductances l_1 and l_2 to windings with number of turns n_1 and n_2, and with resistances r_1 and r_2. The values of l_1 and l_2 depends on the permeability of the material which the

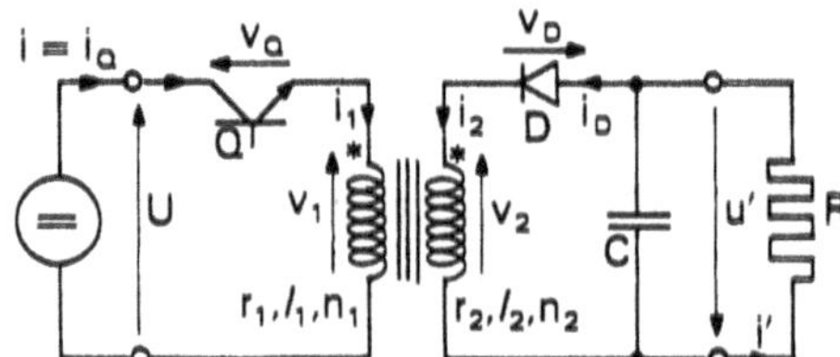

Fig. 6.22

core of the transformer is made of and on the magnetomotive force, equal to $n_1 i_1 + n_2 i_2$. If the width of the hysteresis cycle of the material is ignored and the slope of this cycle taken at the point corresponding to the mean value of $n_1 i_1 + n_2 i_2$, the inductances l_1 and l_2 can be assumed constant.

Transistor Q is saturated from $t = 0$ to $t = \alpha T$, blocked from $t = \alpha T$ to $t = T$. Figure 6.23a shows the waveforms of the different variables for operation with "*incomplete demagnetization*", i.e. when the ampere-turns sum $n_1 i_1 + n_2 i_2$ remains constantly positive.

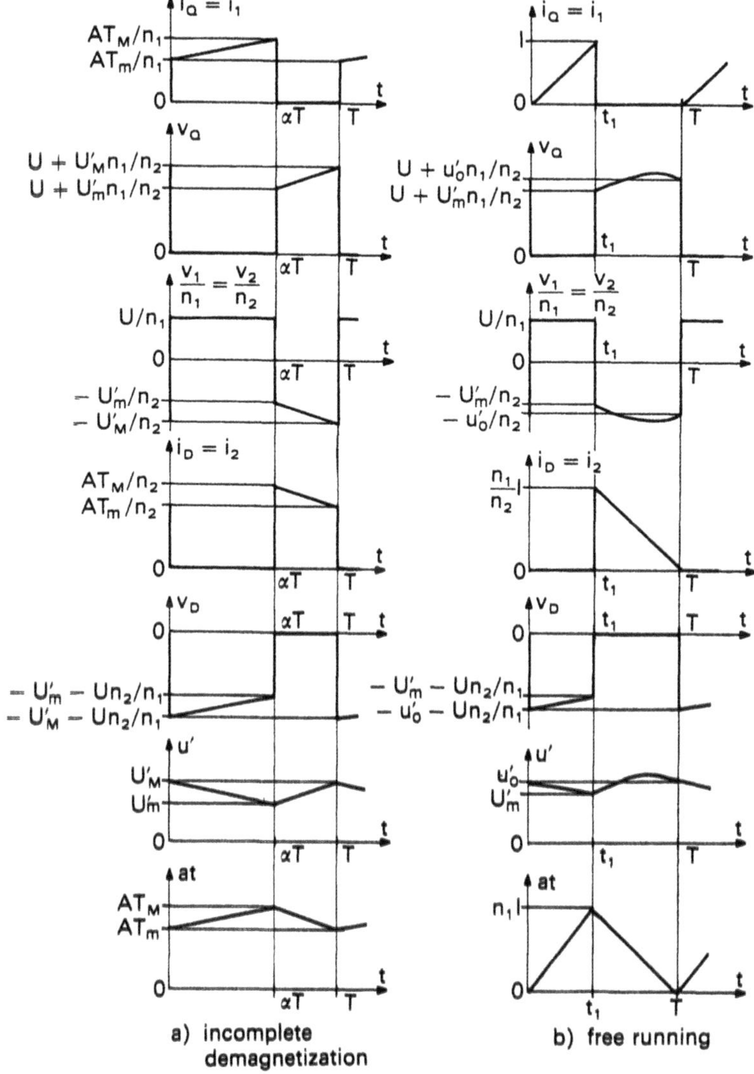

Fig. 6.23

6.3.2.1 Average Values of Voltage u' and Magnetomotive Force at

- Voltages U and u', currents i_1 and i_2, and magnetomotive force at, equal to $n_1 i_1 + n_2 i_2$, are linked by the following equations:
  - From $t = 0$ to $t = \alpha T$ (Q on, D off):

$$U = r_1 i_1 + l_1 \frac{di_1}{dt}, \quad \text{with} \quad i_1 = \frac{\text{at}}{n_1} \quad \text{since} \quad i_2 = 0$$

$$0 = C \frac{du'}{dt} + \frac{u'}{R}$$

or, if the first equation is multiplied by $l_2 n_1$ and the second by n_2,

$$l_2 n_1 U = l_2 r_1 \,\text{at} + l_2 l_1 \frac{d\text{at}}{dt} \tag{6.31a}$$

$$0 = n_2 C \frac{du'}{dt} + n_2 \frac{u'}{R}. \tag{6.31b}$$

  - From $t = \alpha T$ to $t = T$ (Q off, D on):

$$0 = r_2 i_2 + l_2 \frac{di_2}{dt} + u', \quad \text{with} \quad i_2 = \frac{\text{at}}{n_2} \quad \text{since} \quad i_1 = 0$$

$$i_2 = C \frac{du'}{dt} + \frac{u'}{R}$$

or, if the first equation is multiplied by $l_1 n_2$ and the second by n_2,

$$0 = l_1 r_2 \,\text{at} + l_1 l_2 \frac{d\text{at}}{dt} + l_1 n_2 u' \tag{6.32a}$$

$$\text{at} = n_2 C \frac{du'}{dt} + n_2 \frac{u'}{R}. \tag{6.32b}$$

By integrating from 0 to T, by taking the average value of the different terms and by denoting the average values of u' and at as U' and AT, Eqs. (6.31a) and (6.32a) give

$$\alpha l_2 n_1 U = \frac{l_2 r_1}{T} \int_0^{\alpha T} \text{at}\,dt + l_2 l_1 \left(\frac{d\text{at}}{dt}\right)_{\text{mean}} + \frac{l_1 r_2}{T} \int_{\alpha T}^{T} \text{at}\,dt + \frac{l_1 n_2}{T} \int_{\alpha T}^{T} u'\,dt$$

and Eqs. (6.31b) and (6.32b) give

$$\frac{1}{T} \int_{\alpha T}^{T} \text{at}\,dt = n_2 C \left(\frac{du'}{dt}\right)_{\text{mean}} + n_2 \frac{U'}{R}.$$

The mean values of du'/dt and $d\text{at}/dt$ are zero. By taking as the average values of u' and at during the period $(0, \alpha T)$ or the period $(\alpha T, T)$, the same values as for the whole cycle, this gives

$$\alpha l_2 n_1 U = \alpha l_2 r_1 \mathrm{AT} + (1-\alpha) l_1 r_2 \mathrm{AT} + (1-\alpha) l_1 n_2 U'$$

$$(1-\alpha)\mathrm{AT} = n_2 U'/R.$$

From this it can be deduced that

$$U' = \frac{\alpha(1-\alpha)}{\alpha r'_1/R + (1-\alpha) r_2/R + (1-\alpha)^2} \frac{n_2}{n_1} U \tag{6.33}$$

$$\mathrm{AT} = \frac{n_2}{1-\alpha}\frac{U'}{R} = \frac{\alpha}{\alpha r'_1 + (1-\alpha) r_2 + (1-\alpha)^2 R}\frac{n_2^2}{n_1} U \tag{6.34}$$

by denoting r'_1 the resistance of the primary referred to the secondary;

$$r'_1 = r_1 (n_2/n_1)^2 \tag{6.35}$$

and by using the fact that

$$(l_2/l_1) = (n_2/n_1)^2. \tag{6.36}$$

- The *average output voltage* U', which is zero for α equal to zero, increases with α, reaches its maximum and then decreases and falls to zero once again for α equal to 1.

  The derivative of U' with respect to α falls to zero for α equal to α_{max} such that

$$\alpha_{max} = \frac{1}{1 + \sqrt{r'_1/(R + r_2)}}.$$

  U' is then at its maximum, with a value

$$U'_{max} = \frac{R}{r'_1 + r_2 + 2\sqrt{r'_1 (R + r_2)}} \frac{n_2}{n_1} U.$$

- The *average magnetomotive force* AT, which is zero for α equal to zero, increases constantly when α changes from zero to 1. For α equal to 1, it would be equal to $n_1 U/r_1$.

  In practice, α_{max} is not exceeded, the corresponding value of AT is

$$\mathrm{AT}(\alpha_{max}) = \frac{n_2^2 U}{n_1} \frac{1 + \sqrt{(R + r_2)/r'_1}}{r'_1 + r_2 + 2\sqrt{r'_1 (R + r_2)}}.$$

The characteristics which give U' and AT as a function of α for the flyback converter are of the same type as those giving U' and I_l for the chopper with inductive energy storage (see Figs. 6.13 and 6.14).

It can, moreover, be seen that, if $r'_1 = r_2$, Eq. (6.33) is the same as Eq. (6.16), provided only that r'_1 and r_2 are replaced by r and U by $(n_2/n_1)U$.

A comparison of Eqs. (6.17) and (6.34) shows that the relation between

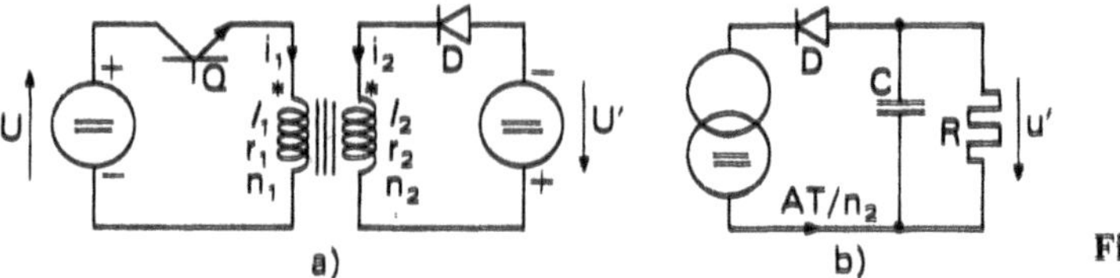

Fig. 6.24

output voltage and currents is the same:

$$I_l = \frac{U'}{R(1-\alpha)}; \quad \frac{\mathrm{AT}}{n_2} = \frac{U'}{R(1-\alpha)}.$$

If $r'_1 = r_2$, the characteristics of Figs. 6.13 and 6.14 can be used to read $U'/(n_2/n_1)U$ and $\mathrm{AT}/(n_2 U/R)$.

6.3.2.2 Magnetomotive Force and Output Voltage Ripples

To provide an estimate of ripple Δat of the MMF, the RC circuit is assumed to be a perfect DC voltage source U' (Fig. 6.24a). To estimate the ripple $\Delta u'$ of the output voltage, the secondary of the transformer is assumed to act as a perfect square-wave current source (Fig. 6.24b).

- *MMF ripple*

From $t = 0$ to $t = \alpha T$, $\quad U = r_1 i_1 + l_1 \dfrac{\mathrm{d}i_1}{\mathrm{d}t}$, $\quad$ with $n_1 i_1 = \mathrm{at}$,
the MMF increases from $\mathrm{AT_m}$ to $\mathrm{AT_M}$ according to the equation

$$\mathrm{at} = \left(\mathrm{AT_m} - n_1 \frac{U}{r_1}\right)\exp(-tr_1/l_1) + n_1 \frac{U}{r_1}.$$

From $t = \alpha T$ to $t = T$, $-U' = r_2 i_2 + l_2 \mathrm{d}i_2/\mathrm{d}t$, with $n_2 i_2 = \mathrm{at}$;
the MMF decreases from $\mathrm{AT_M}$ to $\mathrm{AT_m}$ according to the equation

$$\mathrm{at} = \left(\mathrm{AT_M} + n_2 \frac{U'}{r_2}\right)\exp(-tr_2/l_2) - n_2 \frac{U'}{r_2}.$$

By noting that at takes once more the value $\mathrm{AT_m}$ at instant $t = T$, it is possible to obtain $\mathrm{AT_M}$, $\mathrm{AT_m}$ and their difference Δat:

$$\Delta\mathrm{at} = \left(n_1 \frac{U}{r_1} + n_2 \frac{U'}{r_2}\right) \times \frac{[1 - \exp(-\alpha T r_1/l_1)](1 - \exp[-(1-\alpha)Tr_2/l_2])}{1 - \exp(-\alpha T r_1/l_1)\exp[-(1-\alpha)Tr_2/l_2]}.$$

If $\alpha T r_1/l_1$ and $(1-\alpha)Tr_2/l_2$ are much smaller than 1, the above can be written as

$$\Delta\mathrm{at} \simeq \left(n_1 \frac{U}{r_1} + n_2 \frac{U'}{r_2}\right) \frac{(\alpha T r_1/l_1)[(1-\alpha)Tr_2/l_2]}{\alpha T r_1/l_1 + (1-\alpha)Tr_2/l_2}.$$

By taking as U' the simplified expression obtained by neglecting r'_1 and r_2, i.e.

$$U' \simeq \frac{\alpha}{1-\alpha} \frac{n_2}{n_1} U \tag{6.33'}$$

one gets

$$\Delta\text{at} \simeq n_1 \frac{U}{l_1} \alpha T. \tag{6.37}$$

This expression is analogous to Eq. (6.18) worked out from the same assumptions in the case of the chopper with inductive energy storage. The MMF ripple increases proportionally to α.

• *Output voltage ripple.*

From $t = 0$ to $t = \alpha T$, $0 = C\,\mathrm{d}u'/\mathrm{d}t + u'/R$.

From $t = \alpha T$ to $t = T$, $\mathrm{AT}/n_2 = C\,\mathrm{d}u'/\mathrm{d}t + u'/R$.

These equations are identical to those which give the output voltage ripple of the boost chopper and of the chopper with inductive energy storage. Thus it can be deduced, in the same way, that

$$\Delta u' = R \frac{\mathrm{AT}}{n_2} \frac{1 - \exp[-(1-\alpha)T/RC]}{1 - \exp(-T/RC)} [1 - \exp(-\alpha T/RC)].$$

If RC is much greater than T,

$$\Delta u' \simeq R \frac{\mathrm{AT}}{n_2} (1-\alpha) \frac{\alpha T}{RC}$$

or, since AT equals $n_2 U'/R(1-\alpha)$,

$$\Delta u' \simeq U' \frac{\alpha T}{RC} \tag{6.38}$$

6.3.2.3 Semiconductor Device Ratings

• *Transistor*

For given operating conditions, the average I_Q and peak $i_{\mathrm{Q\,max}}$ current values in the transistor and the maximum value $v_{\mathrm{Q\,max}}$ of the voltage across it are given by

$$I_\mathrm{Q} = \alpha \frac{\mathrm{AT}}{n_1}; \quad i_{\mathrm{Q\,max}} = \frac{1}{n_1}\left(\mathrm{AT} + \frac{\Delta\mathrm{at}}{2}\right)$$

$$v_{\mathrm{Q\,max}} = U + \frac{n_1}{n_2}\left(U' + \frac{\Delta u'}{2}\right).$$

Since AT, Δat, U' and $\Delta u'$ increase with α when the latter moves from zero to the value $\alpha_{\max}$ corresponding to the maximum value of the average output voltage U', the stresses on the transistor must be calculated for $\alpha_{\max}$.

- *Diode*

For given operating conditions,

$$I_D = (1 - \alpha)\frac{AT}{n_2} = \frac{U'}{R}$$

$$i_{D\,max} = \frac{1}{n_2}\left(AT + \frac{\Delta at}{2}\right)$$

$$-v_{D\,max} = U\frac{n_2}{n_1} + U' + \frac{\Delta u'}{2}.$$

The ratings of the diode will also be determined for α equal to α_{max}.

6.3.2.4 Operation in the Free-Running Mode

Instead of operating the flyback circuit at a fixed frequency and regulating the average output voltage U' by means of the switching duty ratio α, this circuit can be operated in a free-running mode.

U' is regulated by imposing the maximum value I of the collector current i_Q of the transistor. The voltage and current waveforms are, in that case, those which are shown on Fig. 6.23b. The transistor is conducting between $t = 0$ and $t = t_1$ while i_Q rises from zero to I. When i_Q reaches I, the transistor is turned off: the diode starts to conduct and ensures the demagnetization of the magnetic core. When i_D falls to zero, for $t = T$, the transistor is once more turned on: a new cycle begins.

This control mode has the following advantage: the transistor is switched on and the diode is turned off when the current is zero. This reduces the interferences inevitably caused by switching.

a) *Currents*

- From $t = 0$ to $t = t_1$, Q is conducting and D in the OFF state:

$$U = r_1 i_1 + l_1\frac{di_1}{dt};$$

As a first approximation, r_1 can be neglected

$$i_1 = \frac{U}{l_1}t; \quad \text{thus, } t_1 = \frac{l_1 I}{U}$$

- From $t = t_1$ to $t = T$, Q is in the OFF state and D is conducting:

$$-u' = r_2 i_2 + l_2\frac{di_2}{dt};$$

As a first approximation, r_2 and the voltage u' ripple can be neglected. Furthermore, the MMF cannot suffer any discontinuity and thus current i_2 takes on value $n_1 I/n_2$ at instant $t = t_1$.

In these conditions,

$$l_2 \frac{di_2}{dt} = -U' \text{ gives}$$

$$i_2 = \frac{n_1}{n_2} I - \frac{U'}{l_2}(t - t_1)$$

Current i_2 falls to zero for $t = T$. From this, the equation linking U', I and cycle T can be deduced:

$$T - t_1 = l_2 \frac{n_1}{n_2} \frac{I}{U'}$$

$$T = \frac{l_1 I}{U} + l_2 \left(\frac{n_1}{n_2}\right)^2 \frac{n_2}{n_1} \frac{I}{U'}$$

$$T = \frac{l_1 I}{U}\left(1 + \frac{n_2}{n_1} \frac{U}{U'}\right). \tag{6.39}$$

- The magnetomotive force $n_1 i_1 + n_2 i_2$ rises linearly from 0 to $n_1 I$ and then falls linearly from $n_1 I$ to 0.

  This gives its average value and its ripple:

$$\mathrm{AT} = \frac{n_1 I}{2}; \quad \Delta \mathrm{at} = n_1 I. \tag{6.40}$$

b) *Output voltage*

- From $t = 0$ to $t = t_1$,

$$0 = C \frac{du'}{dt} + \frac{u'}{R}$$

Voltage u' decreases, starting from its initial value u'_0, in accordance with the equation

$$u' = u'_0 \exp(-t/RC).$$

For $t = t_1$, it reaches its minimum value U'_m:

$$U'_m = u'_0 \exp(-t_1/RC).$$

- From $t = t_1$ to $t = T$,

$$i_2 = C \frac{du'}{dt} + \frac{u'}{R}$$

If i_2 is assumed to decrease linearly,

$$RC \frac{du'}{dt} + u' = R \frac{n_1}{n_2} I - R \frac{n_1}{n_2} I \frac{t - t_1}{T - t_1}.$$

Voltage u' can thus be expressed as

$$u' = R\frac{n_1}{n_2}I\left(\frac{T+RC}{T-t_1} - \frac{t}{T-t_1}\right)$$
$$+\left[U'_{\mathrm{m}} - R\frac{n_1}{n_2}I\frac{T-t_1+RC}{T-t_1}\right]\exp[-(t-t_1)/RC].$$

The voltage increases and reaches a maximum at the instant when, u' becoming equal to Ri_2, the derivative $\mathrm{d}u'/\mathrm{d}t$ falls to zero; u' then decreases and once more takes on value u'_0 at the end of cycle T.

- *The average value* U' of the output voltage can be deduced from the fact that the current in the capacitor has a zero average value. The current in resistance R has the same average value as i_2:

$$\frac{U'}{R} = \frac{1}{2}\frac{n_1}{n_2}I\frac{T-t_1}{T}.$$

If $T - t_1$ and T are replaced by their values as worked out in the preceding section, this results in

$$\frac{U'}{R} = \frac{1}{2}\frac{n_1}{n_2}I\frac{(l_1I/U)(1+(n_2/n_1)(U/U')-1)}{(l_1I/U)(1+(n_2/n_1)U/U')}$$

$$U' = \frac{RI}{2}\frac{1}{(U'/U + n_2/n_1)}.$$

The solution of the resulting second-degree equation in U' gives, since only a positive solution is acceptable:

$$U' = \frac{U}{2}\frac{n_2}{n_1}\left[\sqrt{1+2\left(\frac{n_1}{n_2}\right)^2\frac{RI}{U}} - 1\right]. \tag{6.41}$$

The curve in full line in Fig. 6.25 shows how U' increases as a function of I. U' is referred to Un_2/n_1 and I to U/R', by denoting as R' the load resistance referred to the primary:

$$R' = R(n_1/n_2)^2.$$

- For a given resistance R, I must be increased in order to increase U'.
- To keep U' constant, if R increases, I must be decreased so as to keep the product RI constant.

In the same figure, the curve in dashed lines represents the variations of the operating cycle T as a function of $I/(U/R')$. This cycle is referred to l_1/R'. In plotting this curve, Eq. (6.39) has been used, in the following form:

$$\frac{T}{l_1/R'} = \frac{I}{U/R'}\left(1 + \frac{n_2}{n_1}\frac{U}{U'}\right).$$

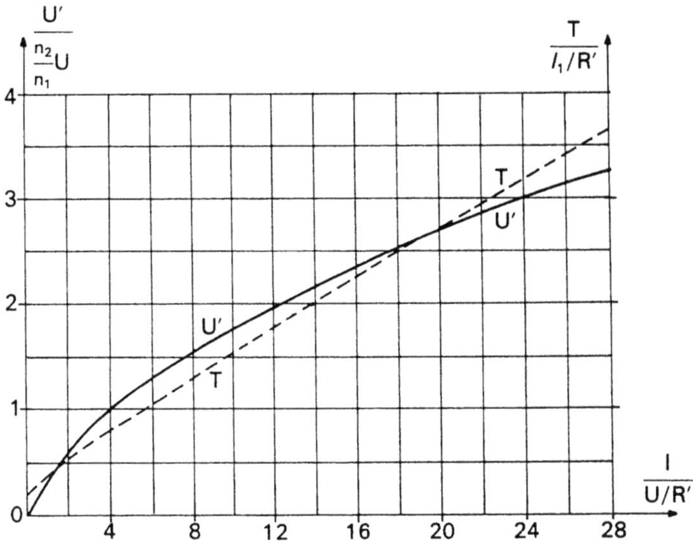

Fig. 6.25

- For a given resistance R, the switching frequency decreases when U' is increased by raising I.
- At a constant voltage U', when R increases, the ratio $R'T/l_1$ remains constant: the switching frequency increases proportionally to R.

- Output voltage *ripple* $\Delta u'$ can be approximately evaluated using period $(0, t_1)$, assuming $U'_M - U'_m$ equal to $u'_0 - U'_m$ and taking for U' the average value of u' during this period.

$$u' = u'_0 \exp(-t/RC)$$

then gives

$$\Delta u' = u'_0[1 - \exp(-t_1/RC)]$$

$$U' = \frac{1}{t_1}\int_0^{t_1} u' \mathrm{d}t = u'_0 \frac{RC}{t_1}[1 - \exp(-t_1/RC)]$$

$$\frac{\Delta u'}{U'} = \frac{t_1}{RC}$$

or, taking the value of t_1 into account,

$$\Delta u' \simeq \frac{U'}{U}\frac{l_1 I}{RC} \tag{6.42}$$

c) *Semiconductor device ratings*

For the transistor,

$$i_{Q\,max} = I\,; \quad I_Q = I\frac{t_1}{2T} = \frac{I}{2}\frac{1}{1 + (n_2/n_1)U/U'}$$

$$v_{Q\,max} = U + \frac{n_1}{n_2}\left(U' + \frac{\Delta u'}{2}\right).$$

For the diode,

$$i_{D\,max} = \frac{n_1}{n_2}I\,; \quad I_D = \frac{n_1}{n_2}I\frac{T - t_1}{2T} = \frac{U'}{R}$$

$$v_{D\,max} = U\frac{n_2}{n_1} + U' + \frac{\Delta u'}{2}.$$

As U' and $\Delta u'$ increase with I, the ratings of the semiconductor devices have to be determined for the maximum value of I, using the previous equations.

6.3.2.5 Remarks

- For operation at constant switching frequency, current i_2 in the diode falls to zero before the end of the cycle when the average output current is low. An analysis of the operation with a magnetomotive force intermittently equal to zero would give results similar to those obtained for the non-isolated buck–boost chopper (see Sect. 6.2.3.4).
- Operation in the free-running mode corresponds to the boundary between the operation with an MMF always positive, and the operation with an MMF intermittently equal to zero. There is also a problem concerning the operation with a low load, since, at a given output voltage U', the switching frequency must increase proportionately to R. When this resistance moves towards infinity (no-load operation), the circuit must operate at I moving towards zero and with a frequency moving towards infinity, if U' is to be kept constant.

6.3.3 Forward Converter

The forward circuit (Fig. 6.26) is deduced from the buck chopper. The transistor is placed at the primary of the transformer, which ensures galvanic isolation and adapts the secondary voltage. A diode D_2 must be added in order to allow for the transistor Q–transformer–diode D_2 set to behave like a controlled unidirectional switch. To avoid the saturation of the core, a third winding of n_3 turns is required; diode D_3 – series-connected to this winding – prevents the latter from conduct when Q is on.

As shown in Fig. 6.2, between the "switch" and the load R which is assumed to be purely resistive, diode D, coil l and capacitor C are found. Only a small residual ripple exists in the output voltage owing to filtering by l, C.

This type of circuit is used for powers between 50 W and 1 kW.

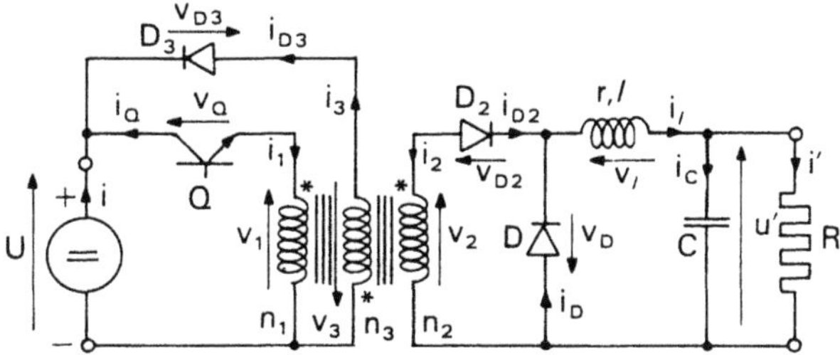

Fig. 6.26

Transistor Q is saturated from $t = 0$ to $t = \alpha T$ and in the OFF state from $t = \alpha T$ to $t = T$; diode D_2 conducts during the same time interval as Q. The off-time of Q must be long enough for the core to have the time to be demagnetized. This is carried out by the current flowing through n_3 and D_3.

Figure 6.27 provides a diagram of the waveforms of the different current and voltages during normal operation.[1] In these diagrams, the ripple in voltage u' has been ignored.

6.3.3.1 Average Values of Voltage u' and Current i_l

Inductance l (see Eq. (6.3)) acts in limiting the voltage u' ripple and, as a result, in limiting current i_C. The load current can thus be increased without creating problems for the capacitor. This circuit is well suited to delivering a high output current under low voltage.

If the output voltage is low, the following must be taken into account:

resistance r_2 of the secondary winding,
resistance r of the smoothing inductor,
internal resistance r_D of diodes D_2 and D,
threshold voltage V_D of these diodes.

Seen from the output terminals, the diagram corresponding to the circuit is as shown in Fig. 6.28. From $t = 0$ to $t = \alpha T$, the inverting switch is in position 1; from $t = \alpha T$ to $t = T$, it is in position 2.

From $t = 0$ to $t = \alpha T$, $i_l = i_{D_2}$,

$$U\frac{n_2}{n_1} = (r_2 + r_D + r)i_l + V_D + l\frac{di_l}{dt} + u'$$

From $t = \alpha T$ to $t = T$, $i_l = i_D$,

$$0 = (r_D + r)i_l + V_D + l\frac{di_l}{dt} + u'$$

By denoting I_l and U' the average values of i_l and u' over a cycle T, and by assuming that i_l has the same average value over interval $(0, \alpha T)$ as over the

[1] The shaded area in the waveform of i_l corresponds to the increase in the current caused by the magnetizing current (see Sect. 6.3.3.3).

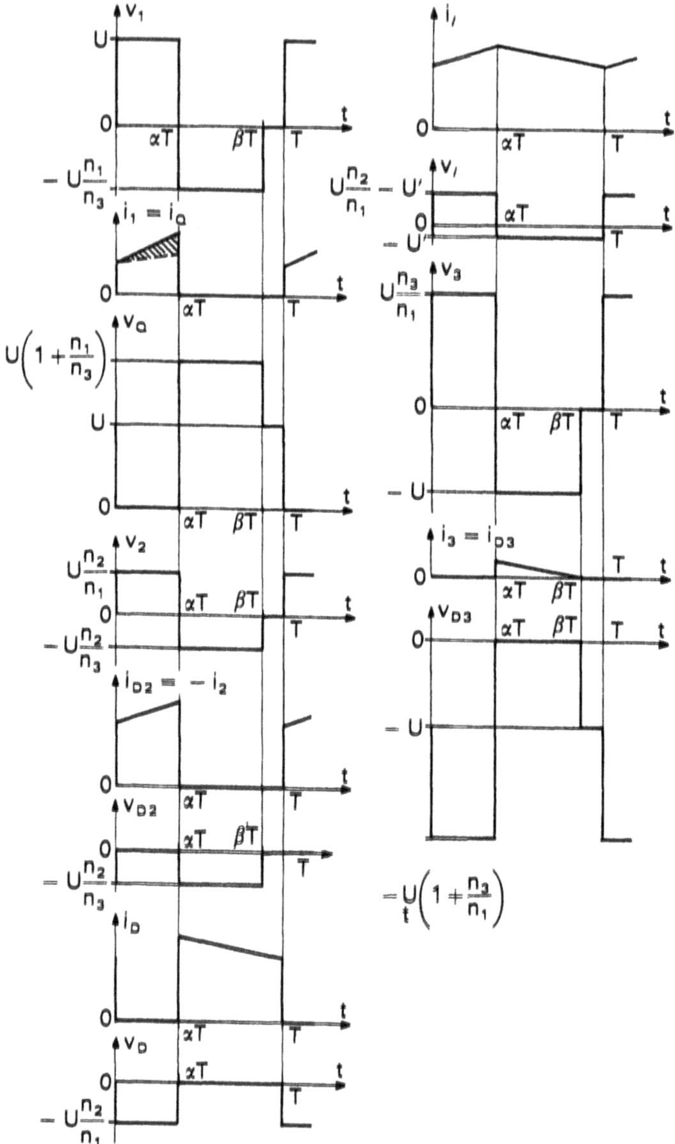

Fig. 6.27

whole cycle, one gets

$$\alpha U \frac{n_2}{n_1} = \alpha r_2 I_l + (r_D + r) I_l + V_D + U'$$

As the average value of i_C is zero, current i_l has the same average value as

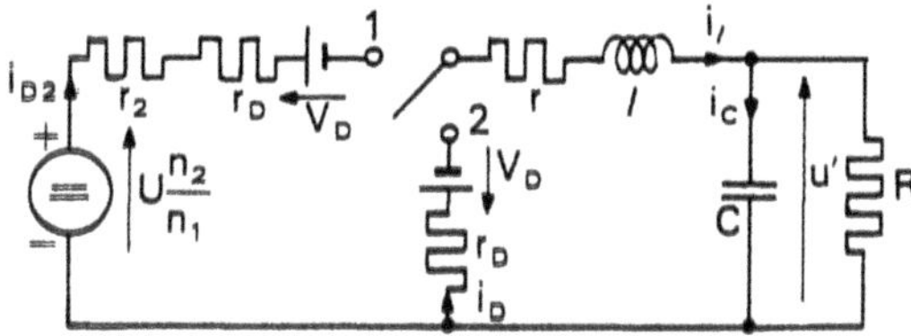

Fig. 6.28

u'/R. If I_l is replaced by U'/R, this results in

$$U' = \frac{\alpha U \dfrac{n_2}{n_1} - V_D}{1 + (r_D + r + \alpha r_2)/R} = RI_l \tag{6.43}$$

As r_2 is low compared to R, the increase in U' as a function of α is virtually linear.

6.3.3.2 Current i_l and Voltage u' Ripples

- In estimating current i_l ripple, voltage u' ripple is assumed to be negligible.
  - From $t = 0$ to $t = \alpha T$,

$$l\frac{di_l}{dt} + (r_2 + r_D + r)i_l = U\frac{n_2}{n_1} - U' - V_D$$

Current i_l increases from I_m to I_M, according to the equation

$$i_l = \frac{U(n_2/n_1) - U' - V_D}{r_2 + r_D + r} + \left[I_m - \frac{U(n_2/n_1) - U' - V_D}{r_2 + r_D + r}\right] \times \exp[-t(r_2 + r_D + r)/l].$$

- From $t = \alpha T$ to $t = T$,

$$l\frac{di_l}{dt} + (r_D + r)i_l = -U' - V_D.$$

Current i_l decreases from I_M to I_m, according to the equation

$$i_l = -\frac{U' + V_D}{r_D + r} + \left[I_M + \frac{U' + V_D}{r_D + r}\right] \times \exp[-(t - \alpha T)(r_D + r)/l].$$

By using the continuity of current i_l at instants $t = \alpha T$ and $t = T$, it is possible to obtain I_m and I_M which give ripple Δi_l equal to $I_M - I_m$. This calculation gives

$$\Delta i_l = \left(\frac{U(n_2/n_1) - U' - V_D}{r_2 + r_D + r} + \frac{U' + V_D}{r_D + r}\right) \cdot \times \frac{1 - \exp[-\alpha T(r_2 + r_D + r)/l]}{1 - \exp[-T(\alpha r_2 + r_D + r)/l]} \cdot (1 - \exp[-(1 - \alpha)T(r_D + r)/l]).$$

If the time constant $l/(r_2 + r_D + r)$ is large enough compared to cycle T, we can write

$$\Delta i_l \simeq \frac{U(n_2/n_1)(r_D + r) + r_2(U' + V_D)}{(r_2 + r_D + r)(r_D + r)}$$

$$\times \frac{((r_2 + r_D + r)/l)\alpha T((r_D + r)/l)(1 - \alpha)T}{((\alpha r_2 + r_D + r)/l)T}$$

$$\Delta i_l \simeq \frac{U(n_2/n_1)(r_D + r) + r_2(U' + V_D)}{\alpha r_2 + r_D + r}(1 - \alpha)\alpha \frac{T}{l} \tag{6.44}$$

If r_2 is neglected, the latter expression gives a result similar to that obtained for the buck chopper:

$$\Delta i_l \simeq U \frac{n_2}{n_1}(1 - \alpha)\alpha \frac{T}{l} \tag{6.44'}$$

- In estimating the *voltage u' ripple*, it can be seen that the configuration of the output stage of the circuit is identical to that of the buck chopper. As in the latter case (Sect. 6.2.1.2), the ripple of u' can be deduced from that of i_l:

$$\Delta u' = \Delta i_l \frac{T}{8C}$$

Given the expression of Δi_l, this gives

$$\Delta u' \simeq \frac{U(n_2/n_1)(r_D + r) + r_2(U' + V_D)}{\alpha r_2 + r_D + r}(1 - \alpha)\alpha \frac{T^2}{8lC} \tag{6.45}$$

Or, if r_2 is ignored,

$$\Delta u' \simeq U \frac{n_2}{n_1}(1 - \alpha)\alpha \frac{T^2}{8lC} \tag{6.45'}$$

6.3.3.3 Demagnetization Period

The flux φ in the cross section of the transformer core is the sum of the fluxes created by each of the three windings. If these three fluxes are denoted as φ_1, φ_2 and φ_3, and the inductances – assumed to be constant – of the three windings as l_1, l_2 and l_3, this leads to the following equations:

$$v_1 = n_1 \frac{d\varphi}{dt}; \quad v_2 = n_2 \frac{d\varphi}{dt}; \quad v_3 = n_3 \frac{d\varphi}{dt}$$

$$\varphi = \varphi_1 + \varphi_2 + \varphi_3, \quad \text{with } \varphi_1 = \frac{l_1}{n_1} i_1; \quad \varphi_2 = \frac{l_2}{n_2} i_2; \quad \varphi_3 = \frac{l_3}{n_3} i_3$$

$$l_1 = n_1^2 \mu \frac{S}{L}; \quad l_2 = n_2^2 \mu \frac{S}{L}; \quad l_3 = n_3^2 \mu \frac{S}{L}$$

In the expressions of the inductances, the following notations are used:
S-the section of the core,

L-the mean length of the field lines,
μ-the permeability of the material used.

- From $t = 0$ to $t = \alpha T$,
the magnetomotive force at equals $n_1 i_1 + n_2 i_2$;
the value of flux is

$$\varphi = \varphi_1 + \varphi_2 = \frac{l_1}{n_1} i_1 + \frac{l_2}{n_2} i_2 = n_1 \mu \frac{S}{L} i_1 + n_2 \mu \frac{S}{L} i_2$$

$$= \mathrm{at}\, \mu \frac{S}{L} = \mathrm{at} \frac{l_1}{n_1^2}$$

Voltage v_1 equals U; thus,

$$U = n_1 \frac{\mathrm{d}\varphi}{\mathrm{d}t} = n_1 \frac{l_1}{n_1^2} \frac{\mathrm{dat}}{\mathrm{d}t}.$$

From this it is deduced that

$$\mathrm{at} = \frac{n_1}{l_1} U t,$$

since the transformer is demagnetized and the MMF is zero for $t = 0$.
At instant $t = \alpha T$, the MMF reaches its maximum:

$$\mathrm{at}(\alpha T) = \frac{n_1}{l_1} U \alpha T.$$

- From $t = \alpha T$ to $t = \beta T$

$$\mathrm{at} = n_3 i_3$$

$$\varphi = \varphi_3 = \frac{l_3}{n_3} i_3 = n_3 i_3 \mu \frac{S}{L} = \mathrm{at} \frac{l_1}{n_1^2}$$

$$v_3 = -U = n_3 \frac{\mathrm{d}\varphi}{\mathrm{d}t} = \frac{n_3}{n_1} \frac{l_1}{n_1} \frac{\mathrm{dat}}{\mathrm{d}t}.$$

From this it is deduced that

$$\mathrm{at} = -U \frac{n_1}{n_3} \frac{n_1}{l_1} (t - \alpha T) + \mathrm{at}(\alpha T)$$

$$\mathrm{at} = \frac{n_1}{l_1} U \left[\alpha T - \frac{n_1}{n_3} (t - \alpha T) \right].$$

The MMF falls back to zero for $t = \beta T$ such that

$$0 = \frac{n_1}{l_1} U \left[\alpha T - \frac{n_1}{n_3} (\beta T - \alpha T) \right].$$

It can be deduced that

$$\beta = \alpha\left(1 + \frac{n_3}{n_1}\right).$$

- For the demagnetization to be complete at the end of each cycle, β must be less than 1. The maximum value of α can be deduced:

$$\alpha_{\max} = \frac{n_1}{n_1 + n_3}. \tag{6.46}$$

The maximum of α is thus limited by the ratio n_3/n_1; for example, if $n_3 = n_1$, α cannot be greater than 0.5. The maximum value of the output voltage corresponds to $\alpha_{\max}$:

$$U'_{\max} = \frac{(n_2/(n_1 + n_3))U - V_D}{1 + (r_D + r)/R + (n_1/(n_1 + n_3))r_2/R} \tag{6.47}$$

6.3.3.4 Semiconductor Device Ratings

- *Transistor* Q

- During the ON-state period $(0, \alpha T)$ of the transistor,

$$at = n_1 i_1 + n_2 i_2\,; \quad i_Q = i_1\,; \quad i_2 = -\, i_l$$

$$i_Q = i_1 = \frac{at}{n_1} - \frac{n_2}{n_1} i_2 = \frac{U}{l_1} t + \frac{n_2}{n_1} i_l$$

Current i_Q has the following average value:

$$I_Q = \frac{1}{T}\int_0^{\alpha T}\left(\frac{U}{l_1} t + \frac{n_2}{n_1} i_l\right) dt$$

$$I_Q = \frac{U}{l_1}\alpha^2\frac{T}{2} + \frac{n_2}{n_1}\alpha I_l$$

and a maximum value:

$$i_{Q\max} = \frac{U}{l_1}\alpha T + \frac{n_2}{n_1}\left(I_l + \frac{\Delta i_l}{2}\right).$$

If i_l and Δi_l are replaced by their expression, this gives

$$I_Q = \frac{U}{l_1}\alpha^2\frac{T}{2} + \frac{\alpha^2(n_2/n_1)^2 U - \alpha(n_2/n_1) V_D}{R + r_D + r + \alpha r_2}$$

$$i_{Q\max} \simeq \frac{U}{l_1}\alpha T + \frac{\alpha(n_2/n_1)^2 U - (n_2/n_1) V_D}{R + r_D + r + \alpha r_2} + \frac{1}{2}\alpha(1-\alpha)\left(\frac{n_2}{n_1}\right)^2\frac{UT}{l}.$$

An analysis of the derivatives of I_Q and $i_{Q\max}$ with respect to α shows that, if the time constant $l/(r + r_D)$ is large enough compared to T, the values of I_Q

and $i_{Q\,max}$ increase continuously when α changes from zero to 1. The current stresses imposed on the transistor must thus be calculated for α equal to α_{max}.

– Voltage v_Q across Q is at its maximum when diode D_3 is conducting:

$$v_{Q\,max} = U\left(1 + \frac{n_1}{n_3}\right).$$

• *Diode* D_2

– The current in diode D_2 has the following average and maximum values:

$$I_{D_2} = \alpha I_l; \quad i_{D_2\,max} = I_l + \frac{\Delta i_l}{2}$$

$$I_{D_2} = \frac{\alpha^2 (n_2/n_1) U - \alpha V_D}{R + r_D + r + \alpha r_2}$$

$$i_{D_2\,max} \simeq \frac{\alpha (n_2/n_1) U - V_D}{R + r_D + r + \alpha r_2} + \frac{1}{2}\alpha(1-\alpha)\left(\frac{n_2}{n_1}\right)\frac{UT}{l}.$$

I_{D_2} increases continuously when α moves from zero to 1; the same applies to $i_{D_2\,max}$ if $l/(r + r_D)$ is noticeably greater than T. The maximum stresses in current are calculated for α_{max}.

– There is negative voltage across D_2 when D_3 is conducting: this voltage has the following value:

$$v_{D_2} = -\frac{n_2}{n_3} U - U' = -\frac{n_2}{n_3} U - \frac{\alpha (n_2/n_1) U - V_D}{1 + (r_D + r + \alpha r_2)/R}.$$

When α varies, the highest reverse voltage is reached for α_{max}

• *Diode* D

– The current in diode D has the following average value:

$$I_D = (1-\alpha) I_l = (1-\alpha)\frac{\alpha (n_2/n_1) U - V_D}{R + r_D + r + \alpha r_2}$$

If α_{max} is above 0.5, when α varies, I_D reaches its maximum for $\alpha = 0.5$. Otherwise, this maximum must be calculated for α_{max}.

The peak value of i_D equals that of i_{D_2}.

– Voltage v_D is negative when Q is conducting and has the following value:

$$v_D = -U\frac{n_2}{n_1}.$$

• *Diode* D_3

– The current in diode D_3 is equal to at/n_3 during the period (αT, βT). It has the following maximum and average values:

$$i_{D_3max} = \frac{at(\alpha T)}{n_3} = \frac{n_1}{n_3}\frac{U}{l_1}\alpha T$$

$$I_{D_3} = \frac{1}{2}(\beta - \alpha) i_{D_3\,max} = \frac{1}{2}\alpha\frac{n_3}{n_1}\frac{n_1}{n_3}\frac{U}{l_1}\alpha T = \frac{U}{2l_1}\alpha^2 T$$

Both values increase with α and are thus to be calculated for α_{max}.

- The reverse voltage across diode D_3 is maximum when Q and D_2 are conducting. It is then equal to:

$$v_{D_3} = -U\left(1 + \frac{n_3}{n_1}\right)$$

6.3.3.5 Remarks on the Half-Bridge Isolated Buck Circuit

The primary winding of the circuit in Fig. 6.26 can carry out the demagnetization, if current i_1 is given a path enabling it to be gradually reduced to zero after the transistor has been turned off. The half-bridge structure with 2 transistors and 2 diodes at the primary (Fig. 6.29) offers this possibility.

- From $t = 0$ to $t = \alpha T$, transistors Q_1, Q'_1 as well as D_2 conduct.
  When Q_1 and Q'_1 are turned off, for $t = \alpha T$, diode D_2 is equally turned off. Current i_l flows through diode D. At the primary, current i_1 goes via diode D_1, voltage source U and diode D'_1; the source applies a voltage equal to $-U$ across the primary; current i_1 becomes zero for $t = \beta T$.
- Seen from the output, the operation is the same as would occur in a forward circuit in which $n_3 = n_1$. The maximum value of α is thus equal to 0.5 in this case.
  At the input, the maximum and average current values in the transistors and diodes are the same as those of the forward circuit in which $n_3 = n_1$. On the other hand, the maximum values of the voltages are divided by 2.
- This configuration uses a simpler transformer but requires two transistors. Their base control circuits must be isolated.

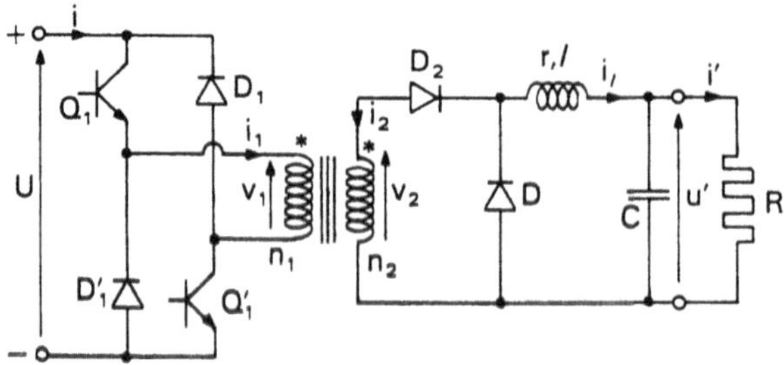

Fig. 6.29

6.4 Multiphase Asymmetrical Circuits

In the switch-mode power supplies, the use of multiphase staggered structures provides the same advantages as with choppers (Sect. 4.3).

- For a given output voltage ripple, the semiconductor devices operate at a lower frequency (conversely, for a given operating frequency, the input and output filter size can be reduced).
- For a given output power, the current stresses are lower on the semiconductor devices (conversely, for given semiconductor devices, the output power can be increased).

6.4.1 Multiphase Flyback Circuits

Figure 6.30 shows the circuit consisting of two staggered flyback converters.

When operating in incomplete demagnetization, transistor Q_1 conducts between $t = 0$ and $t = \alpha T$ and diode D_1 between $t = \alpha T$ and $t = 2T$; transistor Q_2 conducts from $t = T$ to $t = (1 + \alpha)T$, diode D_2 from $t = 0$ to $t = T$ and then from $t = (1 + \alpha)T$ to $t = 2T$.

The current and voltage waveforms show that the output voltage u' and current i'' have a cycle equal to T, while all the other variables have a cycle equal to $2T$.

6.4.1.1 Average Values of Output Voltage and of Magnetomotive Forces

- The MMF of the two transformers are denoted by at_1 and at_2:

$$at_1 = n_1 i_{11} + n_2 i_{21}$$

$$at_2 = n_1 i_{12} + n_2 i_{22},$$

the parameters of the primary coils by l_1, r_1,
the parameters of the secondary coils by l_2, r_2,

$$(l_2/l_1) = (n_2/n_1)^2$$

The two MMFs at_1 and at_2 have a cycle equal to $2T$; as they have the same waveform but are shifted by T, their sum $at_1 + at_2$ has the same cycle T as voltage u'. This enables the analysis period to be reduced to T.

From $t = 0$ to $t = \alpha T$, $i_{21} = 0$, $i_{12} = 0$,

$$n_2 i'' = n_2 i_{22} = (at_1 + at_2) - at_1.$$

From $t = \alpha T$ to $t = T$, $i_{11} = 0$, $i_{12} = 0$,

$$n_2 i'' = n_2 i_{21} + n_2 i_{22} = (at_1 + at_2).$$

The average values are taken over the period $(0, T)$;

$$n_2 \frac{1}{T}\int_0^T i'' \,dt = \frac{1}{T}\int_0^T (at_1 + at_2)\,dt - \frac{1}{T}\int_0^{\alpha T} at_1 \,dt.$$

v_{Q1} i_{11} i_{21} v_{D1}
Q_1 D_1
v_1 n_1 v_2 n_2
C R i''
i_1 U n_1 n_2
v'_1 v'_2 u'
v_{Q2} D_2
Q_2 i_{12} i_{22} v_{D2}
i
i_{11} i_{12}
i_{21} i_{22}
i''
u'
$\frac{v_1}{n_1} = \frac{v_2}{n_2}$
$\frac{v'_1}{n_1} = \frac{v'_2}{n_2}$
v_{Q2} v_{Q1}
v_{D1} v_{D2}
0 αT T (1+α)T 2T t

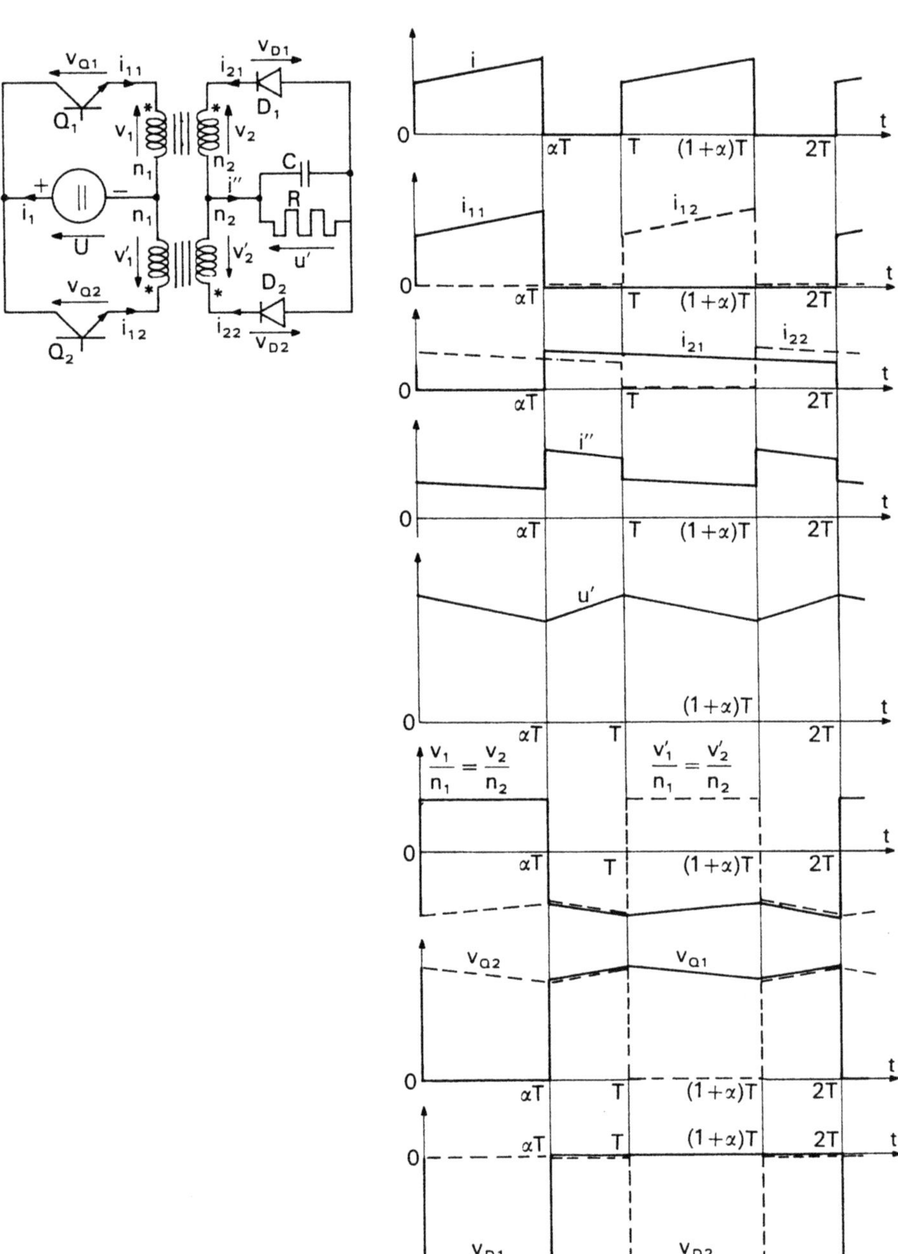

Fig. 6.30

If U' denotes the average value of u' and AT that of at_1 or at_2 and since the current in the capacitor has an average value equal to zero, the above equation gives

$$n_2 \frac{U'}{R} = 2\mathrm{AT} - \frac{1}{T}\int_0^{\alpha T} \mathrm{at}_1 \, \mathrm{d}t$$

If variations of at_1 are assumed to be linear, this MMF has the same average value as for the whole cycle, during its rise period $(0, \alpha T)$. Thus,

$$n_2 \frac{U'}{R} = 2\mathrm{AT} - \alpha \mathrm{AT}$$

$$\mathrm{AT} = \frac{n_2}{2-\alpha} \frac{U'}{R} \tag{6.48}$$

- From $t = 0$ to $t = \alpha T$,

$$U = r_1 i_{11} + l_1 \frac{\mathrm{d}i_{11}}{\mathrm{d}t}$$

$$-u' = r_2 i_{22} + l_2 \frac{\mathrm{d}i_{22}}{\mathrm{d}t}$$

The first equation can be written as

$$n_2 U \frac{n_2}{n_1} = r_1 \left(\frac{n_2}{n_1}\right)^2 \mathrm{at}_1 + l_1 \left(\frac{n_2}{n_1}\right)^2 \frac{\mathrm{dat}_1}{\mathrm{d}t}$$

or, by denoting r'_1 the resistance of the primary referred to the secondary:

$$n_2 U \frac{n_2}{n_1} = r'_1 \mathrm{at}_1 + l_2 \frac{\mathrm{dat}_1}{\mathrm{d}t}.$$

Similarly, for the second equation:

$$-n_2 u' = r_2 \mathrm{at}_2 + l_2 \frac{\mathrm{dat}_2}{\mathrm{d}t}.$$

By adding the equations to each other, this results in

$$n_2 U \frac{n_2}{n_1} - 2n_2 u' + n_2 u' = r_2(\mathrm{at}_1 + \mathrm{at}_2) + (r'_1 - r_2)\mathrm{at}_1 + l_2 \frac{\mathrm{d}(\mathrm{at}_1 + \mathrm{at}_2)}{\mathrm{d}t}. \tag{6.49}$$

From $t = \alpha T$ to $t = T$,

$$-u' = r_2 i_{21} + l_2 \frac{\mathrm{d}i_{12}}{\mathrm{d}t}$$

$$-u' = r_2 i_{22} + l_2 \frac{\mathrm{d}i_{22}}{\mathrm{d}t}.$$

By multiplying by n_2 these two equations and adding them to each other, this gives

$$-2n_2 u' = r_2(\mathrm{at}_1 + \mathrm{at}_2) + l_2 \frac{\mathrm{d}(\mathrm{at}_1 + \mathrm{at}_2)}{\mathrm{d}t}. \tag{6.50}$$

By writing that mean values over period $(0, T)$ of both members of Eqs. (6.49) and (6.50) are equal, and by noting that the mean value of $\mathrm{d}(\mathrm{at}_1 + \mathrm{at}_2)/\mathrm{d}t$ is equal to zero, one gets

$$\alpha n_2 U \frac{n_2}{n_1} - 2n_2 U' + \frac{1}{T} \int_0^{\alpha T} n_2 u' \,\mathrm{d}t = 2r_2 \mathrm{AT} + (r'_1 - r_2) \int_0^{\alpha T} \mathrm{at}_1 \,\mathrm{d}t$$

$$\alpha n_2 U \frac{n_2}{n_1} - n_2(2 - \alpha) U' = 2r_2 \mathrm{AT} + (r'_1 - r_2)\alpha \mathrm{AT}.$$

If AT is replaced by its value given in Eq. (6.48), this leads to

$$U' = \frac{\alpha(n_2/n_1) U}{2 - \alpha + r_2/R + \alpha r'_1/(2 - \alpha) R} \tag{6.51}$$

- If the expressions of AT and U' obtained here are compared with those of the simple flyback circuit, it becomes easy to generalize to a multiphase circuit consisting of k staggered simple converters:

$$\mathrm{AT} = \frac{n_2}{k - \alpha} \frac{U'}{R}$$

$$U' = \frac{\alpha(n_2/n_1) U}{k - \alpha + r_2/R + \alpha r'_1/(k - \alpha) R}$$

If r'_1 and r_2 are negligible compared to R,

$$U' \simeq \frac{\alpha}{k - \alpha} \frac{n_2}{n_1} U, \quad \text{with} \quad 0 < \alpha < \frac{1}{k}$$

6.4.1.2 Magnetomotive Force and Output Voltage Ripples

- The *magnetomotive force* ripple can be calculated using the method outlined in Sect. 6.3.2.2.

When Q_1 conducts, the charge of the first inductance is ruled by the equation

$$U = r_1 i_{11} + l_1 \frac{\mathrm{d}i_{11}}{\mathrm{d}t}.$$

As a first approximation, the following is used:

$$\frac{\mathrm{d}i_{11}}{\mathrm{d}t} \simeq \frac{U}{l_1}; \qquad \frac{\mathrm{dat}_1}{\mathrm{d}t} \simeq n_1 \frac{U}{l_1}$$

$$\Delta \mathrm{at} \simeq n_1 \frac{U}{l_1} \alpha T.$$

- In order to calculate the *output voltage* u' ripple, the MMF ripple is ignored: the MMFs are assumed to be constant and equal to their average value AT.

  From $t = 0$ to $t = \alpha T$, voltage u' changes from its maximum U'_M to its minimum U'_m.

  The equation

$$C\frac{\mathrm{d}u'}{\mathrm{d}t} + \frac{u'}{R} = \frac{\mathrm{AT}}{n_2}$$

gives

$$u' = R\frac{\mathrm{AT}}{n_2} + \left(U'_M - R\frac{\mathrm{AT}}{n_2}\right)\exp(-t/RC).$$

  From $t = \alpha T$ to $t = T$, the increase in u' is ruled by

$$C\frac{\mathrm{d}u'}{\mathrm{d}t} + \frac{u'}{R} = 2\frac{\mathrm{AT}}{n_2}$$

$$u' = 2R\frac{\mathrm{AT}}{n_2} + \left(U'_m - 2R\frac{\mathrm{AT}}{n_2}\right)\exp[-(t-\alpha T)/RC].$$

The expression of U'_M can be found by equating the values of u' provided by the two equations at instant $t = \alpha T$ and by writing that, for $t = T$, u' takes again its value U'_M. From U'_M can be deduced the value of U'_m and ripple $\Delta u'$:

$$\Delta u' = \frac{U'}{2-\alpha}\frac{1-\exp[-(1-\alpha)T/RC]}{1-\exp(-T/RC)}[1-\exp(-\alpha T/RC)].$$

For k paralleled circuits, the following can also be obtained:

$$\Delta u' = \frac{U'}{k-\alpha}\frac{1-\exp[-(1-\alpha)T/RC]}{1-\exp(-T/RC)}[1-\exp(-\alpha T/RC)]$$

or, if RC is much higher than T,

$$\Delta u' \simeq \frac{U'}{k-\alpha}(1-\alpha)\frac{\alpha T}{RC} \tag{6.52}$$

6.4.2 Multiphase Forward Circuits

Figure 6.31 gives the diagram of the multiphase forward circuit comprising two staggered single converters.

- Transistor Q and diode D_2 conduct simultaneously between $t = 0$ and $t = \alpha T$. From $t = \alpha T$ onwards, diode D ensures that current i_l remains continuous, while current i_{D_3} ensures the demagnetization of the first transformer via diode D_3.

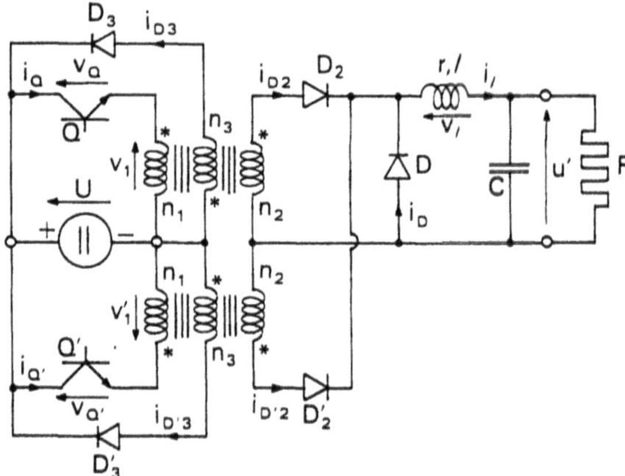

Fig. 6.31

When Q′ is turned on in $t = T$, diode D is turned off, but D_3 continues to conduct as long as i_{D_3} has not fallen to zero; the demagnetization of the first transformer ends, for $t = \beta T$, when i_{D_3} reaches zero.

From $t = T$ to $t = (1 + \alpha)T$, diode D'_2 and transistor Q′ conduct simultaneously. When the latter is turned off, current i_l is transferred from D'_2 to D; the demagnetization of the second transformer is carried out by D'_3 conducting between $t = (1 + \alpha)T$ and $t = (1 + \beta)T$.

The current and voltage waveforms are those shown in Fig. 6.32.

The time period of current and voltage of the two transistors, the two transformers and diodes D_3, D'_3 is twice that of the output voltage. For a given value of U', the stresses on these elements are thus reduced compared to those which correspond to a simple forward circuit.

- All the calculations of average values and ripples carried out in Sect. 6.3.3 remain valid here. The same applies to the duration of the demagnetization period. But the limitation of α which follows from this duration is different; it becomes

$$\beta T = \alpha\left(1 + \frac{n_3}{n_1}\right) < 2T$$

$$\alpha_{\max} = \frac{2n_1}{n_1 + n_3}.$$

If n_3 equals n_1, the theoretical value of α may reach 1.
The maximum value of the output voltage becomes

$$U'_{\max} \simeq \frac{\alpha_{\max} U n_2/n_1}{1 + r/R} = \frac{U}{1 + r/R} \frac{2n_2}{n_1 + n_3}.$$

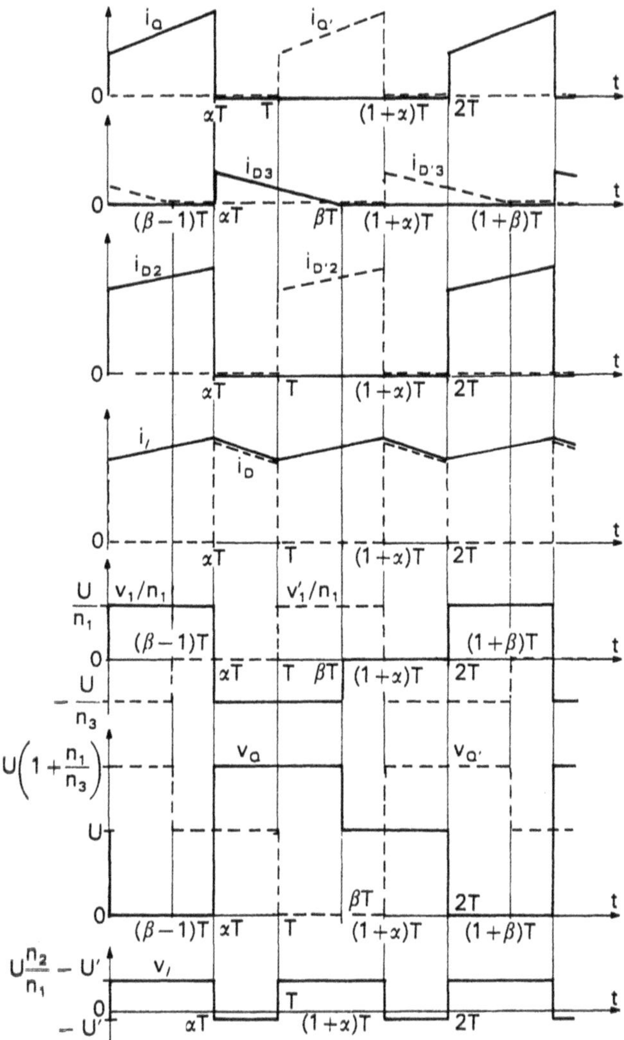

Fig. 6.32

6.5 Asymmetrical Multiple-Output Circuits

The use of a transformer provides the galvanic insulation between output and input. Starting from a single DC input voltage, it also enables several output voltages to be obtained. These are insulated each from other and their values may be different.

6.5.1 Multiple-Output Flyback Circuit

Our analysis will be confined to the converter with two output stages shown in Fig. 6.33. If the resistance of the primary and secondary windings are neglected, this gives simple results which can easily be extended to a larger number of output. If the resistances are taken into account, the calculations below show that the obtained expressions rapidly become more and more complex as the number of outputs increases.

In normal operating conditions, transistor Q conducts from $t = 0$ to $t = \alpha T$; diodes D_1 and D_2 simultaneously conduct during the remainder of the switching cycle.

The notations used are the same as those used in Sects. 6.3.2 and 6.4.1. l_1, l_{21} and l_{22} are used to denote the inductances – assumed to be constant – of the primary, the first secondary and the second secondary:

$$l_1 = n_1^2 \mu \frac{S}{L}; \quad l_{21} = n_{21}^2 \mu \frac{S}{L}; \quad l_{22} = n_{22}^2 \mu \frac{S}{L}$$

φ_1, φ_{21} and φ_{22} are used to denote the flux created throughout a cross section of the core by the currents in the three windings:

$$\varphi_1 = \frac{l_1}{n_1} i_1; \quad \varphi_{21} = \frac{l_{21}}{n_{21}} i_{21}; \quad \varphi_{22} = \frac{l_{22}}{n_{22}} i_{22}$$

The total instantaneous flux through each turn is given by

$$\varphi = \varphi_1 + \varphi_{21} + \varphi_{22}.$$

6.5.1.1 Average Values of the Magnetomotive Force and of Output Voltages

- During the period $(0, \alpha T)$, currents i_{21} and i_{22} are zero:

$$\text{at} = n_1 i_1; \quad \varphi = \varphi_1 = \frac{l_1}{n_1} i_1 = \frac{l_1}{n_1^2} \text{at}.$$

The equation of the primary voltages:

$$U = r_1 i_1 + n_1 \frac{d\varphi}{dt}$$

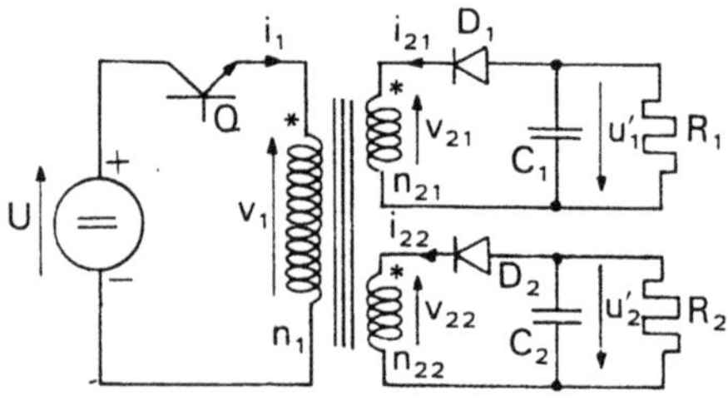

Fig. 6.33

can be written as

$$U = r_1 \frac{\text{at}}{n_1} + \frac{l_1}{n_1}\frac{\text{dat}}{dt} \quad \text{or}$$

$$\frac{U}{n_1} = \frac{r_1}{n_1^2}\text{at} + \frac{l_1}{n_1^2}\frac{\text{dat}}{dt} \tag{6.53}$$

– During the period $(\alpha T, T)$, current i_1 is zero:

$$\text{at} = n_{21} i_{21} + n_{22} i_{22}$$

$$\varphi = \frac{l_{21}}{n_{21}} i_{21} + \frac{l_{22}}{n_{22}} i_{22} = \frac{l_1}{n_1^2}(n_{21} i_{21} + n_{22} i_{22}) = \frac{l_1}{n_1^2}\text{at}$$

The two equations of the secondary circuit voltages

$$-u'_1 = r_{21} i_{21} + n_{21}\frac{d\varphi}{dt}$$

$$-u'_2 = r_{22} i_{22} + n_{22}\frac{d\varphi}{dt}$$

can be written as

$$-\frac{u'_1}{n_{21}} = \frac{r_{21}}{n_{21}} i_{21} + \frac{l_1}{n_1^2}\frac{\text{dat}}{dt} \tag{6.54}$$

$$-\frac{u'_2}{n_{22}} = \frac{r_{22}}{n_{22}} i_{22} + \frac{l_1}{n_1^2}\frac{\text{dat}}{dt}. \tag{6.55}$$

– The average values of the first members of Eqs. (6.53) and (6.54) over a cycle are equated to the average values of the second members.

In order to establish the relations between average values,

- the fact that dat/dt has an average value equal to zero over a cycle is used;
- U'_1, U'_2, AT, I_{21} and I_{22} are used to denote the average values of u'_1, u'_2, at, i_{21} and i_{22} over the whole cycle;
- finally, u'_1, u'_2, at, i_{21} and i_{22} are assumed to have, over one of the periods $(0, \alpha T)$ or $(\alpha T, T)$, the same average value as over the whole cycle.

This gives

$$\alpha\frac{U}{n_1} - (1-\alpha)\frac{U'_1}{n_{21}} = \alpha\frac{r_1}{n_1^2}\text{AT} + (1-\alpha)\frac{r_{21}}{n_{21}} I_{21}. \tag{6.56}$$

By following the same procedure with Eqs (6.53) and (6.55), one gets

$$\alpha\frac{U}{n_1} - (1-\alpha)\frac{U'_2}{n_{22}} = \alpha\frac{r_1}{n_1^2}\text{AT} + (1-\alpha)\frac{r_{22}}{n_{22}} I_{22}. \tag{6.57}$$

- The average value of the magnetomotive force can be deduced from its expression during the second period:

$$\mathrm{AT} = n_{21} I_{21} + n_{22} I_{22}. \tag{6.58}$$

As the average currents in the capacitors are zero, currents u'_1/R_1 and u'_2/R_2 have the same average value as i_{21} and i_{22} over the whole cycle:

$$\frac{U'_1}{R_1} = (1-\alpha) I_{21}; \quad \frac{U'_2}{R_2} = (1-\alpha) I_{22}. \tag{6.59}$$

Transferring the latter into Eq. (6.58), this gives

$$\mathrm{AT} = \frac{1}{1-\alpha}\left(n_{21}\frac{U'_1}{R_1} + n_{22}\frac{U'_2}{R_2}\right). \tag{6.60}$$

Transferring into Eqs. (6.56) and (6.57), this results in

$$\alpha\frac{U}{n_1} - (1-\alpha)\frac{U'_1}{n_{21}} = \frac{\alpha}{1-\alpha}\frac{r_1}{n_1^2}\left(n_{21}\frac{U'_1}{R_1} + n_{22}\frac{U'_2}{R_2}\right) + \frac{r_{21}}{n_{21}}\frac{U'_1}{R_1}$$

$$\alpha\frac{U}{n_1} - (1-\alpha)\frac{U'_2}{n_{22}} = \frac{\alpha}{1-\alpha}\frac{r_1}{n_1^2}\left(n_{21}\frac{U'_1}{R_1} + n_{22}\frac{U'_2}{R_2}\right) + \frac{r_{22}}{n_{22}}\frac{U'_2}{R_2}.$$

From this system of two equations with two unknowns, U'_1 and U'_2, the following can be deduced.

$$U'_1 = \frac{\alpha U (n_{21}/n_1)/(1-\alpha)}{\left[1 + \frac{r_{21}}{(1-\alpha)R_1}\right]\left(1 + \frac{\alpha}{1-\alpha}\frac{r_1}{n_1^2}\left[\frac{n_{21}^2}{r_{21} + (1-\alpha)R_1} + \frac{n_{22}^2}{r_{22} + (1-\alpha)R_2}\right]\right)} \tag{6.61}$$

$$U'_2 = \frac{\alpha U (n_{22}/n_1)/(1-\alpha)}{\left[1 + \frac{r_{22}}{(1-\alpha)R_2}\right]\left(1 + \frac{\alpha}{1-\alpha}\frac{r_1}{n_1^2}\left[\frac{n_{21}^2}{r_{21} + (1-\alpha)R_1} + \frac{n_{22}^2}{r_{22} + (1-\alpha)R_2}\right]\right)} \tag{6.62}$$

- If the values of r_1, r_{21} and r_{22} are small enough to be ignored, we obtain

$$U'_1 \simeq \frac{\alpha}{1-\alpha}\frac{n_{21}}{n_1}U; \quad U'_2 \simeq \frac{\alpha}{1-\alpha}\frac{n_{22}}{n_1}U \tag{6.63}$$

$$\mathrm{AT} \simeq \frac{\alpha}{(1-\alpha)^2}\frac{U}{n_1}\left(\frac{n_{21}^2}{R_1} + \frac{n_{22}^2}{R_2}\right) \tag{6.64}$$

$$I_{21} \simeq \frac{\alpha}{(1-\alpha)^2}\frac{n_{21}}{n_1}\frac{U}{R_1}; \quad I_{22} \simeq \frac{\alpha}{(1-\alpha)^2}\frac{n_{22}}{n_1}\frac{U}{R_2}. \tag{6.65}$$

6.5.1.2 Magnetomotive-Force Ripple

From $t = 0$ to $t = \alpha T$, the expression, as a function of the time, of the MMF at is the same as for a circuit with a single secondary.

From $t = \alpha T$ to $t = T$, the equations of the secondary voltages, (6.54) and (6.55), can be written as

$$-n_{21}\frac{u'_1}{r_{21}} = n_{21}i_{21} + \frac{n_{21}^2}{n_1^2}\frac{l_1}{r_{21}}\frac{\mathrm{dat}}{\mathrm{d}t}$$

$$-n_{22}\frac{u'_2}{r_{22}} = n_{22}i_{22} + \frac{n_{22}^2}{n_1^2}\frac{l_1}{r_{22}}\frac{\mathrm{dat}}{\mathrm{d}t}$$

or, by adding these together,

$$-\left(n_{21}\frac{u'_1}{r_{21}} + n_{22}\frac{u'_2}{r_{22}}\right) = \mathrm{at} + \frac{l_1}{n_1^2}\left(\frac{n_{21}^2}{r_{21}} + \frac{n_{22}^2}{r_{22}}\right)\frac{\mathrm{dat}}{\mathrm{d}t}.$$

The results obtained for a single secondary circuit can thus be used, providing that

$$n_2\frac{u'}{r_2} \text{ is replaced by } n_{21}\frac{u'_1}{r_{21}} + n_{22}\frac{u'_2}{r_{22}}$$

$$\frac{l_2}{r_2} \text{ is replaced by } \left(\frac{n_{21}^2}{r_{21}} + \frac{n_{22}^2}{r_{22}}\right)\frac{l_1}{n_1^2}$$

which gives

$$\Delta\mathrm{at} = \left(n_1\frac{U}{r_1} + n_{21}\frac{U'_1}{r_{21}} + n_{22}\frac{U'_2}{r_{22}}\right)$$
$$\times\frac{1 - \exp(-(n_1^2/l_1)\cdot(1-\alpha)T/(n_{21}^2/r_{21} + n_{22}^2/r_{22}))}{1 - \exp(-\alpha T r_1/l_1)(-(n_1^2/l_1)\cdot(1-\alpha)T/(n_{21}^2/r_{21} + n_{22}^2/r_{22})}$$
$$\times[1 - \exp(-\alpha T r_1/l_1)].$$

If αT and $(1-\alpha)T$ are small compared with the various time constants,

$$\Delta\mathrm{at} \simeq \left(n_1\frac{U}{r_1} + n_{21}\frac{U'_1}{r_{21}} + n_{22}\frac{U'_2}{r_{22}}\right)$$
$$\times\frac{1}{l_1/(r_1\alpha T) + (l_1/n_1^2)(n_{21}^2/r_{21} + n_{22}^2/r_{22})/[(1-\alpha)T]}$$

If U'_1 and U'_2 are replaced by their simplified expressions obtained by ignoring the resistances of the windings, this gives, again,

$$\Delta\mathrm{at} \simeq n_1\frac{U}{l_1}\alpha T.$$

6.5.1.3 Output Voltage Ripples

It is possible to make a calculation similar to that made for a single secondary circuit (see Sect. 6.3.2.2).

For the first secondary, for instance, from

$$0 = C_1 \frac{du'_1}{dt} + \frac{u'_1}{R_1}, \quad \text{for} \quad 0 < t < \alpha T,$$

$$I_{21} = C_1 \frac{du'_1}{dt} + \frac{u'_1}{R_1}, \quad \text{for} \quad \alpha T < t < T,$$

the following can be deduced:

$$\Delta u'_1 = R_1 I_{21} \frac{1 - \exp[-(1-\alpha)T/R_1 C_1]}{1 - \exp(-T/R_1 C_1)} [1 - \exp(-\alpha T/R_1 C_1)]$$

or, if $R_1 C_1$ is much longer than T,

$$\Delta u'_1 \simeq R_1 I_{21} (1 - \alpha) \frac{\alpha T}{R_1 C_1}$$

If I_{21} is replaced by its simplified expression (6.65), this gives

$$\Delta u'_1 \simeq \frac{\alpha^2}{1-\alpha} \frac{T}{R_1 C_1} \frac{n_{21}}{n_1} U. \tag{6.66}$$

Similarly, for the other secondary, one gets

$$\Delta u'_2 \simeq \frac{\alpha^2}{1-\alpha} \frac{T}{R_2 C_2} \frac{n_{22}}{n_1} U. \tag{6.67}$$

6.5.2 Multiple-Output Forward Circuit

Figure 6.34 provides the diagram of a forward circuit with two output stages isolated from the input and isolated between themselves.

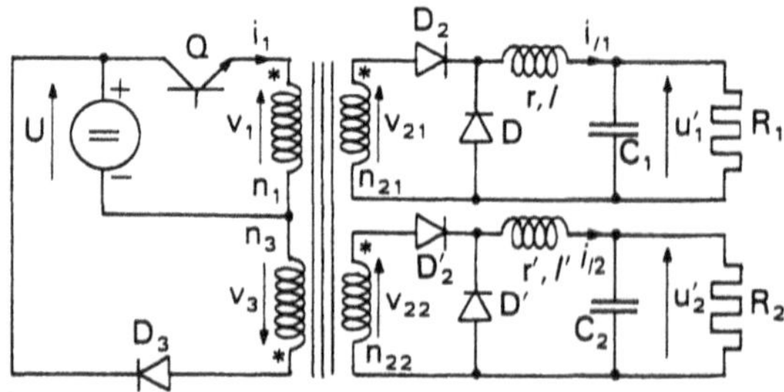

Fig. 6.34

- *In normal operating conditions*, transistor Q and diodes D_2 and D'_2 conduct from $t = 0$ to $t = \alpha T$:

$$v_1 = U; \quad u'_1 = U\frac{n_{21}}{n_1} - ri_{l_1} - l\frac{di_{l_1}}{dt}$$

$$u'_2 = U\frac{n_{22}}{n_1} - r'i_{l_2} - l'\frac{di_{l_2}}{dt}$$

Throughout the whole of the remaining period ($\alpha T < t < T$), diodes D and D′ conduct; one allows i_{l_1} to flow and other i_{l_2}:

$$u'_1 = 0 - ri_{l_1} - l\frac{di_{l_1}}{dt}$$

$$u'_2 = 0 - r'i_{l_2} - l'\frac{di_{l_2}}{dt}$$

From $t = \alpha T$ until $t = \beta T$, winding n_3 and diode D_3 ensure the complete demagnetization of the core.

The average values of the output voltages are close to

$$U'_1 = \alpha U\frac{n_{21}}{n_1}; \quad U'_2 = \alpha U\frac{n_{22}}{n_1}$$

- For all multiple-output circuits, whether they be of the flyback or forward type, there is only one transistor. This latter is thus controlled using an error signal, obtained by comparing *one* of the output voltages, e.g. u'_1, with a reference signal.

  If, at one of the outputs, the current delivered is not sufficient to ensure normal operation, the corresponding voltage becomes closely dependent on the current delivered. This voltage may reach a value which is very different from its value in continuous conduction.

  If, at the output which controls the regulation, the current delivered is not sufficient to ensure normal operation, all the other voltages will be disturbed.

With this type of supply, regulation offers acceptable performances only if the output which controls the regulation (main output) delivers a minimum

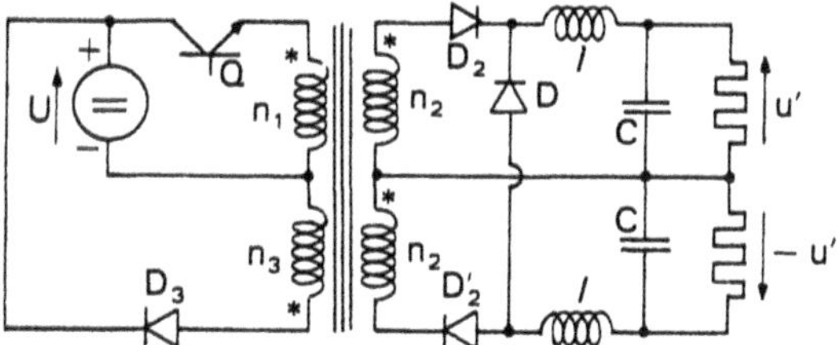

Fig. 6.35

current (often between 10 and 30% of the rated current). The other outputs are then semi-regulated (cross-regulation).

- This disadvantage is, however, greatly reduced when there are two symmetrical voltages in relation to a common point (Fig. 6.35). Because of the single free-wheeling diode D, the currents in the two inductances are both continuous or discontinuous.

6.6 Notes on Symmetrical Circuits

The name of these circuits indicates that an AC magnetomotive force is applied to the core of the transformer producing a bidirectional induction. For a given power, the transformer is used more efficiently and its volume is reduced; but at least two controlled switches must be used.

The converter, which supplies the primary winding with an AC voltage acts as an inverter not as a chopper. The analysis of inverters will be undertaken in the fourth volume of this series which deals with DC–AC conversion. However, the inverter load is of a particular nature in this case, since a rectifier must be placed at the transformer output in order to deliver a DC voltage, after filtering.

The general configuration is the same as before (Fig. 6.36). The DC input voltage U is applied to an inverter which feeds the transformer. The secondary voltage of the latter is rectified; the rectified voltage after filtering makes up the output voltage u'.

But two types of inverters may be used. They can provide

- either an AC signal formed of square-waves (*switching DC power supplies*)
- or an AC signal formed of sine-waves (*resonant DC–DC converters*).

Our analysis will deal only with symmetrical switching DC power supplies.

6.6.1 Various Possible Structures

The square-wave voltage pulses applied to the primary of the transformer can be obtained starting from the three basic inverter structures shown in Fig. 6.37.

a) push-pull circuit, using a centre-tapped transformer,
b) half-bridge circuit, using a capacitive divider,
c) full-bridge circuit.

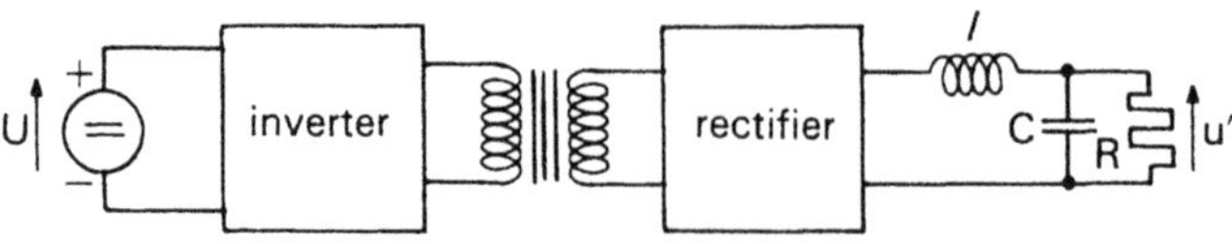

Fig. 6.36

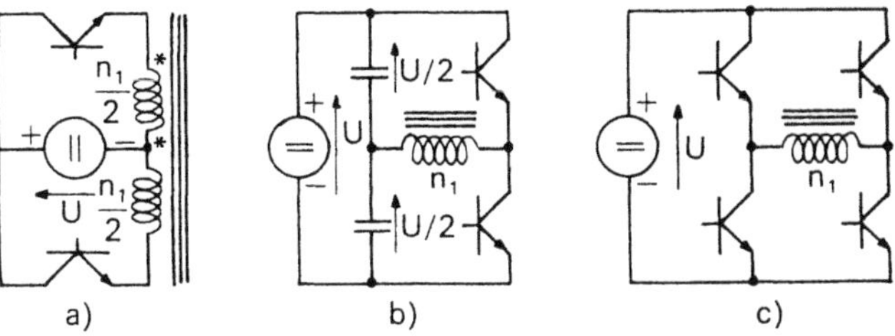

Fig. 6.37

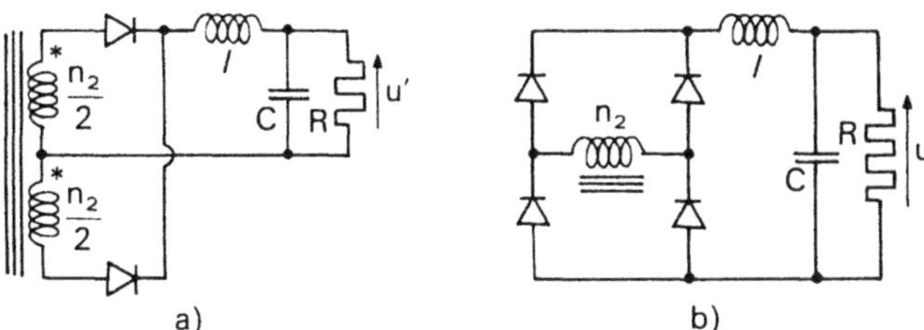

Fig. 6.38

To rectify the secondary voltage, the following can be used (Fig. 6.38):
a) an half-wave rectifier using two diodes and a centre-tapped secondary (P.2)
b) a four-diode bridge rectifier (P.D.2).

Both rectifier structures are compatible with the three inverter structures, thus offering six possibilities. We only analyze the combination of a push–pull inverter with an half-wave rectifier; this structure requires the least number of elements. The analysis of the other structures should be very similar.

6.6.2 The Push-Pull Circuit

The upper part of Fig. 6.39 shows the circuit of the converter under study and indicates the notations used. The waveforms of the different variables have been plotted below.

From $t = 0$ to $t = \alpha T/2$ (with $0 < \alpha < 1$), transistor Q_1 is saturated: $v_{11} = U$. Voltage v_{21} is positive; this causes diode D_1 to conduct.

From $t = \alpha T/2$ to $t = T/2$, Q_1 and Q_2 are off, diodes D_1 and D_2 conduct to ensure the continuity of current i_l in inductance l and of flux φ in the core.

From $t = T/2$, to $t = (1 + \alpha)T/2$, Q_2 is saturated and D_2 conducts.

From $t = (1 + \alpha)T/2$ to $t = T$, Q_1 and Q_2 are off and D_1 and D_2 conduct.

The cycle of output voltage u' and current i_l equals $T/2$, while the triggering cycle of the controlled switches is T.

v_{Q1} i_{11}
i_{21} v_{D1}
i_l
Q_1
D_1
l,r
i_C
U
v_{11} v_{21}
$n_1/2$ $n_2/2$
C R u'
v_{12} v_{22}
Q_2
$n_1/2$ $n_2/2$ D_2
v_{Q2} i_{12}
i_{22} v_{D2}
$v_{11} \simeq v_{12}$
$v_{21} \simeq v_{22}$
i_l
at
i_{21}
i_{22}
v_{D1}
v_{D2}
i_{11}
i_{12}
v_{Q1}
v_{Q2}
U
$-U$
$U\frac{n_2}{n_1}$
$-U\frac{n_2}{n_1}$
$-2U\frac{n_2}{n_1}$
I_M
I_m
$I_M/2$
$I_m/2$
AT_M
$-AT_M$
$2U$
0
$\alpha\frac{T}{2}$
$\frac{T}{2}$
$(1+\alpha)\frac{T}{2}$
T
t

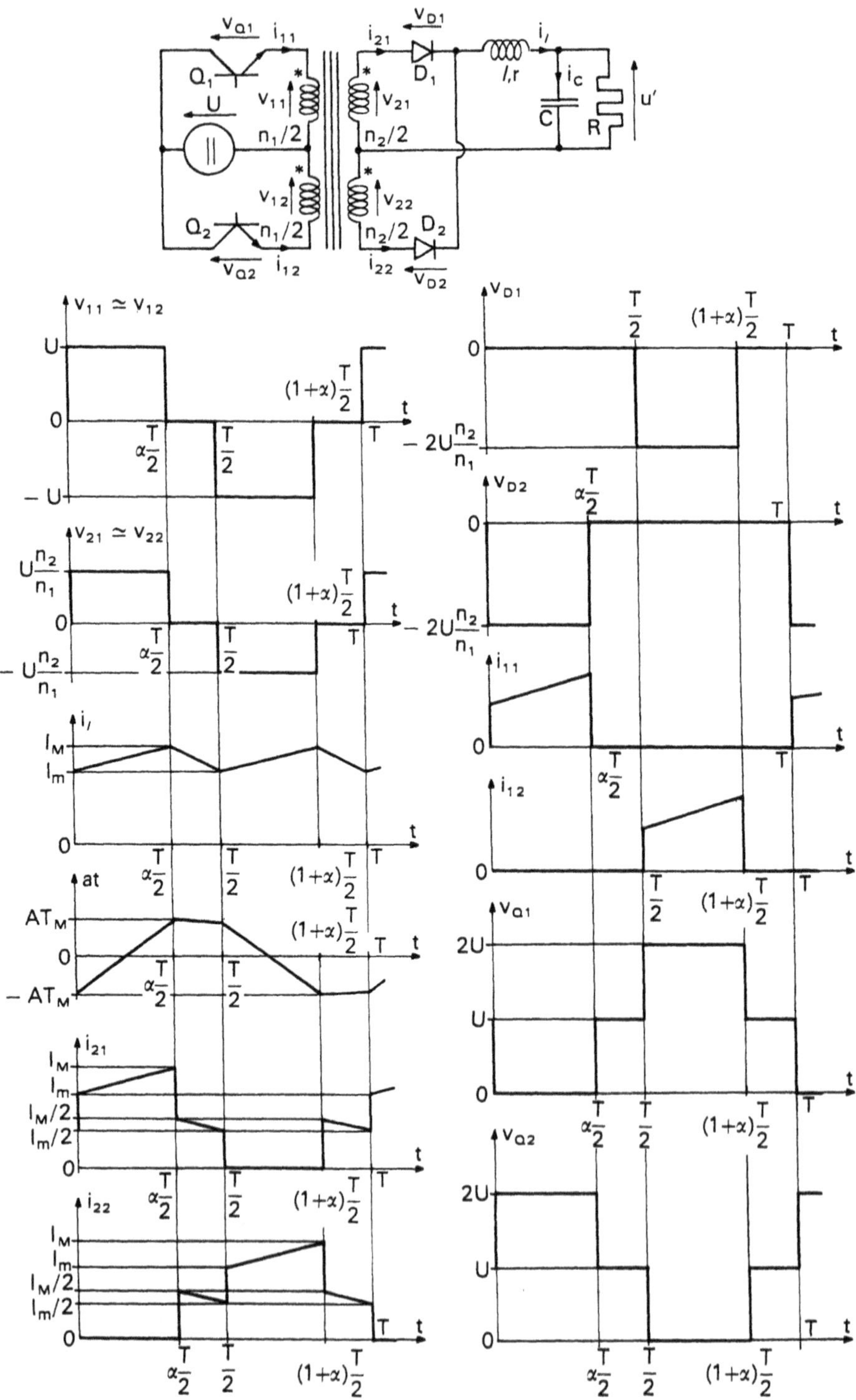

Fig. 6.39

6.6.2.1 Average Value of the Output Voltage

– From $t = 0$ to $t = \alpha T/2$, Q_1 and D_1 conduct:

$$U = r_1 i_{11} + \frac{n_1}{2}\frac{d\varphi}{dt}$$

$$\frac{n_2}{2}\frac{d\varphi}{dt} = r_2 i_{21} + r i_{21} + l\frac{di_l}{dt} + u', \quad \text{with} \quad i_l = i_{21}.$$

If r_1 is neglected, these equations become

$$U = \frac{n_1}{2}\frac{d\varphi}{dt}$$

$$U\frac{n_2}{n_1} = (r + r_2)i_l + l\frac{di_l}{dt} + u'. \tag{6.68}$$

– From $t = \alpha T/2$ to $t = T/2$, D_1 and D_2 conduct:

$$\frac{n_2}{2}\frac{d\varphi}{dt} = r_2 i_{21} + r i_l + l\frac{di_l}{dt} + u' \tag{6.69}$$

$$-\frac{n_2}{2}\frac{d\varphi}{dt} = r_2 i_{22} + r i_l + l\frac{di_l}{dt} + u'. \tag{6.70}$$

Since $i_{21} + i_{22} = i_l$, the half-sum of the previous equations can be written as

$$0 = \left(r + \frac{r_2}{2}\right)i_l + l\frac{di_l}{dt} + u'. \tag{6.71}$$

By taking the average values of i_l and u' over their cycle $T/2$ from the Eqs. (6.68) and (6.71), by assuming that, throughout the period $(\alpha T/2, T/2)$, i_l has the same average value as over the whole cycle $T/2$ and by denoting U' and I_l the average values of u' and i_l, one gets

$$\alpha U\frac{n_2}{n_1} = [r + r_2(1 + \alpha)/2]I_l + U'$$

or, since $I_l = U'/R$

$$U' = RI_l = \frac{\alpha U n_2/n_1}{1 + (1/R)[r + (r_2/2)(1 - \alpha)]} \tag{6.72}$$

If resistances r_2 and r are neglected,

$$U' = RI_l \simeq \alpha U\frac{n_2}{n_1} \tag{6.72'}$$

6.6.2.2 Magnetomotive-Force Amplitude

The MMF cycle is equal to T. Its average value is zero. Its amplitude must be known in order to determine the transformer rating.

- From $t = 0$ to $t = \alpha T/2$, $\text{at} = \frac{n_1}{2} i_{11}$,

$$U = r_1 i_{11} + \frac{n_1}{2}\frac{d\varphi}{dt}$$

If l_1 is used to denote the inductance of one half-primary and if the resistance r_1 of the latter is ignored,

$$U = \frac{n_1}{2}\frac{d\varphi}{dt} = \frac{2l_1}{n_1}\frac{d\text{at}}{dt}$$

The MMF has the following form:

$$\text{at} = \text{at}_0 + \frac{n_1}{2l_1} Ut.$$

It reaches its maximum value for $t = \alpha T/2$:

$$\text{AT}_\text{M} = \text{at}_0 + \frac{n_1}{4l_1} U\alpha T.$$

- From $t = \alpha T/2$ to $t = T/2$, $\text{at} = (n_2/2)(i_{22} - i_{21})$.
 The member-to-member difference between Eqs. (6.69) and (6.70) gives

$$n_2 \frac{d\varphi}{dt} = r_2(i_{21} - i_{22}).$$

If l_2 is used to denote the inductance of each half-secondary, this equation can be written as

$$2\frac{l_2}{n_2}\frac{d\text{at}}{dt} = -2r_2\frac{\text{at}}{n_2}$$

giving

$$\text{at} = \text{AT}_\text{M} \exp[-(t - \alpha T/2) r_2/l_2].$$

- Due to the symmetry in the waveform, the MMF at instant $t = T/2$ must be equal to $-\text{at}_0$:

$$\text{AT}_\text{M} \exp[-(1-\alpha) Tr_2/2l_2] = -\text{at}_0 = -\text{AT}_\text{M} + \frac{n_1}{4l_1} U\alpha T$$

$$\text{AT}_\text{M} = \frac{(n_1/4)(U/l_1)\alpha T}{1 + \exp[-(1-\alpha) Tr_2/2l_2]}.$$

If the influence of r_2 can be neglected,

$$\text{AT}_\text{M} \simeq \frac{n_1}{8}\frac{U}{l_1}\alpha T. \tag{6.73}$$

6.6.2.3 Current i_l and Voltage u' Ripples

- In calculating the current i_l ripple, the voltage u' ripple is ignored. In this case, Eqs. (6.68) and (6.71) can be written as:

$$\text{for } 0 < t < \alpha\frac{T}{2}, \quad l\frac{di_l}{dt} + (r + r_2)i_l = U\frac{n_2}{n_1} - U';$$

$$\text{for } \alpha\frac{T}{2} < t < \frac{T}{2}, \quad l\frac{di_l}{dt} + \left(r + \frac{r_2}{2}\right)i_l = -U'.$$

If the time constants $l/(r + r_2)$ and $l/(r + r_2/2)$ are long compared to the cycle $T/2$ of current i', the effect of resistances r and r_2 can be neglected and Eq. (6.72') used:

$$\Delta i_l \simeq \frac{1}{l}\left(U\frac{n_2}{n_1} - U'\right)\alpha\frac{T}{2}$$

$$\Delta i_l = \frac{1}{l}U\frac{n_2}{n_1}(1 - \alpha)\alpha\frac{T}{2} \tag{6.74}$$

- The output circuit is identical to that of the buck chopper. The voltage u' ripple can be deduced from that of current i_l:

$$\Delta u' \simeq \Delta i_l \frac{1}{8C}\frac{T}{2}$$

$$\Delta u' \simeq \frac{U}{32lC}\frac{n_2}{n_1}(1 - \alpha)\alpha T^2. \tag{6.75}$$

6.6.2.4 Remark on the Simultaneous Diode Conduction

When both transistors are off, flux continuity in the core requires that the magnetizing current flows through the secondaries.

When Q_1 is turned off, if only diode D_1 should conduct, current i_{21} should be equal to i_l, and voltage v_{21} to $-l_2 di_l/dt - r_2 i_l$. However, v_{22} will equal $-l_2 di_l/dt$. Voltage v_{D2} across diode D_2 will be higher than the voltage across diode D_1. Both diodes must conduct simultaneously.

During the period $(\alpha T/2, T/2)$:

$$i_{21} + i_{22} = i_l$$

$$i_{21} - i_{22} = -\frac{2at}{n_2}$$

Thus,

$$i_{21} = \frac{i_l}{2} - \frac{at}{n_2}; \quad i_{22} = \frac{i_l}{2} + \frac{at}{n_2}$$

In normal operating conditions, the magnetizing MMF of a transformer is considerably lower than the secondary or primary MMF. Current at/n_2 can be ignored compared with $i_l/2$ and it can be assumed that, during the simultaneous conduction of both diodes, i_{21} and i_{22} both have a value close to $i_l/2$. This is represented in the waveforms of Fig. 6.39.

Bibliography

Switch-Mode Power Supplies (SMPS) have now replaced almost everywhere the former supplies using a diode rectifier and a ballast semiconductor device, or a rectifier with controlled semiconductor devices, or a diode rectifier and a magnetic stabilizer. The SMPS are cheaper, lighter and less bulky. Moreover, these advantages are increased with the increasingly high operational frequency which has been obtained by means of progress in the semiconductor devices. This increase in frequency also reduces the response time.

After quoting several books [1–6] on SMPSs, we have divided the articles on classical structures, rather artificially, into two groups:

- the first [7–26] essentially concerns the study of operation, characteristics and sizing of the components;
- the remainder [27–56] deals with modeling, control and dynamic behaviour.

The following group of articles [57–68] concerns variations of basic structures, combinations and multiple output converters.

Finally, we have included studies [69–77] concerning the noise of SMPSs and input filters aimed at reducing the current ripple at the supply.

Many recent studies concern resonance supplies. To carry out the DC–DC conversion, these supplies use a resonant inverter followed by a rectifier. We have not quoted the corresponding articles here since resonant inverters will be dealt with in the following volume of this series.

Books

1. Wood P (1981) *Switching Power Converters.* Van Nostrand–Reinhold, New York
2. Sum K (1984) *Switch Mode Power Conversion.* Dekker, New York
3. Chryssis G (1984) *High-Frequency Switching Power Supplies.* Mc Graw-Hill, New York
4. Severns RP, Bloom GF (1985) *Modern DC to DC Switch Mode Power Converter Circuits.* Van Nostrand–Reinhold, New York
5. Ferrieux JP, Forest F (1987) *Alimentations à découpage. Convertisseurs à résonance.* Masson, Paris
6. Mitchell DM (1987) *DC–DC Switching Regulator Analysis.* Mc Graw-Hill, New York

Operation, Characteristics, Design

7. Weischedel HR (1972) An application of frequency entrainment in a DC-to-DC converter. *IEEE Trans. Ind. Appl.*, **8**(4): 437–442

8. Owen HA, Wilson TG, Feng SYM, Lee FCY (1972) A computer-aided design procedure for flyback step-up DC-to-DC converters. *IEEE Trans. Magnetics*, **8**(3): 289–291
9. Judd FF, Wilhart H (1972) Self-oscillating regulated DC-to-DC converter. *IEEE Trans. Ind. Appl.*, **8**(6): 684–689
10. Weischedel HR, Westerman GR (1973) A symmetry correcting pulsewidth modulator for power conditioning applications. *IEEE Trans. Ind. Appl.*, **9**(3): 318–322
11. Chen DY, Owen HA, Wilson TG (1973) Computer-aided design and graphics applied to the study of inductor-energy-storage DC-to-DC electronic power converters. *IEEE Trans. Aerosp. Electron. Syst.*, **9**(4): 585–597
12. Chen DY, Owen HA, Wilson TG (1973) Design of two-winding voltage step-up/current step-up constant frequency DC-to-DC converters. *IEEE Trans. Magnetics*, **9**(3): 252–256
13. Burger P (1975) Analysis of a class of pulse modulated DC/DC power converters. *IEEE Trans. Ind. Electron. Control Instrum.*, **22**(2): 104–116
14. Chen DY, Owen HA, Wilson TG (1976) Table-aided design of energy-storage reactor in DC-to-DC converters. *IEEE Trans. Aerosp. Electron. Syst.*, **12**(3): 374–386
15. Powell DR (1977) Computer-aided design of switching regulators. *IEEE Trans. Magnetics*, **13**(1): 920–928
16. Rao NRM (1980) A unifying principle behind switching converters and some new basic configurations. *IEEE Trans. Consum. Electron.*, **26**(1): 142–180
17. Chen DY, Owen HA, Wilson TG (1980) Design of energy-storage reactors for single-winding constant-frequency DC-to-DC converters operating in the discontinuous-reactor-current mode. *IEEE Trans. Magnetics*, **16**(6): 1422–1426
18. Harada K, Sakamoto H (1981) Pulse synchronising DC-to-DC converter. *IEEE Trans. Aerosp. Electron. Syst.*, **17**(3): 322–332
19. Wu CJ, Lee FC, Balachandran S, Goin HL (1982) Design optimization for a half-bridge DC–DC converter. *IEEE Trans. Aerosp. Electron. Syst.*, **18**(4): 497–507
20. Rahman S, Lee FC (1982) Computer simulations of optimum boost and buck–boost converters. *IEEE Trans. Aerosp. Electron. Syst.*, **18**(5): 598–608
21. Cuk S, Middlebrook RD (1983) Advances in switched-mode power conversion. *IEEE Trans. Ind. Electron.*, **30**(1): 10–19 and 19–29
22. Cuk S (1983) A new zero-ripple switching DC-to-DC converter and integrated magnetics. *IEEE Trans. Magnetics*, **19**(2): 57–75
23. Wilson TG, Whelan EW, Rodriguez R, Dishman JM (1983) DC-to-DC converter power-train optimization for maximum efficiency. *IEEE Trans. Aerosp. Electron. Syst.*, **19**(3): 413–427
24. Shoyama N, Harada K (1984) Steady-state characteristics of the push-pull DC-to-DC converter. *IEEE Trans. Aerosp. Electron. Syst.*, **20**(1): 50–56
25. Kamada A, Mochizuki T, Suzuki K, Torii M (1985) Optimum ferrite core characteristics for a 500 kHz switching mode converter transformer. *IEEE Trans. Magnetics*, **21**(5): 1726–1728
26. Yundt GB (1987) Nonlinear compensation of self-oscillating switching regulators. *IEEE Trans. Power Electron.*, **2**(1): 28–35

Modeling, Control, Dynamic performances

27. Capel A (1972) New control technique in DC/DC regulators for space applications. *IEEE Trans. Aerosp. Electron. Syst.*, **8**(4): 472–480
28. Wester GW, Middlebrook RD (1973) Low-frequency characterization of switched DC/DC converters. *IEEE Trans. Aerosp. Electron. Syst.*, **9**(3): 376–385
29. Calkin ET, Hamilton BH (1976) A conceptually new approach for regulated DC-to-DC converters employing transistor switches and pulsewidth control. *IEEE Trans. Ind. Appl.*, **12**(4): 369–377
30. Clique M, Fossard AJ (1977) A general model for switching converters. *IEEE Trans. Aerosp. Electron. Syst.*, **13**(4): 397–400
31. Burns WW, Wilson TG (1978) A state-trajectory control law for DC-to-DC converters. *IEEE Trans. Aerosp. Electron. Syst.*, **14**(1): 2–20
32. Lee FCY, Iwens RP, Yu Y, Triner JE (1979) Generalized computer-aided discrete time-domain modeling and analysis of DC–DC converters. *IEEE Trans. Ind. Electron. Control Instrum.*, **26**(2): 58–69

33. Lee FCY, Yu Y (1979) Computer-aided analysis and simulation of switched DC–DC converters. *IEEE Trans. Ind. Appl.*, **15**(5): 511–520
34. Wilson TG (1981) Start-up transient of a DC-to-DC converter powered by a current-limited source. *IEEE Trans. Aerosp. Electron. Syst.*, **17**(3): 351–363
35. Philips NJL, Francois GE (1981) Necessary and sufficient conditions for the stability of buck-type switched-mode power supplies. *IEEE Trans. Ind. Electron. Control Instrum.*, **28**(3): 229–234
36. Chetty PRK (1982) Modeling and design of switching regulators. *IEEE Trans. Aerosp. Electron. Syst.*, **18**(3): 334–344
37. Chetty PRK (1982) Current injected equivalent circuit approach to modeling of switching DC–DC converters in discontinuous inductor conduction mode. *IEEE Trans. Ind. Electron.*, **29**(3): 230–234
38. Chetty PRK (1983) Modeling and analysis of CUK converter using current injected equivalent circuit approach. *IEEE Trans. Ind. Electron.*, **30**(1): 56–59
39. Shortt DJ, Lee FC (1983) Improvided switching converter model using discrete and averaging techniques. *IEEE Trans. Aerosp. Electron. Syst.*, **19**(2): 190–202
40. Lee FC, Carter RA, Fang ZD (1983) Investigation of stability and dynamic performance of a current-injected regulator. *IEEE Trans. Aerosp. Electron. Syst.*, **19**(2): 274–287
41. Kislowski AS (1983) Controlled-quantity concept in small-signal analysis of switching power cells. *IEEE Trans. Aerosp. Electron. Syst.*, **19**(3): 438–446
42. Kelkar SS, Lee FC (1984) Stability analysis of a buck regulator employing input filter compensation. *IEEE Trans. Aerosp. Syst.*, **20**(1): 67–77
43. Shortt DJ, Lee FC (1984) Extension of the discrete-average models for converter power stages. *IEEE Trans. Aerosp. Electron. Syst.*, **20**(3): 279–289
44. Kislovski AS (1984) Power switching cell modeling at high-perturbation frequencies. *IEEE Trans. Aerosp. Electron. Syst.*, **20**(5): 521–525
45. Lee FC, Fang AD, Lee TH (1985) Optimal design strategy of switching converters employing current injected control. *IEEE Trans. Aerosp. Electron. Syst.*, **21**(1): 21–35
46. Lee YS (1985) A systematic and unified approach to modeling switches in Switch-Mode Power Supplies. *IEEE Trans. Ind. Electron.*, **32**(4): 445–448
47. Shoyama M, Harada K (1986) Dynamic characteristics of the push–pull DC-to-DC converter. *IEEE Trans. Power Electron.*, **1**(1): 3–8
48. Kelkar SS, Lee FCY (1986) A fast time domain digital technique for power converters: application to a buck converter with feedforward compensation. *IEEE Trans. Power Electron.*, **1**(1): 21–31
49. Mitchell DM (1986) An analytical investigation of current-injected control for constant-frequency switching regulators. *IEEE Trans. Power Electron.*, **1**(3): 167–174
50. Redl R, Sokal NO (1986) Near-optimum dynamic regulation of DC–DC converters using feed-forward of output current and input voltage with current-mode control. *IEEE Trans. Power Electron.*, **1**(3): 181–192
51. McArdle J, Wilson TG, Wong RCS (1987) Effects of power-train parasitic capacitance on the dynamic performance of DC-to-DC converters operating in the discontinuous MMF mode. *IEEE Trans. Power Electron.*, **2**(1): 2–19
52. Birca-Galateanu S (1987) Flyback converter output voltage stabilization. *IEEE Trans. Aerosp. Electron. Syst.*, **23**(2): 146–151
53. Birca-Galateanu S (1987) Frequency-domain analysis of switching closed-loop regulators. *IEEE Trans. Aerosp. Electron. Syst.*, **23**(2): 240–246
54. Middlebrook RD (1987) Topics in multiple-loop regulators and current-mode programming. *IEEE Trans. Power Electron.*, **2**(2): 109–124
55. Wong RC, Owen HA, Wilson TG (1987) An efficient algorithm for the time-domain simulation of regulated energy-storage DC-to-DC converters. *IEEE Trans. Power Electron.*, **2**(2): 154–168
56. Pietkiewicz A, Tollik D (1987) Unified topological modeling method of switching DC–DC converters in duty-ratio programmed-mode. *IEEE Trans. Power Electron.*, **2**(3): 218–226

Variants, Combinations, Multiple outputs

57. Hartignan P (1972) A pulsewidth-modulated DC-to-DC converter that utilizes a small number of components. *IEEE Trans. Aerosp. Electron. Syst.*, **8**(1): 9–12

58. Matsuo H, Harada K (1976) The cascade connection of switching regulators. *IEEE Trans. Ind. Appl.*, **12**(2): 192–198
59. Matsuo H, Harada K (1977) New DC–DC converters with an energy-storage reactor. *IEEE Trans. Magnetics*, **13**(5): 1211–1213
60. Matsuo H, Harada K (1978) New energy-storage DC–DC converter with multiple outputs. *IEEE Trans. Magnetics*, **14**(5): 1005–1007
61. Nakaoka M, Vietson NM, Maruhashi T, Nishimura A (1979) High-power switching regulator type DC–DC converters controlled by time-sharing high-frequency thyristor choppers with energy-storage and transfer reactors. *IEEE Trans. Magnetics*, **15**(6): 1974–1976
62. Matsuo H (1980) Comparison of multiple-output DC–DC converters using cross regulation. *IEEE Trans. Ind. Electron. Control Instrum.*, **27**(3): 176–189
63. Mahmoud FH, Lee FCY (1982) Analysis and design of an adaptative multiloop controlled two-winding buck-boost regulator. *IEEE Trans. Ind. Electron.*, **29**(1): 7–18
64. Kapustra RE, Bush JR, Graves JR, Lanier JR (1986) A programmable transformer coupled converter for high-power space applications. *IEEE Trans. Power Electron*, **1**(1): 32–38
65. Weinberg AH, Schreuders J (1986) A high-power high-voltage DC–DC converter for space applications. *IEEE Trans. Power Electr.*, **1**(3): 148–160
66. Bendien JCh, Fregien G, Van Wyk JD (1986) High-efficiency on-board battery charger with transformer isolation, sinusoidal input current and maximum power factor. *Proc. Inst. Electr. Eng., Part B*, **133**(4): 197–204
67. Shortt DJ, Michael WT, Avant RG, Palma RE (1987) A 600-W four-stage phase-shifted-parallel DC-to-DC converter. *IEEE Trans. Power Electron.*, **2**(2): 101–108
68. Sebastian J, Uceda J (1987) The double converter: a fully regulated two-output DC–DC converter. *IEEE Trans. Power Electron.*, **2**(3): 239–246

Noise, Input filter

69. Harada K, Ninomiya T (1978) Noise generation of a switching regulator. *IEEE Trans. Aerosp. Electron. Syst.*, **14**(1): 178–183
70. Harada K, Ninomiya T, Nishi H (1979) Noise suppression characteristics of a DC-to-DC converter. *IEEE Trans. Aerosp. Electron. Syst.*, **15**(2): 260–266
71. Lee FC, Yu Y (1979) Input-filter design for switching regulators. *IEEE Trans. Aerosp. Electron. Syst.*, **15**(5): 627–634
72. Ninomiya T, Harada K (1980) Common-mode noise generation in a DC-to-DC converter. *IEEE Trans. Aerosp. Electron. Syst.*, **16**(2): 130–137
73. Ninomiya T, Hakihara H, Harada K (1981) Cross noise in multi-output DC-to-DC converters. *IEEE Trans. Aerosp. Electron. Syst.*, **17**(2): 181–189
74. Kelkar SS, Lee FC (1983) A novel feedforward compensation canceling input filter-regulator interaction. *IEEE Trans. Aerosp. Electron. Syst.*, **19**(2): 258–268
75. Kelkar SS, Lee FC (1984) Adaptative input filter compensation for switching regulators. *IEEE Trans. Aerosp. Electron. Syst.*, **20**(1): 57–66
76. Nakahara M, Ninomiya T, Harada K (1985) Surge and noise generation in a forward DC-to-DC converter. *IEEE Trans. Aerosp. Electron. Syst.*, **21**(5): 619–629
77. Ninomiya T, Nakahara M, Tajima H, Harada K (1987) Backward-noise generation in forward DC-to-DC converters. *IEEE Trans. Power Electron.*, **2**(3): 208–217

Appendix

Snubbers

Whenever a semiconductor device operating as a controlled switch is turned on or turned off, the voltage and current across it do not vary instantaneously. The switching periods should be taken into account when calculating the losses in the devices and determining the peak values of the current and of the voltage they must sustain.

Switching problems are particularly important for bipolar transistors and GTO thyristors. In order to improve their switching behaviour, snubbers must often be added to these controlled devices.

A.1 Bipolar Transistors: Switching Losses

Figure A.1 shows a buck converter. It is assumed that the load is sufficiently inductive to allow current I, which flows through it, not to undergo any variations during switching.

• *Preliminary remarks*

The following assumptions have been adopted to make the evaluation of losses during switching and of their reduction by snubbers more easy:

- collector current i_C increases linearly during its rise time t_r;
- current i_C decreases linearly during its fall time t_f;
- the rise and fall times depend solely on current I.

Furthermore the rise and fall times of voltage v_{CE} across the transistor are ignored, and the free-wheeling diode is assumed to be ideal.

Without the snubber, the waveforms are then those shown in Fig. A.1. The importance of these assumptions is clearly apparent when these waveforms are compared with those shown in Figs. 2.25 and 2.30.

Nevertheless, such simplifications are required in order to carry out a rapid, approximate quantitative analysis of the reduction in losses, overvoltages and overcurrents the snubbers provide.

• *Switching losses*

- *At turn-on*, current i_C increases linearly from zero to I when t changes from 0 to t_r. While i_C increases, current i_D in the diode decreases, but voltage v_D

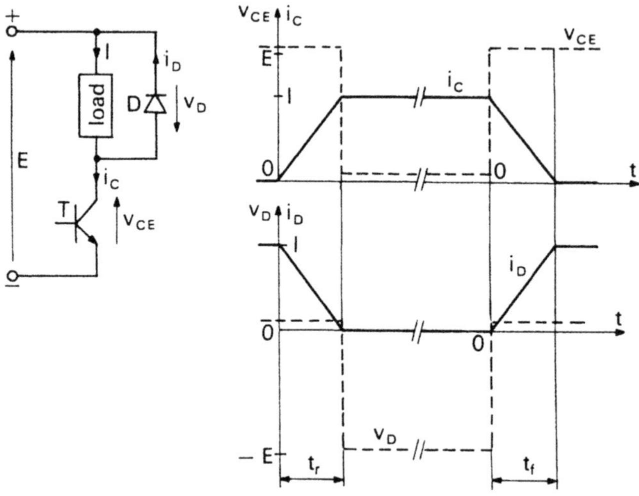

Fig. A.1

across the latter remains negligible as long as i_D is positive: voltage v_{CE} across the transistor remains close to E during the whole of the i_C rising period.

The energy dissipated in the transistor during turn-on thus equals:

$$W_{ON} = \int_0^{t_r} v_{CE} i_C \, dt = \int_0^{t_r} EI \frac{t}{t_r} \, dt$$

$$W_{ON} = \frac{1}{2} EIt_r. \tag{A.1}$$

– *At turn-off*, current i_C changes from I to zero, while t goes from 0 to t_f. As soon as i_C decreases, current i_D in the diode, equal to $I - i_C$, becomes positive: voltage v_D can be neglected and voltage v_{CE} is close to E.

The energy dissipated in the transistor during turn-off thus equals:

$$W_{OFF} = \int_0^{t_f} v_{CE} i_C \, dt = \int_0^{t_f} EI \frac{t_f - t}{t_f} \, dt$$

$$W_{OFF} = \tfrac{1}{2} EIt_f. \tag{A.2}$$

- *Switching trajectory*

In the (v_{CE}, i_C) plane, the operating point follows the path shown in Fig. A.2. It must remain within the forward-bias safe operating area during the turn-on commutation and within the reverse-bias safe operating area during turn-off.

- *Roles of snubbers*

Snubbers play two roles:

- to reduce the switching losses in the transistor so as to enable the switching frequency to be increased;

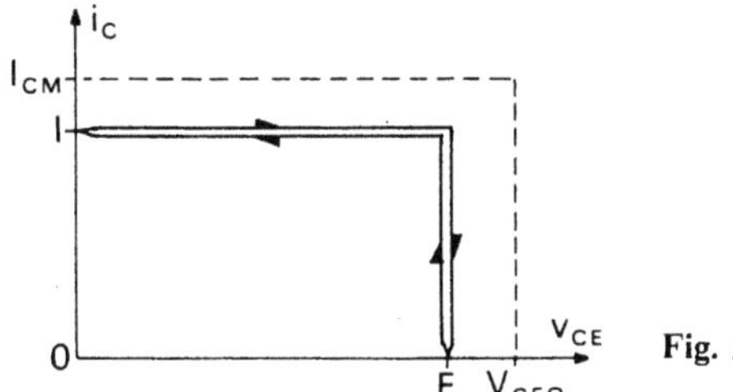

Fig. A.2

- to modify the trajectory of the operating point so to allow, at given E and I, the use of devices which have to withstand lower maximum values of v_{CE} and i_C.

A.2 Turn-on Snubber

In order to reduce voltage v_{CE} during the rise of i_C, an *inductance* l is series-connected to the transistor collector (Fig. A.3). Resistance r and diode D_l enable l to discharge during the following turn-off.

A.2.1 Reduction of Losses in the Transistor

If i_C increases linearly from 0 to I during time period $(0, t_r)$, i_C, v_l and v_{CE} can be written as

$$i_C = i_l = I\frac{t}{t_r}; \quad v_l = l\frac{di_l}{dt} = l\frac{I}{t_r}$$

$$v_{CE} = E + v_D - v_l \simeq E - l\frac{I}{t_r}.$$

Voltage v_{CE} is thus reduced during the rise of i_C. The losses in the transitor during this period are now given by

$$W_T = \int_0^{t_r} v_{CE} i_C \, dt = \int_0^{t_r} \left(E - l\frac{I}{t_r}\right) I \frac{t}{t_r} \, dt$$

$$W_T = \frac{1}{2}\left(E - l\frac{I}{t_r}\right) I \, t_r. \tag{A.3}$$

These losses are less than those, W_{ON}, obtained without l. Theoretically, they would be zero if v_{CE} was equal to zero during the whole rise of i_C. This would happen if

$$l = \frac{E}{I} t_r.$$

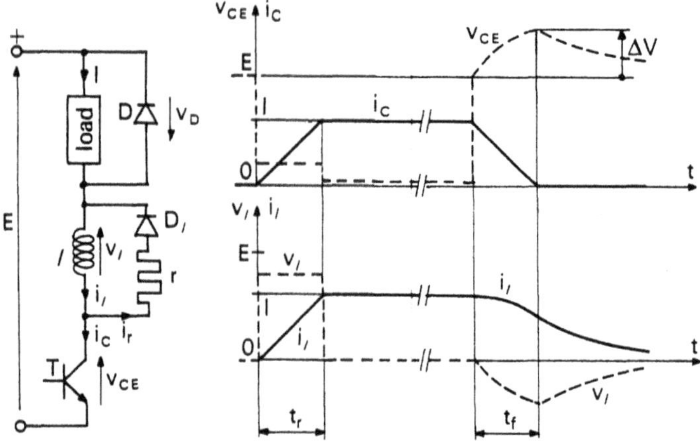

Fig. A.3

A.2.2 Overvoltage, Total Losses

The energy stored in the inductance when i_C is settled, must be eliminated. It can be dissipated in resistance r during the next turn-off of the transistor (diode D_l is intended to avoid current through r when i_C is rising). The discharge of l generates an overvoltage ΔV to appear across the transistor.

- *Current i_l disappears* in two phases:
- *During the decrease of the collector current* i_C, if the beginning of this decrease is taken as the initial time, the following is obtained:

$$i_l = i_r - i_C = i_r - I\,\frac{t_f - t}{t_f}$$

Since $r i_r = -\,l\dfrac{di_l}{dt}$, current i_l is solution of equation

$$i_l + \frac{l}{r}\frac{di_l}{dt} = I\,\frac{t_f - t}{t_f}.$$

From this, the expressions of this current and of voltage v_l can be deduced:

$$i_l = \frac{I}{t_f}\left[\frac{l}{r}\left[1 - \exp\left(-\frac{r}{l}t\right)\right] + t_f - t\right]$$

$$v_l = l\,\frac{di_l}{dt} = -\,l\frac{I}{t_f}\left[1 - \exp\left(-\frac{r}{l}t\right)\right].$$

Voltage v_{CE}, equal to $E - v_l$ is greater than E. This *overvoltage* reaches its maximum for $t = t_f$ and its value is then

$$\Delta V = v_{CE_{max}} - E = l\frac{I}{t_f}\left[1 - \exp\left(-\frac{r}{l}t_f\right)\right]. \tag{A.4}$$

- *After i_C has fallen to zero*, current i_l disappears in the circuit formed by l, r and D_l. If the instant where i_C falls to zero is taken as the new initial time and by denoting I_f the value of i_l at this instant, this gives

$$i_l = i_r = I_f \exp\left(-\frac{r}{l}t\right)$$

$$v_l = -rI_f \exp\left(-\frac{r}{l}t\right); \quad v_{CE} = E - v_l$$

$$\text{with } I_f = \frac{I}{t_f}\frac{l}{r}\left[1 - \exp\left(-\frac{r}{l}t_f\right)\right].$$

Assuming that i_l can be neglected after a damping time equal to three times the time constant l/r, the time interval separating the start of i_C turn-off from the next turn-on has to be such that

$$T_{OFF} > t_f + 3\frac{l}{r}$$

Figure A.4 provides the new switching trajectory in the (v_{CE}, i_C) plane.

- *The energy* stored in inductance l and then dissipated in resistance r is

$$W_l = \frac{1}{2}lI^2.$$

At each turn-on, input DC voltage source must supply

$$W'_{ON} = W_T + W_l = \frac{1}{2}\left(E - l\frac{I}{t_r}\right)It_r + \frac{1}{2}lI^2$$

$$W'_{ON} = \frac{1}{2}EIt_r = W_{ON}.$$

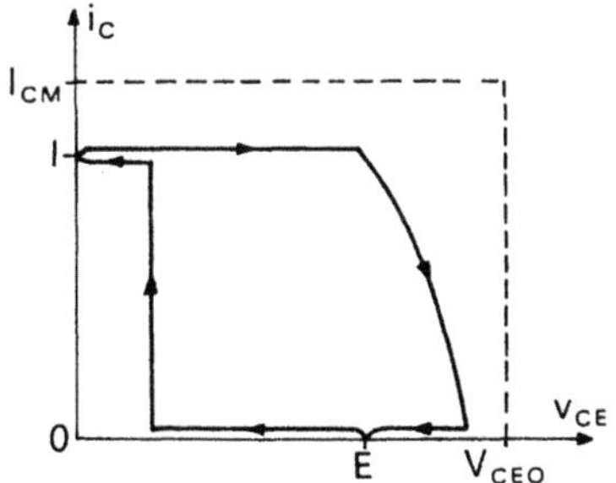

Fig. A.4

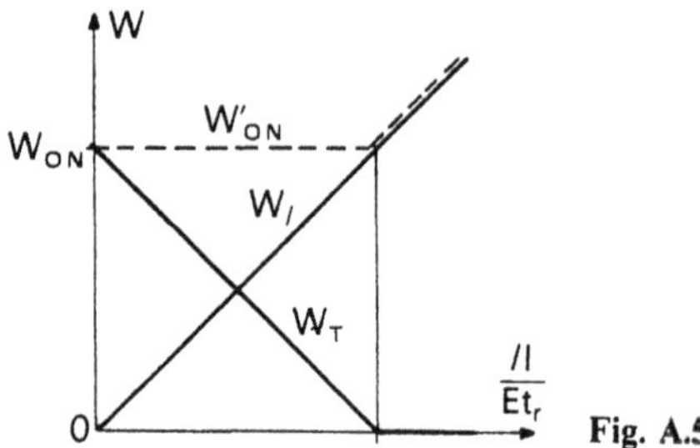

Fig. A.5

The losses in the transistor are reduced, but the total losses remain the same. Figure A.5 shows how W_T, W_l and W'_{ON} vary as a function of inductance l referred to Et_r/I.

If l was equal to Et_r/I, W_T would be zero and W_l would be equal to W_{ON}. If the value of l is still increased, W'_{ON}, equal to W_l, would become greater than W_{ON}, since W_T could no longer decrease.

(Note that adding inductance l slightly increases the losses at turn-off, as voltage v_{CE} is higher than E during the turn-off of i_C.)

- *The choice of values* to be attributed to l and r is the result of an analysis of Eqs. (A.3), (A.4) and (A.5).

  Increasing l reduces W_T but increases the overvoltage ΔV as well as the minimum value to be given to time T_{OFF} between the turn-off starting and the next turn-on.

  Increasing r reduces this value of T_{OFF} but increases overvoltage ΔV.

A.2.3 Using a Saturable Reactor

The constant inductance l can be replaced by a saturable reactor (coil around a ferrite toroid). The latter is calculated in order to have a low magnetizing current and to be saturated when current I flows through the transistor. The waveforms of Fig. A.3 are then modified as shown in Fig. A.6.

- *When conduction begins*, the value of i_C is low and the inductance is not saturated. Its self-induction coefficient is very high and current i_C rises slowly even though voltage v_l is close to E. As i_C rises, the inductance comes into saturation, i_C increases more rapidly, but v_l remains high enough for v_{CE} to be kept to a very low value.

  If, as an initial estimate, it is assumed that the expression of current has the form

$$i_C = I\left(\frac{t}{t_r}\right)^n, \quad \text{with} \quad n > 1,$$

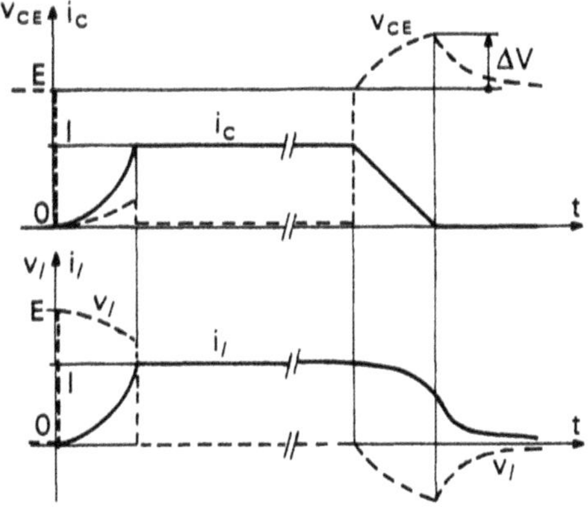

Fig. A.6

the energy delivered by the supply to the commutation circuit is:

$$W'_{ON} = \int_0^{t_r} E i_C \, dt = \int_0^{t_r} E I \left(\frac{t}{t_r}\right)^n dt$$

$$W'_{ON} = \frac{1}{n+1} E I t_r < W_{ON}.$$

The use of a saturable reactor enables the losses in the transistor and the total losses to be simultaneously reduced.

- *At turn-off*, when i_C falls to zero, current i_l falls rapidly at first (saturated inductance), and then much more slowly (unsaturated inductance).

A.2.4 Switching Energy Recovery

If the snubber inductance is the primary of a transformer, the energy stored in this inductance can be recovered, by means of the secondary, and is thus not dissipated in a resistance. Figure A.7a shows the circuit diagram, with a transformer of n_1 turns at the primary and n_2 turns at the secondary; diode D′ enables the current through the secondary to flow back towards the source only.

Outside the commutation periods, voltage v_l is zero, voltage $v_{D'}$, across diode D′ is equal to $-E$ and is thus negative. During the rise of current i_C, v_l is positive and voltage $v_{D'}$, given by $-E-(n_2/n_1)v_l$, is still more negative. Diode D′ can only conduct during i_C turn-off; as the free-wheeling diode is then conducting, the variations of i_C, i_2, v_l and v_{CE} can be analyzed during this period by means of the equivalent diagram in Fig. A.7b.

The operation of the transformer can be accounted for by using the circuit shown in Fig. A.7c, in which

i'_2 is the secondary current referred to the primary,

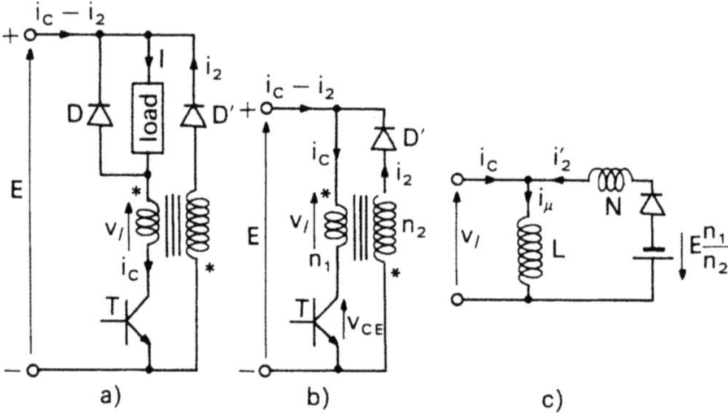

Fig. A.7

L, the magnetizing inductance,
N, the total leakage inductance referred to the primary.

- During the period $(0, t_f)$ when the current i_C is falling, if the fall of this current is assumed to be linear, this gives

  $$i_C = I(1 - t/t_f).$$

  Since $i_C + i'_2 = i_\mu$,

  from $L\mathrm{d}i_\mu/\mathrm{d}t + N\mathrm{d}i'_2/\mathrm{d}t + En_1/n_2 = 0$

  the following can be deduced:

  $$(L + N)\frac{\mathrm{d}i'_2}{\mathrm{d}t} = \frac{LI}{t_f} - E\frac{n_1}{n_2}$$

  giving

  $$i'_2 = \left(\frac{LI}{t_f} - E\frac{n_1}{n_2}\right)\frac{t}{L + N}$$

  Current i'_2 (or i_2) will be positive, i.e. there will be recovery, if

  $$\frac{n_1}{n_2} < \frac{LI}{Et_f}$$

  The primary voltage is then

  $$v_l = L\frac{\mathrm{d}i_\mu}{\mathrm{d}t} = -\frac{L}{L + N}\left(\frac{NI}{t_f} + E\frac{n_1}{n_2}\right); \tag{A.5}$$

  care should be taken that v_{CE}, equal to $E - v_l$, does not exceed the maximum acceptable value V_{CEO}.

- From $t = t_f$ onwards, current i_C is zero; as current i'_2 cannot be discontinuous,

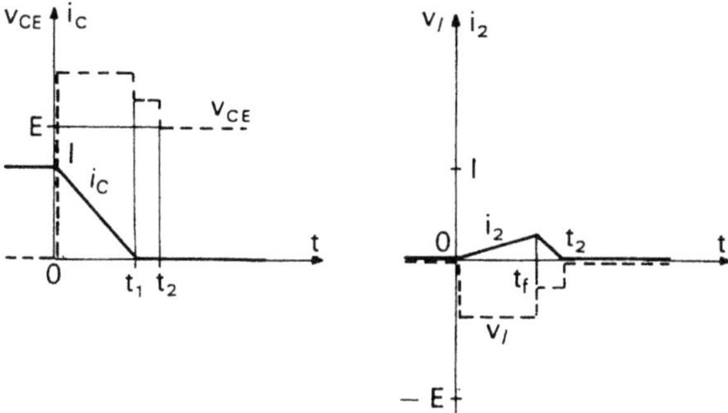

Fig. A.8

the equation

$$(L + N)\frac{di_2'}{dt} = -E\frac{n_1}{n_2}$$

gives

$$i_2' = -\frac{E}{L+N}\frac{n_1}{n_2}(t - t_f) + i_2'(t_f)$$

with

$$i_2'(t_f) = \frac{L}{L+N}I - E\frac{n_1}{n_2}\frac{t_f}{L+N}$$

During this second period,

$$v_l = L\frac{di_2'}{dt} = -\frac{L}{L+N}E\frac{n_1}{n_2} \tag{A.6}$$

Recovery ends when i_2 returns to zero, i.e. at instant $t = t_2$, such that

$$-\frac{E}{L+N}\frac{n_1}{n_2}(t_2 - t_f) + \frac{L}{L+N}I - E\frac{n_1}{n_2}\frac{t_f}{L+N} = 0$$

$$t_2 = \frac{LI}{E}\frac{n_2}{n_1}$$

– Figure A.8 shows the waveforms of voltages v_{CE} and v_l and of currents i_C and i_2.

A.3 Turn-off Snubber

In order to reduce voltage v_{CE} during the fall of i_C, a *capacitor* γ is connected between the transistor collector and emitter (Fig. A.9). This capacitor will be

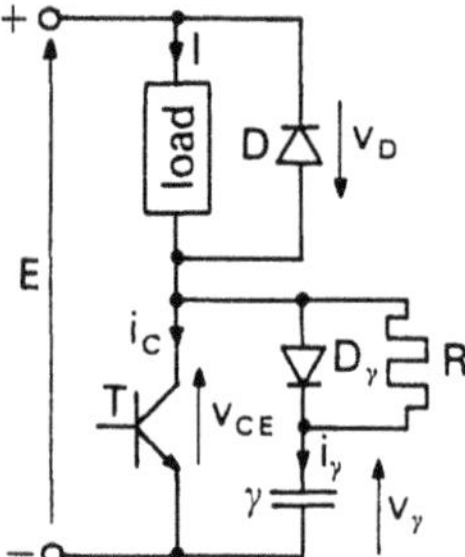

Fig. A.9

discharged via R at the start of the next conduction period. A diode D_γ is added to prevent R from slowing down the charging of γ, and voltage $R(I - i_C)$ from being added to v_γ during this same charging.

A.3.1 Reduction of Losses in the Transistor

If it is assumed that the collector current falls linearly from I to zero during the time interval t_f and if the beginning of this fall is taken as the initial time,

$$i_C = I \frac{t_f - t}{t_f}$$

The difference between I and i_C is supplied to the capacitor via diode D_γ:

$$i_\gamma = I - i_C = I \frac{t}{t_f} = \gamma \frac{dv_\gamma}{dt}$$

If the low value of voltage v_{CE} across the transistor during its conduction is ignored, v_γ and v_{CE} start from zero and are expressed as

$$v_{CE} = v_\gamma = \frac{I}{\gamma} \frac{t^2}{2t_f}$$

Two cases may arise, depending on whether γ is greater or less than C_0. C_0 denotes the value of γ for which the rise of v_γ from 0 to E has a duration equal to t_f.

If $\gamma = C_0$,

$$E = \frac{I}{C_0} \frac{t_f^2}{2t_f} \quad \text{or} \quad C_0 = \frac{I}{E} \frac{t_f}{2} \tag{A.7}$$

A.3.1.1 First Case: $\gamma > C_0$ (Fig. A.10a)

- If capacitor γ is greater than C_0, voltage v_γ has not reached value E when i_C becomes zero. Its value V_F is then

$$V_F = \frac{I}{2\gamma} t_f < E$$

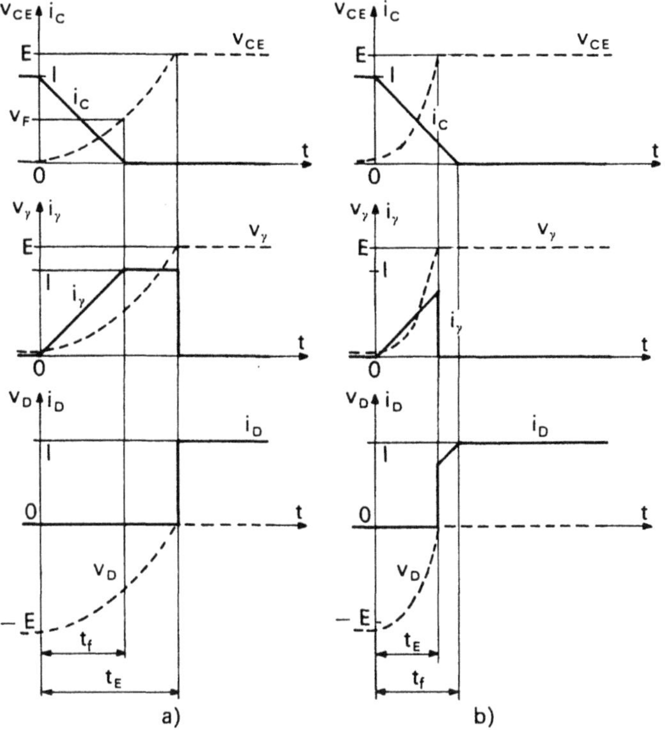

Fig. A.10

Voltage v_D, equal to $v_{CE} - E$ or to $v_\gamma - E$, is negative; current i_D remains at zero and the whole current flows through the capacitor, which is charged at a constant current:

$$i_\gamma = I\,; \; v_\gamma = v_{CE} = v_F + \frac{I}{\gamma}(t - t_f).$$

Voltage v_γ reaches value E for $t = t_E$, such that

$$V_F + \frac{I}{\gamma}(t_E - t_f) = E$$

Since V_F is known, the following can be deduced:

$$t_E = \frac{t_f}{2} + \frac{\gamma E}{I}. \tag{A.8}$$

At the instant $t = t_E$, v_D is zero and diode D begins conducting: current i_D rises very quickly (see remark at the end of this section) from 0 to I, current i_γ falls to zero and the capacitor remains charged at voltage E.

– The energy dissipated in the transistor during the turn-off is:

$$W_T = \int_0^{t_f} v_{CE} i_C \, dt = \int_0^{t_f} \frac{I}{\gamma} \frac{t^2}{2t_f} I \frac{t_f - t}{t_f} \, dt$$

$$W_T = \frac{I^2}{\gamma} \frac{t_f^2}{24}. \tag{A.9}$$

By writing

$$\gamma = k C_0, \tag{A.10}$$

ratio k is, in this case, greater than 1, since t_E is greater than t_F.

W_T can be expressed as a function of k and of energy W_{OFF} which would be dissipated in the transistor without turn-off snubber.

$$W_T = \frac{I^2}{kC_0} \frac{t_f^2}{24} = \frac{I^2}{k} \frac{2E}{It_f} \frac{t_f^2}{24} = \frac{1}{12k} E I t_f$$

$$W_T = \frac{1}{6k} W_{OFF} < W_{OFF}. \tag{A.10'}$$

– *Remark.* At instant $t = t_E$, current i_D does not change instantaneously from 0 to I; current i_γ does not fall instantaneously to zero. This brings v_γ and v_{CE} beyond E.

This phenomenon cannot be accounted for by the simplified diagram used in this analysis. The simultaneous conduction of D and D_γ once again makes v_γ and v_{CE} equal to E, and does not show the possibility of overvoltage.

In fact, the stray inductances of the connection and, possibly, of the turn-on snubber must be taken into account. When D begins conducting, these inductances cause a brief oscillation before voltage v_γ and v_{CE} stabilize at value E. This overvoltage must be taken into account in determining the maximum value of v_{CE} the transistor must sustain.

A.3.1.2 Second Case: $\gamma < C_0$ (Fig. A.10b)

If γ is less than C_0, voltage v_γ reaches value E, for $t = t_E$, before the instant, $t = t_f$, when i_C falls to zero.

– During the whole time interval $(0, t_E)$, voltage v_γ is given by

$$v_\gamma = \frac{I}{\gamma} \frac{t^2}{2t_f}$$

From this it can be deduced that

$$t_E = \sqrt{\frac{2\gamma E t_f}{I}} \tag{A.11}$$

or, if Eqs (A.7) and (A.10) are taken into account,

$$t_E = t_f\sqrt{k}\,. \tag{A.11'}$$

At instant $t = t_E$, diode D becomes conducting and current i_γ falls suddenly to zero (the remark made at the end of the analysis of the first case can once more be applied here).

– The energy dissipated in the transistor during turn-off is:

$$W_T = \int_0^{t_f} v_{CE} i_C \,dt = \int_0^{t_E} \frac{I}{\gamma}\frac{t^2}{2t_f} I \frac{t_f - t}{t_f}\,dt + \int_{t_E}^{t_f} EI \frac{t_f - t}{t_f}\,dt$$

$$= \frac{I^2}{2\gamma t_f}\left(\frac{t_E^3 t_f}{3} - \frac{t_E^4}{4}\right) + \frac{EI}{t_f}\left(t_f^2 - t_f t_E - \frac{t_f^2}{2} + \frac{t_E^2}{2}\right)$$

or, if Eqs. (A.7) and (A.11′) are taken into account, after simplification,

$$W_T = EIt_f\left(\frac{\sqrt{k}}{3} - \frac{k}{4} + \frac{1}{2} - \sqrt{k} + \frac{k}{2}\right)$$

$$W_T = W_{OFF}\left(1 - \frac{4\sqrt{k}}{3} + \frac{k}{2}\right) < W_{OFF}\,. \tag{A.12}$$

A.3.2 Overcurrent: Total Losses

The energy stored in the capacitor during turn-off is dissipated when the next turn-on begins. For the diagram in Fig. A.9, this energy is dissipated in resistance R.

- As long as the diode is conducting (interval $(0, t_r)$ in Fig. A.11), its flow imposes a value $+E$ on v_{CE} and on v_γ. The capacitor discharge, via R and the transistor, only begins at the instant when i_D falls to zero. By taking this instant as the initial time, this gives

$$v_\gamma = E\exp\left(-\frac{t}{R\gamma}\right); \quad i_\gamma = \gamma\frac{dv_\gamma}{dt} = -\frac{E}{R}\exp\left(-\frac{t}{R\gamma}\right)$$

$$i_C = I - i_\gamma$$

The overcurrent, caused by $-i_\gamma$ flowing through the transistor, is at its maximum at the beginning of the discharge; its value is then

$$\Delta I = E/R. \tag{A.13}$$

The discharge can be considered as over after a time equal to three times the time constant γR; the transistor conduction period T_{ON} must thus be such that

$$T_{ON} > t_r + 3R\gamma. \tag{A.14}$$

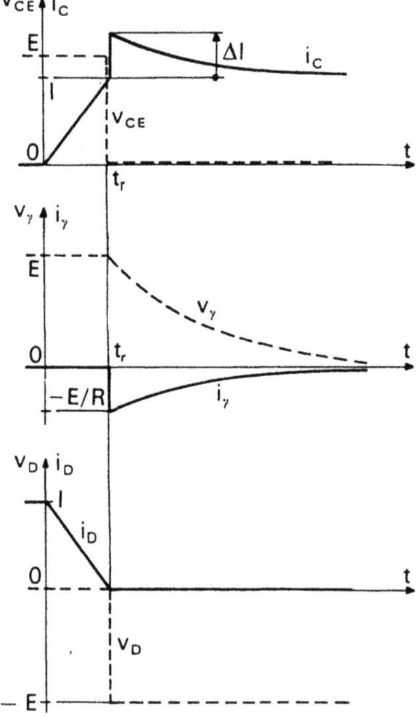

Fig. A.11

Figure A.12 shows the switching loci in the two cases which may arise for the charging of $\gamma (t_E > t_f$ and $t_f > t_E)$

- *The energy* stored in the capacitor during turn-off is

$$W_\gamma = \frac{1}{2}\gamma E^2 = \frac{1}{2}kC_0E^2 = \frac{k}{4}EIt_f = \frac{k}{2}W_{OFF}.$$

This energy is then dissipated in R. The turn-off snubber thus reduces the losses in the transistor but brings in additional losses. The supply must provide

$$W'_{OFF} = W_T + W_\gamma$$

$$W'_{OFF} = \begin{cases} W_{OFF}\left(\dfrac{1}{6k} + \dfrac{k}{2}\right), & \text{if } k > 1; \\ W_{OFF}\left(1 - \dfrac{4\sqrt{k}}{3} + k\right), & \text{if } k < 1. \end{cases}$$

These expressions take no account of additional losses caused by the discharge of γ via the transistor.

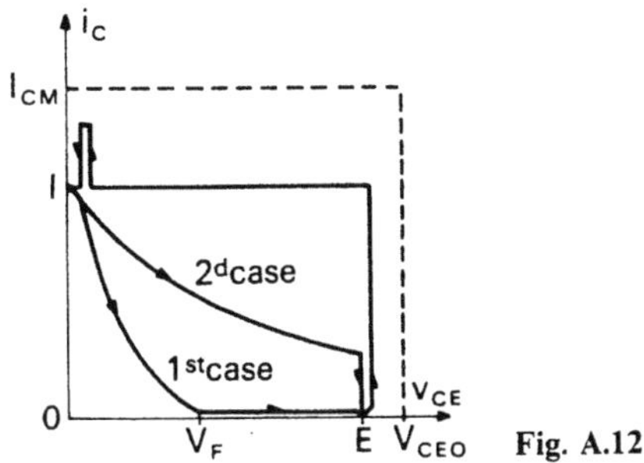

Fig. A.12

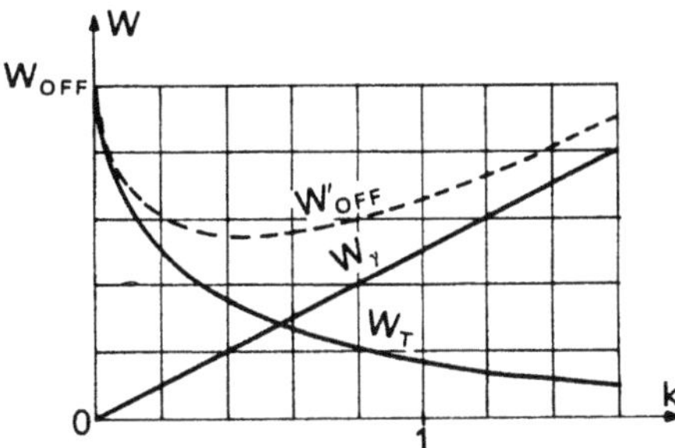

Fig. A.13

Figure A.13 gives the variations of W_T, W_γ and W'_{OFF} as a function of k. All these energies are referred to W_{OFF}.

- *The choice of* γ is linked to that of k. An analysis of Figs. A.12 and A.13 shows that k should be chosen greater than 1 in order to limit the losses in the transistor, even though the minimum of W'_{OFF} corresponds to k less than 1. For given values of E and I, the stresses imposed on the transistor are lower: when v_{CE} goes from V_F to E the transistor is already blocked.

  The choice of R is the result of a compromise: an increase in R decreases the overcurrent ΔI, but increases the minimum of T_{ON}.

A.3.3 Switching Energy Recovery

As in the case of the turn-on snubber, part of the energy stored in the turn-off snubber capacitor can be recovered.

- *A first solution* (Fig. A.14a) is to divide the capacitor into two parts γ_1 and γ_2. At the beginning of each conduction period, the energy stored in γ_2 is used to increase the rate of rise of the base current and to bring about the current peak necessary to reduce the current i_C rise time (see Sect. 2.2.4.4). Diode D_2, parallel-connected across γ_2, prevents the base control from modifying the transistor collector-emitter voltage v_{CE}.

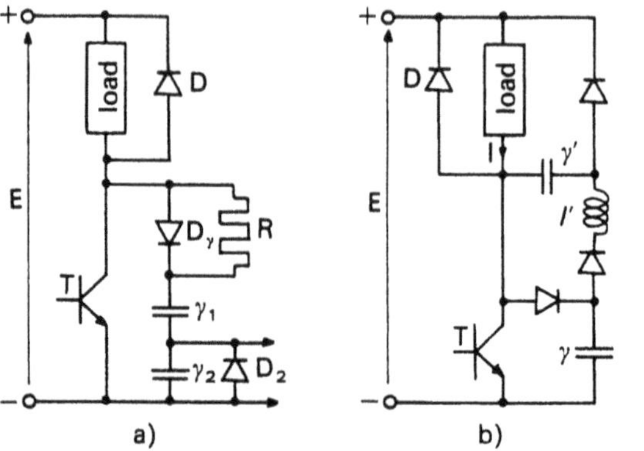

Fig. A.14

– *A second solution* (Fig. A.14b), known as the KNOLL circuit, needs additional elements: an inductance l', a capacitor γ' and a diode. When T begins to conduct, the energy of γ is used to charge γ' via l'. When T is turned off, the discharge of γ' supplies a part of current I in the load. A more detailed analysis of such a type of circuit will be given in Sect. A.5. It is slightly more complex, since it includes both turn-off snubber and turn-on snubber.

A.4 Combination of the Two Snubbers Without Energy Recovery

Both snubbers can be used simultaneously. During switchings, overvoltages and overcurrents can best be controlled if each snubber comprises a resistance and a diode (Fig. A.15a). A variant consists of series-connecting r with R during the discharge of γ (connection 1–2 replaced by connection 1–3 shown in broken line).

But if the transistor is operating well below its repetitive maximum peak values, the diagram can be simplified by using only one diode D′ and one resistance R (Fig. A.15b).

We will describe the switchings on the first diagram, indicating how the results obtained can be adapted to the circuit with only one resistance.

A.4.1 Turn-on: Discharge of γ (Fig. A.16)

At the start of the transistor turn-on period, current i_C, equal to i_l, rises from 0 to I while the free-wheeling diode current changes from I to 0. Voltage v_l is positive and diode D_l is blocked. The capacitor, charged at voltage E, cannot be discharged. The currents in resistances R and r are zero: the operation is the same as that described in Sect. A.2.1.

• *Start of capacitor γ discharge*

As soon as i_C reaches value I, diode D turns off and γ can be discharged via l, R and the transistor. As long as i_C continues to increase, diode D_l remains blocked and the circuit diagram is equivalent to that shown in Fig. A.16.

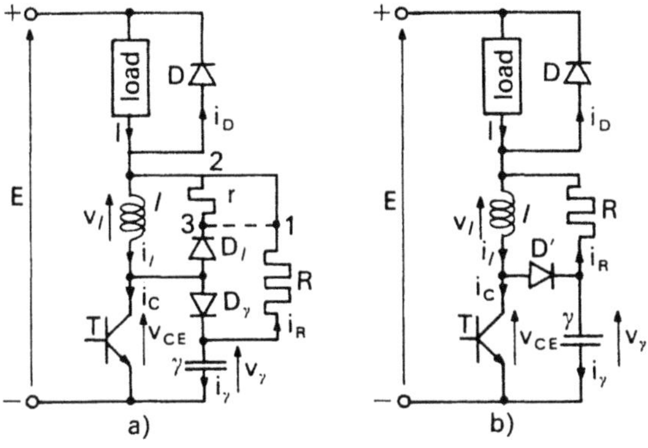

Fig. A.15

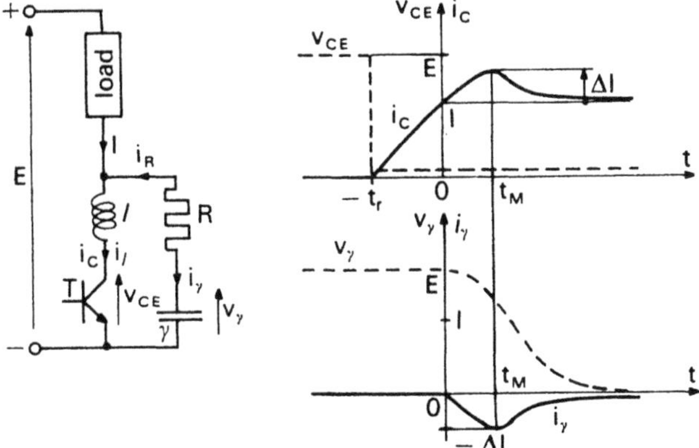

Fig. A.16

From

$$l\frac{di_l}{dt} - v_\gamma + Ri_R = 0$$

with

$$i_l = I + i_R; \quad i_R = -i_\gamma \quad \text{and} \quad i_\gamma = \gamma\frac{dv_\gamma}{dt}$$

the differential equation giving v_γ can be deduced:

$$l\gamma\frac{d^2v_\gamma}{dt^2} + R\gamma\frac{dv_\gamma}{dt} + v_\gamma = 0. \tag{A.15}$$

We will use

$$\alpha = \frac{R}{2l}; \quad \omega_0 = \frac{1}{\sqrt{l\gamma}}; \quad z = \frac{\alpha}{\omega_0} = \frac{R}{2}\sqrt{\frac{\gamma}{l}} \tag{A.16}$$

– *If there is low damping* ($z < 1$):

By taking the beginning of discharge as the initial time, v_γ and i_γ can be expressed as

$$v_\gamma = \frac{E}{\sqrt{1 - z^2}} \exp(-\alpha t) \cos[(\omega_0 \sqrt{1 - z^2})t - \varphi]$$

$$i_\gamma = -\frac{\gamma \omega_0 E}{\sqrt{1 - z^2}} \exp(-\alpha t) \sin(\omega_0 \sqrt{1 - z^2})t$$

with $\varphi = \arcsin z$.

Current $-i_\gamma$ reaches its maximum ΔI, and thus current i_C reaches its maximum $I + \Delta I$, for

$$t = t_M = \frac{1}{\omega_0 \sqrt{1 - z^2}} \left(\frac{\pi}{2} - \varphi \right) \tag{A.17}$$

and the value of the *overcurrent* is

$$\Delta I = \gamma \omega_0 E \exp(-\alpha t_M). \tag{A.18}$$

– *If there is high damping* ($z > 1$):

Voltage v_γ and current i_γ are given by

$$v_\gamma = \frac{E}{\alpha_2 - \alpha_1} [\alpha_2 \exp(\alpha_1 t) - \alpha_1 \exp(\alpha_2 t)]$$

$$i_\gamma = \gamma E \frac{\alpha_1 \alpha_2}{\alpha_1 - \alpha_2} [\exp(\alpha_1 t) - \exp(\alpha_2 t)]$$

with

$$\alpha_1 = -\alpha + \sqrt{\alpha^2 - \omega_0^2}; \quad \alpha_2 = -\alpha - \sqrt{\alpha^2 - \omega_0^2}.$$

The maximum $I + \Delta I$ of current i_C is obtained for

$$t = t_M = \frac{1}{\alpha_1 - \alpha_2} \ln\left(\frac{\alpha_2}{\alpha_1} \right) \tag{A.17'}$$

and the value of the *overcurrent* is

$$\Delta I = \frac{\gamma \omega_0^2 E}{\alpha_1 - \alpha_2} [\exp(\alpha_1 t_M) - \exp(\alpha_2 t_M)]. \tag{A.18'}$$

– Equations (A.17) and (A.17′), (A.18) and (A.18′) can be used directly for the single resistance circuit in Fig. A.15b. For the variant, shown in dotted line, of the diagram in Figure A.15a, R must be replaced by $R + r$.

• *End of capacitor γ discharge*

The end of the discharge ($t > t_M$) depends on the circuit used.

- With the circuit in Fig. A.15a, v_l becomes negative and diode D_l starts conducting, as soon as i_C begins to decrease. The capacitor is discharged by R, l and r in parallel, and by the transistor. The flow of a current through r makes the potential of point 2 more negative than that of the collector and speeds up the discharge of γ via R. The waveforms in Fig. A.16 apply to this case.
- With the variant of the circuit shown in Fig. A.15a, the state corresponding to the start of discharge continues slightly beyond $t = t_M$, until $-l(di_C/dt)$ becomes equal to ri_C.
- With the single resistance circuit (Fig. A.15b), if z is greater than 1, the state corresponding to the first stage of the discharge continues until v_γ and i_γ become negligible. If z is less than 1, the initial state continues until v_γ reaches zero, for $t > t_M$; diode D′ then starts to conduct, v_γ remains zero and l discharges into R.

A.4.2 Turn-off: Discharge of l (Fig. A.17)

The study will be limited to the case in which t_E is greater than t_f and current i_C falls to zero before voltage v_{CE} has reached value $+E$.

As long as v_{CE} remains below E, diode D is blocked. Current I, first equal to $i_C + i_\gamma$ and then to i_γ, flows through inductance l and voltage v_l is zero, as are the currents in resistances r and R. The operational mode is the same as that described in Sect. A.3.1.1.

• *Start of l discharging*

As soon as voltage v_{CE}, equal to v_γ, exceeds the value $+E$, diode D starts to conduct. As long as v_γ increases, i_γ is positive. Voltage v_l, equal to $E - v_\gamma$,

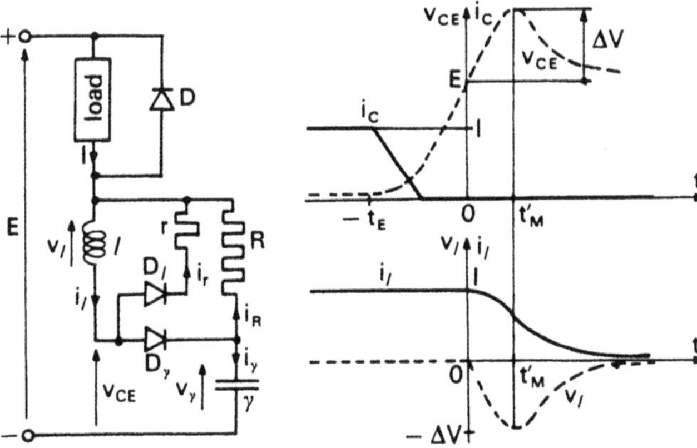

Fig. A.17

becomes negative; this leads to a current i_r flowing through resistance r, via D_l, and a current i_R flowing, via D_γ, through resistance R. The active part of the circuit is that shown in Fig. A.17.

From

$$l\frac{di_l}{dt} + ri_r = 0,$$

with

$$i_r = i_l - i_\gamma - i_R; \quad i_\gamma = \gamma\frac{dv_\gamma}{dt}; \quad v_\gamma = E - l\frac{di_l}{dt} \text{ and } Ri_R = ri_r$$

the differential equation giving i_l can be deduced:

$$l\gamma\frac{d^2 i_l}{dt^2} + l\frac{r+R}{rR}\frac{di_l}{dt} + i_l = 0. \tag{A.19}$$

The following notations are introduced:

$$\alpha' = \frac{1}{2\gamma}\frac{r+R}{rR}; \quad \omega_0 = \frac{1}{\sqrt{l\gamma}}; \quad z' = \frac{\alpha'}{\omega_0} = \frac{1}{2}\frac{r+R}{rR}\sqrt{\frac{l}{\gamma}} \tag{A.20}$$

– *If there is low damping* ($z' < 1$):

By taking the start of the inductor discharge as the initial time, the following can be obtained for i_l and v_l:

$$i_l = \frac{I}{\sqrt{1-z'^2}}\exp(-\alpha' t)\cos[(\omega\sqrt{1-z'^2})t - \varphi']$$

$$v_l = -\frac{l\omega_0 I}{\sqrt{1-z'^2}}\exp(-\alpha' t)\sin(\omega_0\sqrt{1-z'^2})t,$$

with $\varphi' = \arcsin z'$.

Voltage v_l reaches its minimum $-\Delta V$ and, thus, voltage v_{CE} its maximum $E + \Delta V$, for

$$t = t'_M = \frac{1}{\omega_0\sqrt{1-z'^2}}\left(\frac{\pi}{2} - \varphi'\right) \tag{A.21}$$

and *overvoltage* ΔV is given by

$$\Delta V = l\omega_0 I\exp(-\alpha' t'_M). \tag{A.22}$$

– *If there is high damping* ($z' > 1$):

Current i_l and voltage v_l are given by

$$i_l = \frac{I}{\alpha'_2 - \alpha'_1}[\alpha'_2\exp(\alpha'_1 t) - \alpha'_1\exp(\alpha'_2 t)],$$

$$v_l = -lI\frac{\alpha'_1\alpha'_2}{\alpha'_1 - \alpha'_2}[\exp(\alpha'_1 t) - \exp(\alpha'_2 t)]$$

with

$$\alpha'_1 = -\alpha' + \sqrt{\alpha'^2 - \omega_0^2}; \quad \alpha'_2 = -\alpha' - \sqrt{\alpha'^2 - \omega_0^2}.$$

Voltage v_{CE}, equal to $E - v_l$ reaches its maximum for

$$t = t'_M = \frac{1}{\alpha'_1 - \alpha'_2} \ln\left(\frac{\alpha'_2}{\alpha'_1}\right) \tag{A.21'}$$

and *overvoltage* ΔV is:

$$\Delta V = \frac{l\omega_0^2 I}{\alpha'_1 - \alpha'_2} [\exp(\alpha'_1 t'_M) - \exp(\alpha'_2 t'_M)]. \tag{A.22'}$$

- Equations (A.21), (A.22), (A.21′) and (A.22′) can be used for the variant shown in Fig. A.15a; resistance r has simply to replace $rR/(r + R)$ in the expressions of α' and z' in Eq. (A.20). For the single resistance circuit in Fig. A.16, R has merely to replace $rR/(r + R)$.

- *End of l discharging*

The instant at which the previous state ends and the end of discharge depend on the circuit used:

With the circuit shown in Fig. A.15a, current i_r flows through diode D_l and current $i_R + i_\gamma$ through diode D_γ, at the beginning of discharge. Current i_r and i_R are positive throughout the decrease of i_l. But current i_γ, which is positive when v_{CE} is increasing ($t < t'_M$), then becomes negative. If the sum $i_R + i_\gamma$ falls to zero, diode D_γ turns off, inductance l ends discharging into r via D_l; the excess charge of the capacitor is dissipated into R.

With the single resistance circuit shown in Fig. A.15b, the entire discharge takes place according to the same equations, if there is high damping. If the damping is low, current i_l falls to zero shortly after instant $t = t'_M$, and diode D′ turns off: i_l remains at zero and v_γ moves towards E as the capacitor excess charge is dissipated into R.

With the variant shown in Fig. A.15a, the end of discharge is similar to that of the single resistance circuit. However, after i_l has fallen to zero, the excess charge of γ is dissipated into $R + r$.

A.4.3 Characteristics

The similarity of the equations which give overcurrent ΔI, when γ is discharged and overvoltage ΔV, when l is discharged, enables the variations of stresses imposed on the transistor to be represented by a single curve.

This curve (Fig. A.18) gives

$$\frac{\Delta I}{\gamma\omega_0 E} \quad \text{or} \quad \frac{\Delta I}{E\sqrt{\gamma/l}} \quad \text{as a function of } z \text{, equal to} \quad \frac{R}{2}\sqrt{\frac{\gamma}{l}}$$

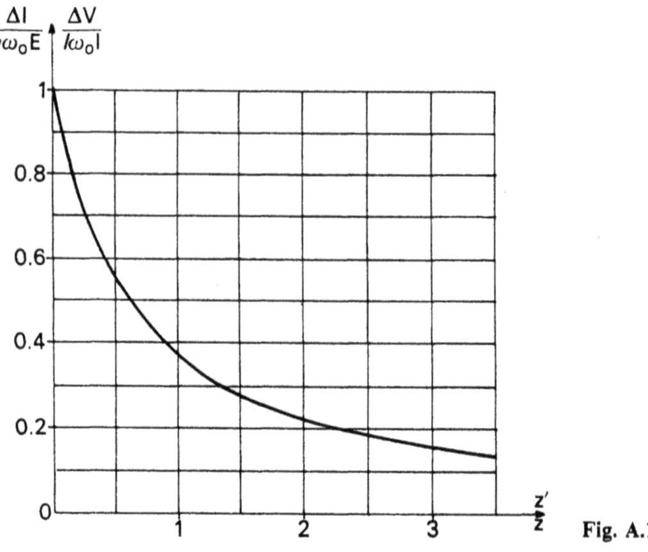

Fig. A.18

$$\frac{\Delta V}{l\omega_0 I} \quad \text{or} \quad \frac{\Delta V}{I\sqrt{l/\gamma}} \text{ as a function of } z', \text{ equal to } \frac{1}{2}\frac{r+R}{rR}\sqrt{\frac{l}{\gamma}}$$

- When the values of l and γ have been selected, according to the required reduction of the losses in the transistor at turn-on and at turn-off, the above curve enables R and r to be chosen in order to limit ΔI and ΔV.

  This curve shows the advantages resulting from increasing R and reducing r: the former increases z and thus reduces ΔI; the latter increases z' and thus reduces ΔV.

  The increase of R and the decrease of r are limited by their influence on the increase in the minimum values to be given to T_{ON} and T_{OFF}.
- By multiplying the two ratios of which the curve in Fig. A.18 shows the variations,

$$\frac{\Delta I}{\gamma\omega_0 E}\frac{\Delta V}{l\omega_0 I} = \frac{\Delta I}{I}\frac{\Delta V}{E},$$

it can be seen that this product is equal to the relative values of the overcurrent and the overvoltage. The product zz' must be about 5 in order that $\Delta I/I$ and $\Delta V/E$ be both about 0.2.

- *Remarks*

- With the single resistance circuit in Fig. A.15b, the product zz' equals 1/4. The overcurrent and the overvoltage cannot be reduced simultaneously. The presence of only a single resistance leads to the latter having a value too small to effectively limit ΔI and too high to effectively limit ΔV.

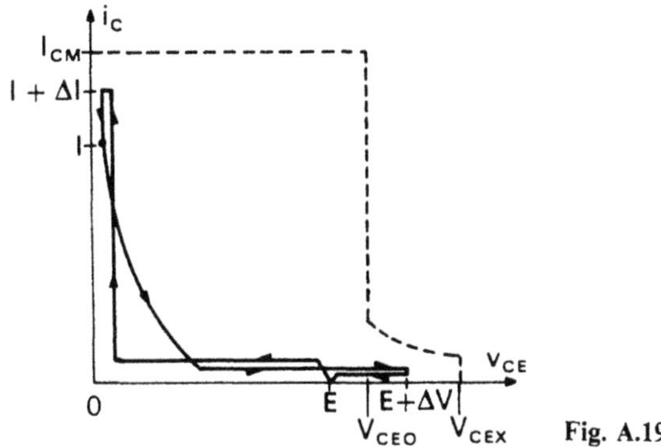

Fig. A.19

Figure A.19 shows the switching loci when the two circuits are used simultaneously. Since overvoltage ΔV occurs in the off state after current i_C has fallen to zero, this enables the reverse-bias safe operating area to be used. The maximum value of v_{CE} can go beyond V_{CEO} but must remain less than V_{CEX}.

A.5 Combination of the Two Snubbers with Energy Recovery

By using additional components, the energy stored in the snubber inductor or capacitor can be recovered, and its dissipation in resistances can be avoided. Among the various possible circuits we will analyze the example shown in Fig. A.20.

In this circuit still can be found inductor l, capacitor γ and diode D_γ; diode D_l and resistances R and r have been removed; diodes D_2, D_3, D_4, inductor l_2 and capacitor γ_2 have been added.

We will denote

$$\gamma_2 = k\gamma \,; \quad L = l + l_2 \tag{A.23}$$

A.5.1 Turn-on (Fig. A.21)

As will be shown by the analysis of the turn-off, capacitor γ is charged at voltage E when conduction begins while capacitor γ_2 is completely discharged.

The turn-on process can be divided into three phases:

- *First phase*

During the rise of current i_C from zero to I, the free-wheeling diode D is conducting. Voltages v_γ across γ and v_2 across γ_2 cannot vary.

- *Second phase*

- As soon as diode D turns off, the voltage v_l across l can decrease. This enables γ to be discharged via D_2, l_2, γ_2, l and the transistor.

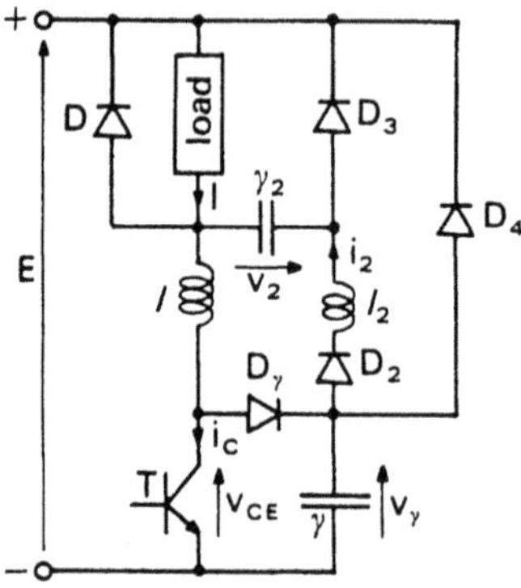

Fig. A.20

From $I = i_C - i_2$

$$v_\gamma - v_2 = l\frac{di_C}{dt} + l_2\frac{di_2}{dt} = L\frac{di_2}{dt}$$

$$i_2 = -\gamma\frac{dv_\gamma}{dt} = \gamma_2\frac{dv_2}{dt} = k\gamma\frac{dv_2}{dt}$$

it can be deduced that

$$L\gamma\frac{k}{1+k}\frac{d^2 i_2}{dt^2} + i_2 = 0.$$

If the beginning of this phase is taken as the initial time, this gives

$$i_2 = \frac{E}{L\omega_1}\sin\omega_1 t, \quad \text{with} \tag{A.24}$$

$$\omega_1 = \frac{1}{\sqrt{L\gamma k/(1+k)}}$$

it can be deduced that

$$i_C = I + \frac{E}{L\omega_1}\sin\omega_1 t; \quad v_\gamma = \frac{E}{1+k}(1 + k\cos\omega_1 t);$$

$$v_2 = \frac{E}{1+k}(1 - \cos\omega_1 t). \tag{A.25}$$

During this second phase, current i_C can reach a maximum value equal to $I + (E/L\omega_1)$.

Capacitor γ can be completely discharged only if the cosine term of the expression of v_γ has an amplitude $kE/(1+k)$ greater than $E/(1+k)$. It is

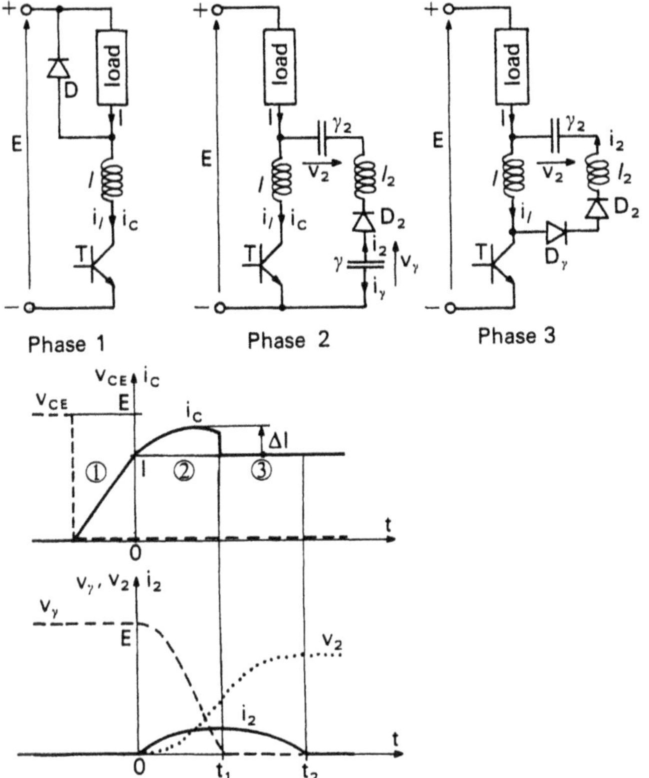

Fig. A.21

therefore necessary that

$$k > 1 \text{ or } \gamma_2 > \gamma. \tag{A.26}$$

– This operational mode ends, for $t = t_1$, when v_γ reaches zero:

$$t_1 = \frac{1}{\omega_1} \arccos(-1/k) \tag{A.27}$$

At this instant, i_2 and v_2 have the following values:

$$i_2(t_1) = \frac{E}{L\omega_1} \frac{\sqrt{k^2 - 1}}{k}; \quad v_2(t_1) = \frac{E}{k}$$

• *Third phase*

– As soon as v_γ falls to zero, diode D_γ becomes conducting. Current i_C reverts to value I and keeps that value. Current i_2 flows through D_γ, D_2, l_2, γ_2 and l; in l it is added to I. During this operational mode, which ends when i_2 falls to zero, the following can be written:

$$v_2 + l\frac{di_l}{dt} + l_2\frac{di_2}{dt} = 0$$

$$i_l = I + i_2$$

$$i_2 = \gamma_2\frac{dv_2}{dt} = k\gamma\frac{dv_2}{dt}$$

Voltage v_2 is thus the solution of the equation

$$kL\gamma\frac{d^2v_2}{dt^2} + v_2 = 0;$$

which gives

$$v_2 = \frac{E}{\sqrt{k}}\sin[\omega_2(t - t_1) + \varphi]$$

$$i_2 = \sqrt{k}\gamma\omega_2 E\cos[\omega_2(t - t_1) + \varphi] \tag{A.28}$$

with

$$\omega_2 = \frac{1}{\sqrt{kL\gamma}}; \quad \varphi = \arctan\frac{1}{\sqrt{k-1}}.$$

- Current i_2 falls to zero for $t = t_2$ given by

$$t_2 = t_1 + \frac{1}{\omega_2}\left(\frac{\pi}{2} - \varphi\right) \tag{A.29}$$

and voltage v_2 is then:

$$v_2(t_2) = E/\sqrt{k}. \tag{A.30}$$

- It can be seen that the energy stored in γ has been used to charge capacitor γ_2.

In order to reduce the overcurrent ΔI of current i_C during the second phase, the values of additional elements (especially that of inductor l_2) must be chosen in order to give to $L\,\omega_1$ a sufficiently high value. As a result, phases 2 and 3 last longer. Once again, a compromise has to be found between reducing ΔI and increasing the minimum value of T_{ON}.

A.5.2 **Turn-off** (Fig. A.22)

When current i_C begins to decrease,

$$v_\gamma = 0; \quad v_2 = E/\sqrt{k}; \quad i_l = I.$$

The turn-off procedure can be divided into five phases:

load
E
I
l
i_l
D_γ
i_C
T
v_{CE}
γ
v_γ
i_γ
D_3
γ_2
v_2
D_4
D
Phase 1
Phase 2
Phase 3
Phase 4
Phase 5
v_{CE}, i_C
v_2, i_l
t
0
t_1
t_f
t_3
t_4
①
②
③
④
⑤

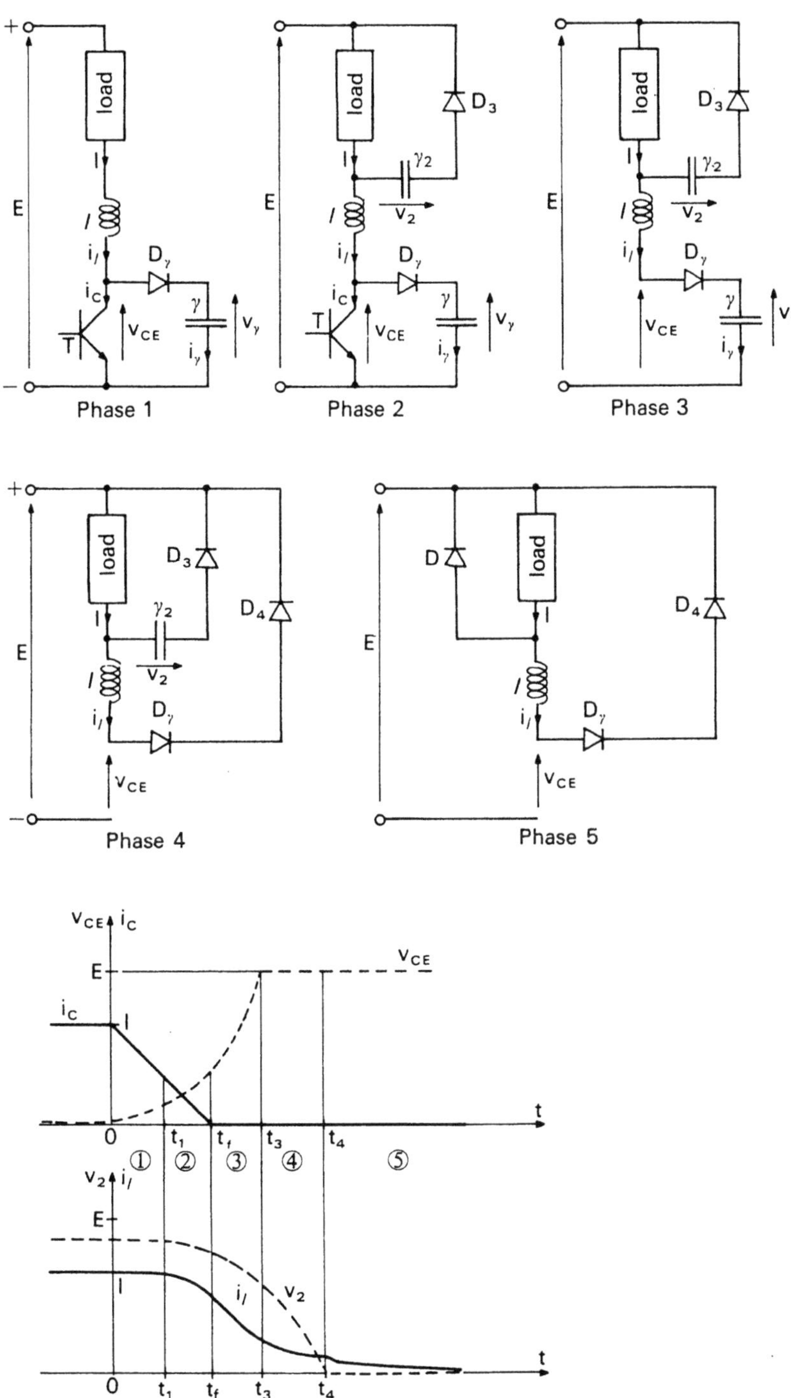

Fig. A.22

- *First phase*

- Starting from I, current i_C in the transistor initially decreases linearly. Current i_l in inductor l remains constant and equal to I. The difference between I and i_C charges capacitor γ via diode D_γ; the voltage across γ is given by

$$v_\gamma = \frac{1}{\gamma}\int_0^t (I - i_C)\,dt = \frac{I}{\gamma}\frac{t^2}{2t_f} \tag{A.31}$$

- This phase ends, for $t = t_1$, when diode D_3 becomes conducting.

  Since $l\,di_l/dt$ is zero, the voltage across D_3 is given by

$$v_{D_3} = -E + v_\gamma + v_2, \quad \text{with} \quad v_2 = E/\sqrt{k}.$$

  This gives the following value of t_1:

$$t_1 = \sqrt{\frac{2\gamma t_f}{I} E\left(1 - \frac{1}{\sqrt{k}}\right)} \tag{A.32}$$

  and the value of v_γ at the end of this phase:

$$v_\gamma(t_1) = E\left(1 - \frac{1}{\sqrt{k}}\right).$$

- *Second phase*

- The conduction of D_3 enables γ_2 to begin discharging and to supply the load with a part of current I.

  Current i_l is therefore no longer equal to I and begins to decrease; current i_C continues to fall, and voltages v_{CE} and v_γ are still rising as $i_l - i_C$ charges capacitor γ.

  From equation

$$v_\gamma + l\frac{di_l}{dt} + v_2 = E$$

  with

$$I - i_l = -\gamma_2\frac{dv_2}{dt}; \quad i_C = I\left(1 - \frac{t}{t_f}\right); \quad i_l - i_C = \gamma\frac{dv_\gamma}{dt}$$

  the following can be deduced:

$$\frac{k}{1+k} lC\frac{d^2 i_l}{dt^2} + i_l = I\left(1 - \frac{k}{1+k}\frac{t}{t_f}\right).$$

This gives the expressions of current i_l, and of voltages v_γ and v_2:

$$i_l = I\left[1 - \frac{k}{1+k}\frac{1}{t_f}\left(t + \frac{\sin[\omega_0(t-t_1)+\varphi_1]}{\omega_0\cos\varphi_1}\right)\right]$$

$$v_\gamma = E\left(1 - \frac{1}{\sqrt{k}}\right) + \frac{I}{(1+k)\gamma t_f}\left[\frac{t^2-t_1^2}{2} + \frac{k}{\omega_0^2}\right.$$

$$\left.\times\left(1 - \frac{\cos[\omega_0(t-t_1)+\varphi_1]}{\cos\varphi_1}\right)\right] \tag{A.33}$$

$$v_2 = \frac{E}{\sqrt{k}} - \frac{I}{(1+k)\gamma t_f}\left[\frac{t^2-t_1^2}{2} + \frac{1}{\omega_0^2}\left(\frac{\cos[\omega_0(t-t_1)+\varphi_1]}{\cos\varphi_1} - 1\right)\right].$$

with

$$\omega_0 = \frac{1}{\sqrt{l\gamma k/(1+k)}}, \quad \tan\varphi_1 = \omega_0 t_1. \tag{A.34}$$

– If the value of capacitor γ has been chosen in order to greatly reduce the losses in the transistor, voltage v_γ will not have reached value E when, for $t = t_f$, current i_C falls to zero. This moment marks the end of the second phase. At the end of this period, values I_{lF}, $V_{\gamma F}$, V_{2F} of current i_l and of voltages v_γ and v_2 are obtained by making $t = t_f$ in Eqs. (A.33).

- *Third phase*

– Current i_C is zero. Capacitor γ_2 still delivers part of current I. Current i_l charges capacitor γ via inductor l.

Since

$$l\frac{di_l}{dt} + v_\gamma + v_2 = E, \quad \text{with} \quad i_l = \gamma\frac{dv_\gamma}{dt} \quad \text{and} \quad I - i_l = -\gamma_2\frac{dv_2}{dt},$$

current i_l is now a solution of

$$\frac{k}{1+k}l\gamma\frac{d^2 i_l}{dt^2} + i_l = \frac{I}{1+k}$$

From this, the expression of i_l can be deduced and, from the latter, those of v_γ and v_2:

$$i_l = \frac{I}{1+k}[1 - \cos\omega_0(t-t_f)] - \frac{V_{\gamma F} + V_{2F} - E}{l\omega_0} \times \sin\omega_0(t-t_f)$$

$$+ I_{lF}\cos\omega_0(t-t_f)$$

$$v_\gamma = \frac{1}{1+k}\left\{V_{\gamma F} - k(V_{2F} - E) + k(V_{\gamma F} + V_{2F} - E)\cos\omega_0(t-t_f)\right.$$

$$+\frac{1}{\gamma\omega_0}(I[\omega_0(t-t_f)-\sin\omega_0(t-t_f)]+(1+k)I_{lF}\sin\omega_0(t-t_f))\Big\}$$

$$v_2=\frac{1}{1+k}\Big\{kV_{2F}-V_{\gamma F}+E+(V_{\gamma F}+V_{2F}-E)\cos\omega_0(t-t_f)$$

$$-\frac{1}{k\gamma\omega_0}(I[k\omega_0(t-t_f)-\sin\omega_0(t-t_f)]+(1+k)I_{lF}\sin\omega_0(t-t_f))\Big\}. \tag{A.35}$$

- This phase ends, for $t = t_3$, when voltage v_γ across γ reaches value E. Diode D_4 then starts to conduct and prevents v_γ (and hence v_{CE}) from exceeding E.
  By making $t = t_3$ in Eqs. (A.35) values I_{l3} and V_{23} of current i_l and voltage v_2 at the end of this period can be obtained.

- *Fourth phase*

- Capacitor γ_2 discharges completely via diode D_3. Current i_l continues to decrease in the circuit made up of l, D_γ, D_4 and the load.
  From equations

$$l\frac{di_l}{dt}+v_2=0;\quad I-i_l=-\gamma_2\frac{dv_2}{dt}$$

the following can be deduced:

$$kL\gamma\frac{d^2i_l}{dt^2}+i_l=I$$

giving

$$i_l=I-\frac{V_{23}}{l\omega_0'}\sin\omega_0'(t-t_3)-(I-I_{l3})\cos\omega_0'(t-t_3)$$

$$v_2=V_{23}\cos\omega_0'(t-t_3)-l_1\omega_0'(I-I_{l3})\sin\omega_0'(t-t_3) \tag{A.36}$$

with

$$\omega_0'=\frac{1}{\sqrt{kl\gamma}}$$

- This phase ends, for $t = t_4$, when voltage v_2 across γ_2 becomes equal to zero. The free-wheeling diode D starts to conduct and keeps v_2 at zero.
  The second of Eqs. (A.36) gives the value of t_4:

$$t_4=t_3+\frac{1}{\omega_0'}\arctan\frac{V_{23}}{l\omega_0'(I-I_{l3})} \tag{A.37}$$

- *Fifth phase*

From $t = t_4$, current I in the load flows partly through the free-wheeling diode and partly via l, D_γ and D_4.

Theoretically, if the voltage drops in the diodes and in the resistance of l are ignored, the voltage across l is zero and current i_l keeps the value which it has at instant $t = t_4$. In practice, current i_l gradually disappears during the conduction of D.

The values i_2, v_γ and v_2 ($i_2 = 0$, $v_\gamma = E$, $v_2 = 0$) are those considered as initial values for the analysis of the turn-on process.

- This circuit is designed to avoid the systematic dissipation, in the resistances, of the energy stored in the inductor l and the snubber capacitor γ. This energy is recovered by the load when capacitor γ_2 discharges.

  During the period (t_1, t_f), current i_l is less than I. This has a slight slowing down effect on the rise of v_{CE}.

  A further advantage of this circuit lies in the suppression of the overvoltage across the transitor.

  The minimum off-time of the transistor must be greater than t_4.

A.6 GTO Thyristor, Turn-off Snubber

The essential difference between the GTO thyristor and the conventional thyristor lies in the turn-off process.

An inductor enables the di/dt to be limited when conduction starts. This inductor can have a very low value since the GTO thyristor can withstand very high di/dt on account of its highly interdigitated structure.

The turn-off deserves a closer analysis: the turn-off snubber must limit the dv/dt and the overvoltage at the end of the voltage rise.

A.6.1 Turn-off

Application of a negative gate current brings about a fast and uncontrollable decrease in the main current i_T. A turn-off snubber circuit must be added in order to avoid unacceptable overvoltage. We have already described (Sect. 2.5.2.2) how voltage is restored at the terminals when the simplest type of snubber (Fig. A.23) is used. This snubber consists of a diode, a resistance and a capacitor.

- During the decrease in i_T, part of the load current is transferred to the capacitor, via the diode. However, the latter does not begin conducting instantaneously and there is a stray inductance in the loop formed by D_C, C and the GTO which prevents the capacitor from playing its full role. Towards the end of the fall time, voltage v_D shows a *voltage spike* V_S.
- When only the tail current remains in the GTO, the diode is conducting. It enables the capacitor to limit the rise slope dv/dt of v_D, but this latter exceeds supply voltage E to reach V_{DM}. The *overvoltage* at the end of the voltage rise should not bring V_{DM} beyond the repetitive peak forward voltage V_{DRM}.

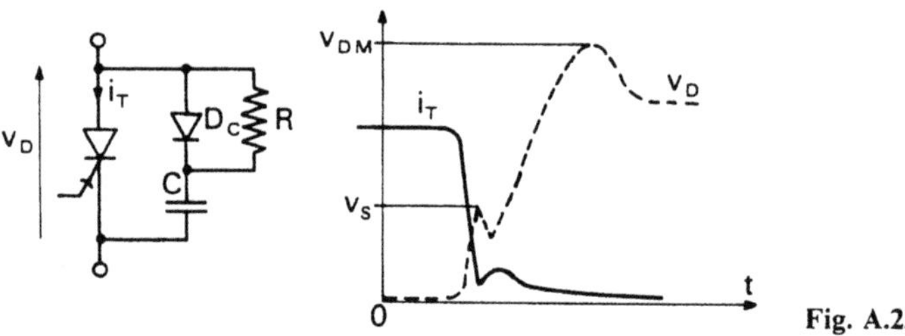

Fig. A.23

A.6.2 Approximate Calculation of the Overvoltage

In order to estimate the overvoltage, it is assumed that voltage v_C across C (Fig. A.24) only begins to increase when current i_T has fallen to zero. Three successive phases can thus be defined.

- *First phase*

First of all, the capacitor charges up at a constant current via the di/dt limiting inductor l and diode D_C. If the voltage drop in the latter is ignored, this gives

$$v_C = v_D = \frac{I}{C}t.$$

For $t = t_E$, this phase ends when v_C reaches value E:

$$t_E = CE/I. \tag{A.38}$$

- *Second phase*

– As soon as voltage v_C reaches E, the free-wheeling diode D starts to conduct and applies voltage E across the series-connection of l, D_C and C:

$$v_C + l\frac{\mathrm{d}i_l}{\mathrm{d}t} = E, \quad \text{with} \quad i_l = C\frac{\mathrm{d}v_C}{\mathrm{d}t}$$

gives

$$lC\frac{\mathrm{d}^2 v_C}{\mathrm{d}t^2} + v_C = E.$$

From this it can be deduced that

$$v_C = E + \frac{I}{C\omega}\sin\omega(t - t_E)$$

$$i_l = I\cos\omega(t - t_E) \tag{A.39}$$

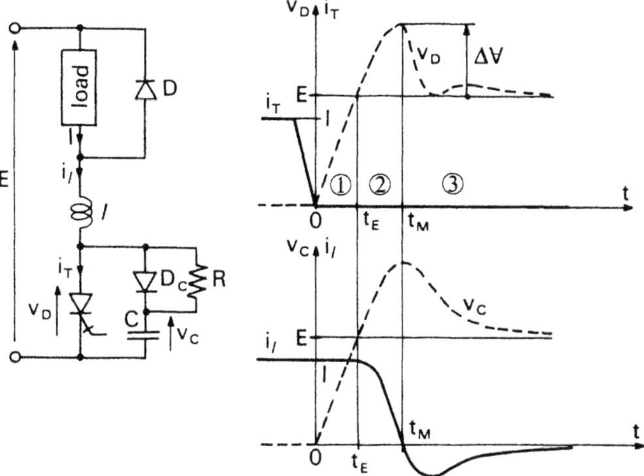

Fig. A.24

with

$$\omega = 1/\sqrt{lC}$$

– The second phase ends, for $t = t_M$, when i_l falls to zero:

$$t_M = t_E + \frac{\pi}{2\omega} \tag{A.40}$$

Voltage v_C, equal to v_D, is then maximum and has the following value

$$v_{C_{max}} = E + \Delta V = E + \frac{I}{C\omega} = E + I\sqrt{\frac{l}{C}} \tag{A.41}$$

– Instant t_M and voltage $v_{C_{max}}$ can be expressed as a function of the maximum value of di/dt and dv/dt, which appear during the GTO operation:

at turn-on, $\left(\frac{di}{dt}\right)_{max} = \frac{E}{l}$

at turn-off, $\left(\frac{dv}{dt}\right)_{max} = \frac{I}{C}$

As a result, this gives

$$t_M = t_E + \frac{\pi}{2\omega} = \frac{EC}{I} + \frac{\pi}{2}\sqrt{lC} = \frac{E}{(dv/dt)_{max}} + \frac{\pi}{2}\sqrt{\frac{EI}{(di/dt)_{max}(dv/dt)_{max}}} \tag{A.40'}$$

$$v_{C\,max} = E + I\sqrt{\frac{l}{C}} = E + I\sqrt{\frac{E/(di/dt)_{max}}{I/(dv/dt)_{max}}} = E + \sqrt{\frac{EI\,(dv/dt)_{max}}{(di/dt)_{max}}}. \tag{A.41'}$$

The latter relation shows that the high value of di/dt which is characteristic of the GTO thyristor must be used to reduce the overvoltage.

The expression (A.40′) of t_M shows that this time interval can be reduced by getting close to the maximum values of di/dt and dv/dt. Time interval t_M plays a role in the minimum value of T_{OFF} and thus in the limitation of the maximum operating frequency. In practice, however, this limitation is mainly linked to the duration of the tail current.

• *Third phase*

As diode D_C is blocked, C discharges into voltage supply E via R, l and D. Voltage v_D is now equal to $E - l\,(di_l/dt)$ and no more to v_C.

The standard equations of a series RLC circuit can be used here. Figure A.24 gives the waveforms of v_D, v_C and i_l in the case of an aperiodic response with a high damping ($R > 2\sqrt{l/C}$).

In order to decrease the overvoltage ΔV at instant t_M, C can be increased; by doing this, however, the durations of the three phases are similarly increased and the maximum operating frequency is reduced.

A.6.3 Overvoltage Reduction

The circuit shown in Fig. A.25 enables the overvoltage to be limited without increasing the commutation time of the GTO.

- *The first phase* is exactly the same as that described in Sect A.6.2. (Placing inductance l in front of the load in no way alters its contribution during the turn-on process.) Capacitor C is charged by a constant current I, via D_C, the load and l. Diodes D and D_2 are blocked. Capacitor C_2 has been charged via R_2 at voltage E.
- *The second phase* begins, for $t = t_E$, when v_C reaches value E, thus causing diodes D and D_2 to turn-on. If the low current flowing through resistance R_2 is ignored, current i_l can be divided into two parts:

  one part charges C_2 via D_2,
  the other continues to charge C via the load and D_C.
  This corresponds to the following equations:

$$C\frac{dv_C}{dt} + C_2\frac{dv_2}{dt} = i_l$$

$$l\frac{di_l}{dt} + v_2 = E\,; \quad v_2 = v_C\,.$$

  The differential equation giving v_C is thus

$$l(C + C_2)\frac{d^2v_C}{dt^2} = E$$

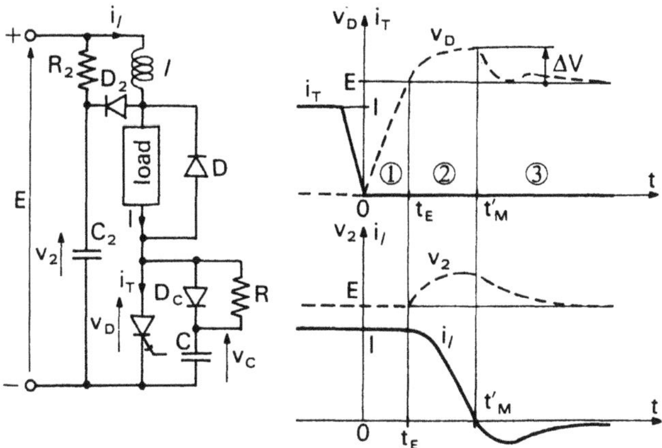

Fig. A.25

and v_C and i_l are expressed as

$$v_C = E + \frac{I}{(C + C_2)\omega'} \sin \omega'(t - t_E)$$

$$i_l = I \cos \omega'(t - t_E) \qquad \text{(A.42)}$$

with $\omega' = 1/\sqrt{l(C + C_2)}$.

Current i_l and the currents charging C and C_2 fall to zero for $t = t'_M$:

$$t'_M = t_E + \frac{\pi}{2\omega'} = \frac{EC}{I} + \frac{\pi}{2}\sqrt{l(C + C_2)}.$$

Voltage v_D is then at its maximum value and is

$$v_{D_{max}} = v_{C_{max}} = E + \frac{I}{(C + C_2)\omega'} = E + I\sqrt{\frac{l}{C + C_2}} \qquad \text{(A.43)}$$

By choosing C_2 high enough compared to C, the overvoltage at the end of turn-off is considerably reduced. The value of t'_M is greater than that of t_M but the third phase is shorter since $v_C(t'_M)$ is less than $v_C(t_M)$. The minimum value of T_{OFF} is very little affected.

- *The third phase* is identical to that described in Sect. A.6.2, as far as v_C is concerned. The capacitor C_2 discharges into voltage supply E via R_2, and voltage v_2 reverts to value E and keeps this value until the next GTO turn-off.

A.6.4 Energy Recovery

The type of circuit described in Sect. A.5 can be adapted to the case of GTOs. However, the power concerned is noticeably greater, both for the GTO and for snubber components.

For example, it becomes difficult to find fast diodes which can sustain the necessary voltages. Similarly, the kind of assembly which could be needed to recover the capacitor energy are often bulky and expensive: they usually limit the maximum operating frequency. Lastly, the smallest stray inductances are no more negligible in circuits where the di/dt may reach 10^9 A/S.

Bibliography

1. Locher RE (1970) On switching inductive loads with power transistors. *IEEE Trans. Ind. Electron. Control Instrum.*, **17**(4): 256–262
2. Hall JK (1971) Improvement of turn-off performance of series-connected thyristors by magnetic components. *IEEE Trans. Magnetics*, **7**(2): 297–304
3. Mc Murray W (1972) Optimum snubbers for power semiconductors. *IEEE Trans. Ind. Appl.*, **8**(5): 593–600
4. Calkin ET, Hamilton BH (1976) Circuit techniques for improving the switching loci of transitor switches in switching regulators. *IEEE Trans. Ind. Appl.*, **12**(4): 364–369
5. Harada K, Ninomiya T, Kohno M (1979) Optimum design of RC snubbers for switching regulators. *IEEE Trans. Aerosp. Electron. Syst.*, **15**(2): 209–218
6. Mc Murray W (1980) Selection of snubbers and clamps to optimize the design of transitor switching converters. *IEEE Trans. Ind. Appl.*, **16**(4): 513–523
7. Ohashi H (1983) Snubber circuit for high-power Gate Turn-Off thyristors. *IEEE Trans. Ind. Appl.*, **19**(4): 655–664
8. Williams BW (1984) High-voltage high-frequency power-switching transistor module with switching-aid-circuit energy recovery. *Proc. Inst. Electr. Eng., Part B*, **131**(1): 7–12
9. Zach FC, Kaiser KH, Kolar JW, Haselsteiner FJ (1986) New lossless turn-on and turn-off snubber. Network for inverters, including circuits for blocking voltage limitation. *IEEE Trans. Power Electron.*, **1**(2): 65–75
10. Steyn CG, Van Wyk JD (1986) Study of application of nonlinear turn-off snubber for power electronic switches. *IEEE Trans. Ind. Appl.*, **22**(3): 471–477
11. Pong MH, Jackson RD (1987) Computer-aided design of power transistor inverter snubber circuits. *Proc. Inst. Electr. Eng., Part B*, **134**(2): 69–77
12. Mc Murray W (1987) Efficient snubbers for voltage-source GTO inverters. *IEEE Trans. Power Electron.*, **2**(3): 264–272

Subject Index